Springer Series in Statistics

Springer
New York
Berlin
Heidelberg
Barcelona
Hong Kong
London
Milan
Paris
Singapore
Tokyo

Springer Series in Statistics

Andersen/Borgan/Gill/Keiding: Statistical Models Based on Counting Processes.
Andrews/Herzberg: Data: A Collection of Problems from Many Fields for the Student and Research Worker.
Anscombe: Computing in Statistical Science through APL.
Berger: Statistical Decision Theory and Bayesian Analysis, 2nd edition.
Bolfarine/Zacks: Prediction Theory for Finite Populations.
Borg/Groenen: Modern Multidimensional Scaling: Theory and Applications
Brémaud: Point Processes and Queues: Martingale Dynamics.
Brockwell/Davis: Time Series: Theory and Methods, 2nd edition.
Daley/Vere-Jones: An Introduction to the Theory of Point Processes.
Dzhaparidze: Parameter Estimation and Hypothesis Testing in Spectral Analysis of Stationary Time Series.
Fahrmeir/Tutz: Multivariate Statistical Modelling Based on Generalized Linear Models.
Farebrother: Fitting Linear Relationships: A History of the Calculus of Observations 1750 - 1900.
Farrell: Multivariate Calculation.
Federer: Statistical Design and Analysis for Intercropping Experiments, Volume I: Two Crops.
Federer: Statistical Design and Analysis for Intercropping Experiments, Volume II: Three or More Crops.
Fienberg/Hoaglin/Kruskal/Tanur (Eds.): A Statistical Model: Frederick Mosteller's Contributions to Statistics, Science and Public Policy.
Fisher/Sen: The Collected Works of Wassily Hoeffding.
Good: Permutation Tests: A Practical Guide to Resampling Methods for Testing Hypotheses.
Goodman/Kruskal: Measures of Association for Cross Classifications.
Gouriéroux: ARCH Models and Financial Applications.
Grandell: Aspects of Risk Theory.
Haberman: Advanced Statistics, Volume I: Description of Populations.
Hall: The Bootstrap and Edgeworth Expansion.
Härdle: Smoothing Techniques: With Implementation in S.
Hart: Nonparametric Smoothing and Lack-of-Fit Tests.
Hartigan: Bayes Theory.
Hedayat/Sloane/Stufken: Orthogonal Arrays: Theory and Applications.
Heyde: Quasi-Likelihood and its Application: A General Approach to Optimal Parameter Estimation.
Heyer: Theory of Statistical Experiments.
Huet/Bouvier/Gruet/Jolivet: Statistical Tools for Nonlinear Regression: A Practical Guide with S-PLUS Examples.
Jolliffe: Principal Component Analysis.
Kolen/Brennan: Test Equating: Methods and Practices.
Kotz/Johnson (Eds.): Breakthroughs in Statistics Volume I.

(continued after index)

A.S. Hedayat N.J.A. Sloane
John Stufken

Orthogonal Arrays

Theory and Applications

Springer

A.S. Hedayat
Department of Mathematics, Statistics, and Computer Science
University of Illinois
Chicago, IL 60607-7045
USA
hedayat@uic.edu

N.J.A. Sloane
AT&T Labs Research
180 Park Avenue
Florham Park, NJ 07932-0971
USA
njas@research.att.com

John Stufken
Department of Statistics
Iowa State University
Ames, IA 50011
USA
jstufken@iastate.edu

Library of Congress Cataloging-in-Publication Data
Hedayat, A.
Orthogonal arrays : theory and applications / A.S. Hedayat, N.J.A. Sloane, John Stufken.
p. cm. — (Springer series in statistics)
Includes bibliographical references and indexes.
ISBN 0-387-98766-5 (hc. : alk. paper)
1. Orthogonal arrays. I. Sloane, N. J. A. (Neil James Alexander), 1939– . II. Stufken, John. III. Title. IV. Series.
QA165.H43 1999
511′.6—dc21 99-30377

Printed on acid-free paper.

Production managed by A. Orrantia; manufacturing supervised by Joe Quatela.
Photocomposed copy prepared from the authors' PostScript file.
Printed and bound by Maple-Vail Book Manufacturing Group, York, PA.
Printed in the United States of America.

9 8 7 6 5 4 3 2 1

ISBN 0-387-98766-5 Springer-Verlag New York Berlin Heidelberg SPIN 10712489

Dedicated to

Batool, Leyla and Yashar Hedayat,

Susanna Cuyler Sloane,

Lili, Sharon and Byron Stufken

Preface

Who should read this book? Anyone who is running experiments, whether in a chemistry lab or a manufacturing plant (trying to make those alloys stronger), or in agricultural or medical research. Anyone interested in one of the most fascinating areas of discrete mathematics, connected to statistics and coding theory, with applications to computer science and cryptography. This is the first book on the subject since its introduction more than fifty years ago, and can be used as a graduate text or as a reference work. It features all of the key results, many very useful tables, and a large number of exercises and research problems. Most of the arrays that can be obtained by the methods in this book are available electronically.

Orthogonal arrays are beautiful and useful. They are essential in statistics and they are used in computer science and cryptography. In statistics they are primarily used in designing experiments, which simply means that they are immensely important in all areas of human investigation: for example in medicine, agriculture and manufacturing.

Your automobile lasts longer today because of orthogonal arrays.[1]

The mathematical theory is extremely beautiful: orthogonal arrays are related to combinatorics, finite fields, geometry and error-correcting codes. The definition of an orthogonal array is simple and natural, and we know many elegant constructions — yet there are at least as many unsolved problems.

[1] "The new mantra: MVT", *Forbes*, Mar. 11, 1996, pp. 114–118. (See also Section 11.8.)

Here is an example of an orthogonal array of strength 2:

```
0 0 0 0 0 0 0 0 0 0 0
1 1 1 0 1 1 0 1 0 0 0
0 1 1 1 0 1 1 0 1 0 0
0 0 1 1 1 0 1 1 0 1 0
0 0 0 1 1 1 0 1 1 0 1
1 0 0 0 1 1 1 0 1 1 0
0 1 0 0 0 1 1 1 0 1 1
1 0 1 0 0 0 1 1 1 0 1
1 1 0 1 0 0 0 1 1 1 0
0 1 1 0 1 0 0 0 1 1 1
1 0 1 1 0 1 0 0 0 1 1
1 1 0 1 1 0 1 0 0 0 1
```

Pick any two columns, say the first and the last:

```
0 0
1 0
0 0
0 0
0 1
1 0
0 1
1 1
1 0
0 1
1 1
1 1
```

Each of the four possible rows we might see there,

0 0, 0 1, 1 0, 1 1,

does appear, and they all appear *the same number of times* (three times, in fact). That's the property that makes it an orthogonal array.

Only 0's and 1's appear in that array, but for use in statistics

0 or 1

in the first column might be replaced by

"butter" or "margarine" ,

and in the second column by

"sugar" or "no sugar" ,

and so on. Or

"slow cooling" or "fast cooling" ,

"catalyst" or "no catalyst" ,

etc., depending on the application.

Since only 0's and 1's appear, this is called a *2-level array.* There are 11 columns, which means we can vary the levels of up to *11 different variables*, and 12 rows, which means we are going to bake 12 different cakes, or produce *12 different samples* of the alloy. In short, we call this array an

$$OA(12, 11, 2, 2) .$$

The first "2" indicates the number of levels, and the second "2" the *strength*, which is the number of columns where we are guaranteed to see all the possibilities an equal number of times. In an orthogonal array of strength 3 (with two levels), in any *three* columns we would see each of the eight possibilities

$$000, 001, 010, 011, 100, 101, 110, 111$$

equally often. (The formal definition is given in the first chapter.)

As already mentioned, the main applications of orthogonal arrays are in planning experiments. The rows of the array represent the experiments or tests to be performed— cakes to be baked, samples of alloy to be produced, integrated circuits to be etched, test plots of crops to be grown, and so on.

The columns of the orthogonal array correspond to the different variables whose effects are being analyzed. The entries in the array specify the levels at which the variables are to be applied. If a row of the orthogonal array reads

$$110100\ldots$$

this could mean that in that test the first, second, fourth variables (where the 1's occur) are to be set at their "high" levels, and the third, fifth, sixth variables (where the 0's occur) at their "low" levels.

By basing the experiment on an orthogonal array of strength t we ensure that all possible combinations of up to t of the variables occur together equally often.

The aim here is to investigate not only the effects of the individual variables (or factors) on the outcome, but also how the variables interact. Obviously, even with a moderate number of factors and a small number of levels for each factor, the number of possible level combinations for the factors increases rapidly. It may therefore not be feasible to make even one observation at each of the level combinations. In such cases observations are made at only some of the level combinations, and the purpose of the orthogonal array is to specify which level combinations are to be used. Such experiments are called "fractional factorial" experiments. While there are nowadays other applications of orthogonal arrays in statistics (for example in computer experiments[2] and survey sampling[3]), the principal application is in the selection of level combinations for fractional factorial experiments.

Since the rows of an orthogonal array represent runs (or tests or samples) — which require money, time, and other resources — there are always practical constraints on the number of rows that can be used in an experiment. Finding the smallest possible number of rows is a problem of eminent importance. On the other hand, for a given number of runs we may want to know the largest number of columns that can be used in an orthogonal array, since this will tell us how many variables can be studied. We also want the strength to be large, though in many real-life applications this is set at 2, 3 or 4.

Then the main questions we ask are:

- for which values of the numbers of rows, columns, strength and levels does an orthogonal array exist?
- how can we construct the array, if it exists?

We have already mentioned the use of orthogonal arrays in automobile design. The following are some typical applications in medicine:

- Pharmaceutical companies use orthogonal arrays to investigate the stability and shelf life of drugs, which usually involve many different factors. The use of orthogonal arrays can lead to more economical tests and provide better statistical information.
- Multiple drug therapy is widely practiced in order to treat either a single medical disorder or to treat several concurrently existing ailments in the same patient. But many drugs interact with each other. Bioavailability studies are used to evaluate the combinations of various doses. Orthogonal arrays can help to identify the medically safe and effective regions so that physicians can adjust dosage levels to avoid or minimize interactions when using multiple drugs.
- Orthogonal and related arrays have been used in clinical trials which study how drugs are absorbed, distributed, metabolized and eliminated from the body. Orthogonal arrays can be useful for studying the effects of multiple factors on these drug characteristics. Orthogonal arrays or closely related

---

[2]Owen (1992).
[3]McCarthy (1969), Wu (1991).

arrays[4] can also be useful when using crossover designs, where the N rows of an array can be thought of as specifying N treatment sequences, one for each of the N subjects in the trial.

Although half a century has passed since orthogonal arrays were introduced by C. R. Rao, this is the first book entirely devoted to them. While many books on experimental design (and some books on combinatorial mathematics) devote a chapter to certain types of orthogonal arrays, it is astounding that after so many years this is the *first* book exclusively about orthogonal arrays. Perhaps it is the breadth of the subject that has discouraged others: although the main applications are in factorial experiments, orthogonal arrays are closely connected with finite geometries, Latin squares, Hadamard matrices and above all with error-correcting codes. A multidisciplinary effort—as presented in the present work—is required to do justice to the subject.

The book is arranged as follows. After an introductory chapter, Chapters 2 and 3 consider two of the fundamental questions concerning orthogonal arrays. The first is the existence question. Rao established inequalities (Theorem 2.1) that the parameters of an orthogonal array must satisfy, although these are in general not sufficient to guarantee the existence of the array. Improvements to Rao's inequalities are studied in Chapters 2 and 4. The second is the construction question: given that an orthogonal array exists, how do we find it? Galois fields turn out to be a powerful tool for the construction of orthogonal arrays, and various construction methods using such fields are presented in Chapter 3.

Chapters 4 and 5 establish the connection between orthogonal arrays and coding theory. The subject of error-correcting codes is also about 50 years old, and the two topics have a great deal in common, although for the most part they have developed separately. There are however many similar bounds and constructions, and in Chapters 4 and 5 we consider the fundamental questions of the previous two chapters from this new point of view. Chapter 4 includes a detailed discussion of Delsarte's linear programming bound, which in most cases improves on Rao's inequalities. In Chapter 5 we give a number of constructions for orthogonal arrays that are based on error-correcting codes.

One of the earliest methods for constructing certain orthogonal arrays is via "difference schemes": this is the subject of Chapter 6.

Orthogonal arrays are related to various other combinatorial structures. Chapters 7 and 8 focus on two of these. In Chapter 7 we explore the relationship between orthogonal arrays and Hadamard matrices, while Chapter 8 deals with orthogonal arrays, pairwise orthogonal Latin squares and pairwise orthogonal F-squares.

All of Chapters 1 through 8 restrict attention to orthogonal arrays in which

[4] Stufken (1991)

each factor has the same number of levels (each column contains the same set of numbers). In statistical applications this is often too restrictive. The extension of orthogonal arrays to so-called mixed-level orthogonal arrays is the subject of Chapter 9.

Chapter 10 contains a smorgasbord of topics which, while extremely interesting and important, did not have a natural home in the earlier chapters. The topics include some methods for constructing orthogonal arrays inspired by coding theory, bounds on large orthogonal arrays, brief introductions to compound orthogonal arrays, balanced orthogonal multi-arrays, transversal designs, resilient functions and nets, and connections with association schemes.

The focus of Chapter 11 is on applications of orthogonal arrays in statistics. In particular, we give a detailed discussion of how orthogonal arrays are used in factorial experiments. Special consideration is given to 2-level orthogonal arrays with a defining relation and to blocking. We discuss orthogonal main-effects plans and characterize orthogonal arrays in terms of information matrices. The final sections deal with the use of orthogonal arrays in robust design and with other families of designs (balanced arrays, central composite designs, supersaturated designs, covering designs and computer-generated designs).

Chapter 12 contains two sets of tables of orthogonal arrays. The first set (Tables 12.1–12.3) gives tables showing the smallest possible index (and hence the smallest number of runs) in 2-, 3- and 4-level orthogonal arrays with at most 32 factors and strengths between 2 and 10. The second set (Tables 12.6 (a)–(g) and Table 12.7) summarizes most of the arrays constructed in this book, and includes a table of both mixed- and fixed-level orthogonal arrays of strength 2 with up to 100 runs. These tables serve as a road-map to the whole book. This chapter also contains a table summarizing the connections between orthogonal arrays and other combinatorial structures. The chapter concludes by briefly discussing what to do when the orthogonal array you want doesn't exist (or is not yet known to exist).

Galois fields are an important tool for the construction of orthogonal arrays. They play a role in all the constructions in Chapter 3 and in many later chapters. Appendix A provides an introduction to Galois fields, covering the results needed for our constructions.

It is always hard to know when to stop. There are many further topics we could have included, such as other related combinatorial structures, connections with numerical analysis, other types of statistical designs, and data analysis techniques. But our primary goal has been to write about *orthogonal arrays.* In many chapters we have provided references on additional topics, and we have also created an extensive bibliography to assist in finding key references for further reading—both on orthogonal arrays and on related topics.

We have tried to write both a graduate text-book and a reference work.

With its many examples and problems, the book can serve as the sole graduate text-book for a special topics course, or as one of the main references in a broader graduate course on designs. With its many results, research problems and extensive bibliography, it should also be a valuable reference for research workers. The extensive tables of orthogonal arrays will make it useful for those just looking for an orthogonal array to use in their experiments.

This book began when one of us (A.S.H.) gave a course on orthogonal arrays in Chicago in the winter of 1985/86. J.S. took this course, but left Chicago in 1986; these two authors then worked on the manuscript on and off for the next few years. In the mean time N.J.A.S. had always been interested in the connections between orthogonal arrays and codes (cf. MacWilliams and Sloane, 1977), and had computed extensive tables of lower and upper bounds on orthogonal arrays. The authors joined forces in the mid 1990's. What you hold in your hand is therefore the product of many years of work.

Comments and corrections will be welcomed. The authors' email addresses are `hedayat@uic.edu`, `njas@research.att.com` and `jstufken@iastate.edu`. We plan to maintain a list of corrections, updates and improvements on a web site devoted to the book that will be found on our home pages. At the present time our home pages have the following addresses:

A.S.H.:	`www.uic.edu/depts/statlab/sam.htm`
N.J.A.S.:	`www.research.att.com/~njas/`
J.S.:	`www.public.iastate.edu/~jstufken/`

Electronic data-bases containing a large number of the orthogonal arrays and Hadamard matrices constructed in the book can be found at

`www.research.att.com/~njas/oadir/`

and

`www.research.att.com/~njas/hadamard/` .

Acknowledgments

Samad Hedayat and John Stufken have benefited greatly from comments by various students. The comments of Guoqin Su, who took a reading course from John Stufken when he was visiting the University of Illinois at Chicago, have been especially helpful. We are also grateful to Don Kreher, Arnold Neumaier and Eric Rains for comments on the manuscript.

Neil Sloane thanks his colleagues Rob Calderbank, John Conway, Anne

Freeny, Ron Hardin, Colin Mallows and Eric Rains for many helpful discussions about orthogonal arrays over the past fifteen years.

While we were writing this book, Neil Sloane offered two bottles of wine to the first person who could construct an $OA(45, 22, 3, 2)$, or one bottle for a proof of its nonexistence. He now owes Vladimir Tonchev and Noam Elkies one bottle each. The argument that establishes the nonexistence will be found in Chapter 4.

Samad Hedayat and John Stufken gratefully acknowledge the research support that they have received during the preparation of the book. Most recently, Samad Hedayat's research was supported through NSF grant DMS-9803596 and NIH grant PO1-CA48112, and that of John Stufken through NSF grants DMS-9504882 and DMS-9803684.

We thank Brenda Hewitt, who typed the preliminary versions of several chapters, and above all Sue Pope, who has typed and retyped all the chapters many times in the past five years.

The end of a proof or an example is indicated by the symbol: ■

Foreword

It gives me great pleasure to write a foreword to the remarkable, timely and most comprehensive treatise on "Orthogonal Arrays: Theory and Applications", written by Hedayat, Sloane and Stufken, who are all well known for their outstanding contributions to Combinatorics and Design of Experiments. They have designed this book with great skill to meet multiple objectives. It will be ideal as a textbook for students who want to make a systematic study of and specialize in this area, as a research monograph for those who want to know the current state of the art, and as a practical manual for consultants and technicians involved in designing experiments to determine the optimum mix of factors.

It may be of interest to the reader to learn something about the historical background of and my involvement with orthogonal arrays. When I joined the Indian Statistical Institute in January 1941 to study statistics, I was surprised to find that combinatorics, based on the concepts of Galois fields and finite projective geometries, was the main field of research of many faculty members under the leadership of R. C. Bose. With my knowledge of mathematics it was not difficult for me to read the relevant literature and jump on the bandwagon. I started writing papers on design of experiments in collaboration with K. R. Nair, an associate of R. C. Bose, the first of which was published in 1941 a few months after I joined the Institute. In 1943 I took my master's degree in statistics, and as a part of the final examination I chose to write a thesis in lieu of two practical papers. One of the chapters in the thesis was on asymmetrical factorial designs where I introduced the concept of "hypercubes of strength d", a special kind of orthogonal array. The concept was later exploited in a series of papers (Rao, 1945, 1946a, 1947) for the construction of confounded factorial designs with full or fractional replication.

In 1947, during my stay in Cambridge, U.K., I wrote a paper giving the general definition of an $OA(N, k, s, t)$ as outlined in the present book and discussing various applications. Since my paper provided a generalization of the multifactorial designs of Plackett and Burman (1946) which appeared in *Biometrika*, I submitted it to the same journal for publication. I was disappointed when

I received a letter from the editor, E. S. Pearson, stating that the paper was too mathematical for *Biometrika* and the applications discussed were not significant enough for publication. I decided to split the paper into two parts. The part dealing with the general theory of orthogonal arrays I sent to the *Proceedings of the Edinburgh Mathematical Society.* The editor commented on the paper as "highly original" and published it (Rao, 1949). The part dealing with applications was sent to the *Journal of the Royal Statistical Society* where it was accepted without any revision and published (Rao, 1947). This is a brief account of the introduction of orthogonal arrays into the literature.

I was glad to see some early applications of orthogonal arrays in the construction of response surface designs by Box, error correcting codes by Joshi and Hamming, and later in industrial experimentation by Taguchi. In another direction, upon my advice Chakravarti (1956, 1961, 1963) studied partially balanced arrays in his Ph.D. dissertation, relaxing the restriction on the index of orthogonal arrays. Some of the current developments are described in a special issue of the *Journal of Statistical Planning and Inference* (Vol. 56, 1996) edited by N. M. Singhi.

Orthogonal arrays are likely to receive further applications in the future, and the book by Hedayat, Sloane and Stufken, which has many attractive features, will indeed play a valuable role in promoting future research.

C. R. Rao
Eberly Professor of Statistics
Pennsylvania State University

Contents

List of Symbols

Symbol	Meaning	Section
$D(r,c,s)$	difference scheme with r rows, c columns and s levels	6.1
$D_t(r,c,s)$	difference scheme with r rows, c columns, s levels and strength t	6.5
$f(N,s,t)$	maximal number of factors in an $OA(N,k,s,t)$	2.1
$F(k,s,t)$	minimal number of runs in an $OA(N,k,s,t)$	2.1
$GF(s)$	Galois field of order s	3.1, Appendix
$GF(s)^n$	n-tuples from $GF(s)$	3.1
I_n	$n \times n$ identity matrix	1
J_n	$n \times n$ matrix of 1's	1
k	number of factors	1
l.c.m.	least common multiple	
$MOLS(s,t)$	same as $POL(s,t)$	8.1
N	number of runs	1
$OA(N,k,s,t)$	orthogonal array with N runs, k factors, s levels and strength t	1
$OA(N,s^k,t)$	same as $OA(N,k,s,t)$	9.1
$OA_I(N,k,s,t)$	orthogonal array of Type I with N runs, k factors, s levels and strength t	6.5
$OA_{II}(N,k,s,t)$	orthogonal array of Type II with N runs, k factors, s levels and strength t	6.5
$POF(n,s,t)$	A set of t pairwise orthogonal $n \times n$ F-squares with s levels	8.2
$POL(s,t)$	A set of t pairwise orthogonal Latin squares of order s	8.1
s	number of levels	1
S	set of symbols or levels	1
t	strength of array	1
T	transpose of a matrix	1
$\mathbb{Z}_m$	integers modulo m	6.1
λ	index of array	1
1_n	column vector of n 1's	2.2
$\lceil x \rceil$	least integer not less than x	
$\lfloor x \rfloor$	greatest integer not exceeding x	
$\langle\ \rangle$	cyclically permute the enclosed symbols	5.2
$\langle\ \rangle_\beta$	shift and multiply by constant	5.2
$\otimes$	tensor product	6.2
$-$	complex conjugate	5.1

Chapter 1

Introduction

In the 1940's, in a series of seminal papers (Rao, 1946a, 1947, 1949), C. R. Rao introduced certain combinatorial arrangements with applications to statistics. Although Rao (1946a) at first considered only a subclass of these arrangements, the entire class became quickly known by their current name of orthogonal arrays (or OA's). Since their introduction, many prominent researchers have found a source of inspiration in this fascinating subject. Both statisticians and mathematicians can be credited with significant contributions to this field.

The diversity in background of the contributors to this subject shows its richness. On the negative side, this diversity has also caused the contributions to be published in numerous journals and to be presented in many different styles. With this book, we hope to create some order out of this chaos. A unified presentation of the available results and connections with related areas of research should be beneficial not only to those that have had some exposure to the subject, but especially to those who have not yet encountered the many mysteries that surround this area of research.

The prominent role that orthogonal arrays have played and continue to play in the design of experiments provided us with more than enough motivation to write this book. Its success will be complete if it manages to inspire some of its readers to tackle the many unsolved problems.

In this chapter we present some basic definitions and terminology and list some elementary properties of orthogonal arrays.

Let S be a set of s symbols or levels. The term "level" is used because in the chief application of these arrays, in the design of experiments, the symbols typically indicate the levels or settings of the factors or variables whose effects on a response of interest are to be studied. Usually we will denote the possible levels by $0, 1, \ldots, s-1$, and often we will interpret the levels as the elements of some special structure, such as a group or Galois field. But for the time being no such interpretation is required. Throughout the book, by an $a \times b$ array (or matrix) with entries from S we shall mean a collection of ab elements of S arranged in a rows and b columns with one element per row-column pair. Formally, we can then define an orthogonal array as follows.

Definition 1.1. An $N \times k$ array A with entries from S is said to be an *orthogonal array with s levels, strength t and index* λ (for some t in the range $0 \le t \le k$) if every $N \times t$ subarray of A contains each t-tuple based on S exactly λ times as a row.

In the important case when $\lambda = 1$ it is customary to say that the orthogonal array has *index unity*. The integers N, k, s, t and λ may be referred to as the parameters of the orthogonal array. We will denote such an array by $OA(N, k, s, t)$. It is unnecessary to mention λ in this notation, since it is determined by the other parameters — see Equation (1.1).

A less formal way of defining an orthogonal array is to say that it is an array with the property that in any t columns you see all of the t-tuples that it is possible to see there, and furthermore you see each of them equally often. In an array of index unity you see every possibility exactly once.

Some authors prefer to represent an orthogonal array as a $k \times N$ array rather than an $N \times k$ array. This requires fewer lines, but is less natural for statisticians who are accustomed — especially when using statistical software packages — to having the rows of the array represent the level combinations and the columns the factors. Our illustrations will sometimes show the transposed array, in order to save space, and when this is the case — as in Table 2.3 for instance — the word "transposed" will appear in the caption.

The notation $OA(N, k, s, t)$ for an orthogonal array is not uncommon, but neither is it universal. A number of other symbols have been used in the literature, among which the notation $L_N(s^k)$ occurs perhaps most frequently, for example in Taguchi and Wu (1985). Although compact, this notation is incomplete (since it omits the parameter t), and uses subscripts and superscripts which is awkward in complicated situations. Among other symbols found in the literature are L_N, $L_N(\lambda \times s^k)$, $L_{\lambda \times s^t}(k)$, s-$OA(t, k, \lambda)$, $OA(N, k, s, t) : \lambda$, $OA(N, s^k, t)$, $OA_\lambda(t, k, s)$ and $OA(s, k, \lambda)$. We recommend that $OA(N, k, s, t)$ be used henceforth as the standard abbreviation.

The words "strength" and "index" are commonly accepted for referring to the parameters t and λ. However, various names appear in the literature for the other three parameters. The number of rows N is also known as the size of the array, the number of runs (or observations), the number of assemblies, or the number of level or treatment combinations. The number of columns k is also called the number of constraints, or the number of factors or variables, while s is called the number of symbols or the number of levels. The terminology "an N-run orthogonal array of strength t and index λ with k factors each at s levels" is also used. We shall endeavor to be consistent and refer to the parameters N, k and s as the number of runs, the number of factors and the number of levels, respectively.

Example 1.2. The array in Table 1.3 is an orthogonal array based on two levels, with strength three, of index unity, with eight runs and with four factors. It is an $OA(8, 4, 2, 3)$. The reader is invited to verify this. ■

Table 1.3. An $OA(8, 4, 2, 3)$.

$$\begin{array}{cccc} 0&0&0&0\\ 0&0&1&1\\ 0&1&0&1\\ 0&1&1&0\\ 1&0&0&1\\ 1&0&1&0\\ 1&1&0&0\\ 1&1&1&1 \end{array}$$

Example 1.4. Every $N \times k$ array has strength 0. Orthogonal arrays of strength 1 can be constructed trivially, as this example of an $OA(2, 12, 2, 1)$ shows:

$$\begin{array}{cccccccccccc} 0&0&0&0&0&0&0&0&0&0&0&0\\ 1&1&1&1&1&1&1&1&1&1&1&1 \end{array}$$

■

Example 1.5. Only slightly less trivial is this $OA(4, 3, 2, 2)$.

$$\begin{array}{ccc} 0&0&0\\ 0&1&1\\ 1&0&1\\ 1&1&0 \end{array}$$

■

Generally, the larger the desired strength, the harder it is to construct the array.

Note. The adjective "orthogonal" has many meanings. In common usage it means "perpendicular": one says that two vectors $u = (u_1, \ldots, u_m)$ and $v = (v_1, \ldots, v_m)$ are orthogonal if their inner product

$$u \cdot v = uv^T = u_1v_1 + \cdots + u_mv_m$$

is zero (T denotes transposition). However, an "orthogonal array" is *not* an array whose columns or rows are orthogonal in the above sense: there may not even be an appropriate definition of inner product. The adjective "orthogonal" has another interpretation in statistics that explains why Bush (1950) introduced the term "orthogonal array" — we will return to this in Chapter 11. Problems 1.8 and 1.9 give some indication of this interpretation.

In subsequent chapters we will derive various properties of orthogonal arrays, some of which are very general and some of which are only applicable under additional assumptions. The following properties, however, follow immediately from Definition 1.1. We leave their verification to the reader (Problem 1.1).

1. The parameters of an orthogonal array satisfy the equality

$$\lambda = N/s^t \ . \tag{1.1}$$

2. Any orthogonal array of strength t is also an orthogonal array of strength t', $0 \le t' < t$. The index of the array when considered as an array of strength t' is $\lambda s^{t-t'}$, where λ denotes the index of the array when considered to have strength t.

   Describing an array of strength 4 (say) as having strength 2 is rather like describing a 300 horse-power automobile as having 10 horse-power. It is technically correct, but possibly misleading!

3. If A_i, $i = 1, \ldots, r$, is an $OA(N_i, k, s, t_i)$ then the array A obtained from the juxtaposition of these r arrays,

$$A = \begin{bmatrix} A_1 \\ A_2 \\ \vdots \\ A_r \end{bmatrix} ,$$

   is an $OA(N, k, s, t)$, where $N = N_1 + \cdots + N_r$ and the strength is t for some $t \ge \min\{t_1, \ldots, t_r\}$. Further, when $r = s$ and each A_i is an $OA(N, k, s, t)$, after appending a 1 to each row of A_1, a 2 to each row of A_2, and so on, we obtain an $OA(sN, k+1, s, t)$.

4. A permutation of the runs or factors in an orthogonal array results in an orthogonal array with the same parameters.

5. A permutation of the levels of any factor in an orthogonal array results in an orthogonal array with the same parameters.

6. Any $N \times k'$ subarray of an $OA(N, k, s, t)$ is an $OA(N, k', s, t')$, where $t' = \min\{k', t\}$.

7. Taking the runs in an $OA(N, k, s, t)$ that begin with 0 (or any other particular symbol) and omitting the first column yields an $OA(N/s, k-1, s, t-1)$. If we assume that these are the initial runs, the process can be represented by the following diagram:

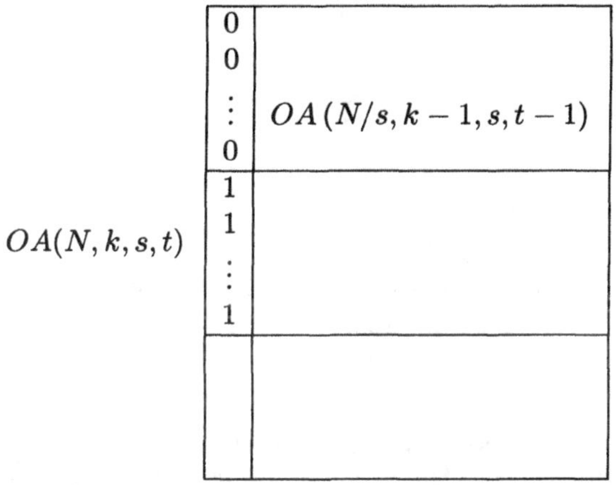

8. Let $\mathcal{C}$ be the set of all possible runs that could have occurred in a particular orthogonal array A, and for $c \in \mathcal{C}$ let f_c be the frequency of c in A. Let f denote the maximal f_c over all $c \in \mathcal{C}$. Then the array which contains run c with frequency $f - f_c$, for all $c \in \mathcal{C}$, is said to be the *set-theoretic complement*, or simply the *complement*, of A. The complement of an $OA(N, k, s, t)$ is an $OA(fs^k - N, k, s, t)$.

9. More generally, suppose $A = \begin{bmatrix} A_1 \\ A_2 \end{bmatrix}$ is an $OA(N, k, s, t)$, where A_1 is itself an $OA(N_1, k, s, t_1)$. Then A_2 is an $OA(N - N_1, k, s, t_2)$ with $t_2 \geq \min\{t, t_1\}$.

Properties 7 and 9 are both converses to Property 3.

We now turn to an issue where mathematical and statistical interests seem to be in conflict. One of the first questions a mathematician will consider, when investigating a new combinatorial structure, is the number of essentially distinct structures of a given type. In light of this, we make the following definition.

Definition 1.6. Two orthogonal arrays are said to be *isomorphic* if one can be obtained from the other by a sequence of permutations of the columns, the rows, and the levels of each factor.

Example 1.7. The arrays in Tables 1.3 and 1.8 are isomorphic. In Problem 2.15 we will see that any two orthogonal arrays with these parameters are isomorphic. ■

Table 1.8. Another $OA(8,4,2,3)$.

1	0	0	0
0	1	0	0
0	0	1	0
0	0	0	1
0	1	1	1
1	0	1	1
1	1	0	1
1	1	1	0

However, the main question of interest to a statistician is whether two arrays consist of the same runs, possibly in a different order. Only then are the arrays equivalent when used in an experiment, as will be explained in Chapter 11. Therefore we make a second definition.

Definition 1.9. Two orthogonal arrays are said to be *statistically equivalent* if one can be obtained from the other by a permutation of the runs.

Obviously the arrays in Tables 1.3 and 1.8 are not statistically equivalent.

Orthogonal arrays may be regarded as special cases of some more general classes of arrays that are also "orthogonal" in the statistical sense. Three such generalizations are given in the following examples, and others will appear in later chapters.

Example 1.10. Let A be the 8×5 array in Table 1.11. This is a *mixed* or *asymmetrical orthogonal array* of strength 2.

Table 1.11. A mixed orthogonal array.

0	0	0	0	0
0	1	1	1	1
1	0	1	0	1
1	1	0	1	0
2	0	0	1	1
2	1	1	0	0
3	0	1	1	0
3	1	0	0	1

Although not all factors have the same number of levels, the array still has the property that in any two columns all possible pairs occur as rows with the same frequency. However, the number of possible pairs depends on which columns we are considering. We will return to these arrays in Chapter 9. ■

Example 1.12. Let A be the 9×4 array in Table 1.13. All four factors are now at 2 levels, but since N is odd this cannot even be an orthogonal array of strength 1.

Table 1.13. An orthogonal main-effects plan.

0	0	0	0
0	0	0	0
0	1	0	0
0	0	1	0
0	0	0	1
0	1	1	1
1	1	0	0
1	0	1	0
1	0	0	1

However, this array does have a slightly different orthogonality property. For example, in the first two factors the combinations $(0,0)$, $(0,1)$, $(1,0)$ and $(1,1)$ occur respectively $4, 2, 2$ and 1 times as a row. These numbers are proportional to the products of the frequencies of the corresponding levels for the individual factors, which in this case are respectively $36, 18, 18$ and 9 (since level 0 occurs with frequency 6 in each factor and level 1 with frequency 3). The same property holds for any other pair of columns. Such an array may be called an *orthogonal effects plan of strength* 2, and is also known as an *orthogonal main-effects plan.* We return to these arrays in Chapter 11. ■

Example 1.14. The array in Table 1.15 combines the features of the arrays in Examples 1.10 and 1.12. It may be called a *mixed orthogonal effects plan of strength* 2 or a *mixed orthogonal main-effects plan.*

Table 1.15. A mixed orthogonal main-effects plan.

0	0	0	0	0
0	1	1	1	1
0	0	0	1	1
0	1	1	0	0
1	0	1	0	1
1	1	0	1	0
2	0	1	1	0
2	1	0	0	1

More about these arrays will be found in Chapter 11. ■

Although the arrays in these three examples are interesting in their own right, the main body of this book will treat orthogonal arrays as defined in

Definition 1.1. We will be concerned with the more general arrays because of their statistical applications.

1.1 Problems

1.1. Verify the validity of the nine properties of orthogonal arrays listed in this chapter.

1.2. To save paper we have transposed the following arrays. Complete each of them, if possible, to an $OA(16, 4, 4, 2)$.

a.

$$\begin{array}{cccccccccccccc}
0 & 0 & 3 & 2 & 2 & 3 & 3 & 2 & 1 & 0 & 0 & 1 & 1 & 2 \\
2 & 1 & 3 & 1 & 3 & 0 & 2 & 2 & 3 & 3 & 0 & 1 & 0 & 0 \\
2 & 1 & 0 & 3 & 1 & 3 & 1 & 0 & 2 & 3 & 0 & 0 & 1 & 2 \\
2 & 1 & 2 & 2 & 0 & 1 & 3 & 1 & 1 & 3 & 0 & 3 & 2 & 3
\end{array}$$

b.

$$\begin{array}{cccccccccccccc}
2 & 2 & 0 & 0 & 1 & 3 & 3 & 2 & 3 & 2 & 0 & 0 & 1 & 1 \\
0 & 3 & 1 & 0 & 0 & 3 & 2 & 1 & 0 & 2 & 2 & 3 & 1 & 3 \\
1 & 2 & 2 & 0 & 2 & 0 & 2 & 3 & 3 & 0 & 1 & 3 & 0 & 1 \\
3 & 2 & 1 & 2 & 0 & 0 & 3 & 0 & 1 & 1 & 0 & 3 & 3 & 2
\end{array}$$

c.

$$\begin{array}{cccccccccccccccc}
1 & 1 & 0 & 0 & 3 & 2 & 0 & 0 & 2 & 2 & 3 & 3 & 3 & 2 & 1 & 1 \\
1 & 3 & 3 & 2 & 0 & 0 & 1 & 0 & 1 & 3 & 3 & 2 & 1 & 2 & 0 & 2 \\
2 & 3 & 1 & 3 & 3 & 1 & 0 & 2 & 3 & 2 & 0 & 2 & 1 & 0 & 0 & 1
\end{array}$$

d.

$$\begin{array}{cccccccccccccccc}
0 & 1 & 2 & 1 & 1 & 0 & 3 & 2 & 3 & 2 & 2 & 1 & 0 & 3 & 3 & 0 \\
2 & 3 & 0 & 2 & 1 & 0 & 0 & 3 & 1 & 2 & 1 & 0 & 3 & 2 & 3 & 1 \\
1 & 0 & 1 & 3 & 1 & 0 & 3 & 2 & 0 & 0 & 3 & 2 & 3 & 2 & 1 & 2
\end{array}$$

e.

$$\begin{array}{ccccccccccccccc}
1 & 2 & 0 & 0 & 0 & 0 & 2 & 3 & 1 & 1 & 1 &   & 3 & 3 &   \\
2 & 1 & 1 & 0 &   & 2 & 3 & 1 & 0 &   & 1 & 0 & 3 & 2 & 0 \\
1 & 0 & 3 & 1 & 2 & 0 & 1 & 1 & 0 & 3 & 2 &   & 0 & 2 &   \\
0 & 3 & 0 & 3 &   & 1 & 1 & 2 & 2 &   & 1 & 0 & 0 & 3 & 1
\end{array}$$

1.3. Among the arrays in Problem 1.2 that can be completed to an orthogonal array $OA(16, 4, 4, 2)$, which can be completed to an $OA(16, 5, 4, 2)$? (Perhaps you should consider the more general question of whether every $OA(16, 4, 4, 2)$ can be completed to an $OA(16, 5, 4, 2)$.)

1.4. Identify the parameters of the following orthogonal array.

$$\begin{array}{ccccc}
0 & 0 & 0 & 0 & 0 \\
0 & 0 & 1 & 1 & 1 \\
0 & 1 & 0 & 1 & 0 \\
0 & 1 & 1 & 0 & 1 \\
1 & 0 & 0 & 1 & 1 \\
1 & 0 & 1 & 0 & 0 \\
1 & 1 & 0 & 0 & 1 \\
1 & 1 & 1 & 1 & 0
\end{array}$$

1.5. Does the orthogonal array in Problem 1.4 contain a subarray that forms an orthogonal array of higher strength?

1.6. Give an example of an $OA(16, 4, 2, 2)$.

1.7. (i) Give an example of an $OA(16, 4, 2, 3)$. (ii) Give an example of an $OA(16, 4, 2, 1)$ with a subarray $OA(8, 4, 2, 1)$ which when removed leaves an $OA(8, 4, 2, 3)$. This shows that in Property 9 the strength of A_2 may exceed $\min\{t, t_1\}$.

1.8. This problem introduces some notation that will be used throughout the book. Let 1_N denote a column vector of N 1's, let I_N denote an $N \times N$ identity matrix, and let J_N denote an $N \times N$ matrix of 1's.

Suppose A is an $N \times k$ matrix with entries $+1$ or -1. Show that the following are equivalent.

(i) A is an $OA(N, k, 2, 2)$.

(ii) $X^T X = N I_{k+1}$, where $X = [1_N \; A]$.

(iii)

$$A^T \left(I_N - \frac{1}{N} J_N \right) A = N I_k \ .$$

1.9. Let A be an $N \times k$ array with entries ± 1, and let $A^{(2)}$ be the $N \times \binom{k}{2}$ matrix whose columns are the componentwise products of pairs of columns of A. If $A = (a_{ij})$, then $A^{(2)}$ has a column

$$\begin{bmatrix} a_{1u} a_{1v} \\ \vdots \\ a_{Nu} a_{Nv} \end{bmatrix}$$

for every pair u, v with $1 \le u < v \le k$. Let $Y = [1_N \; A^{(2)}]$ and let $P_Y = Y(Y^T Y)^- Y^T$, where $-$ denotes a generalized inverse. Show that A is an $OA(N, k, 2, 3)$ if and only if

$$A^T (I - P_Y) A = N \; I_k \ .$$

Chapter 2

Rao's Inequalities and Improvements

If the other parameters of an orthogonal array are specified, there is a limit on the number of possible factors, imposed by the defining conditions. We shall discuss these restrictions in this chapter and in Chapter 4. Section 2.2 presents the celebrated inequalities found by Rao (1947). Section 2.3 discusses improvements on Rao's bounds for orthogonal arrays of strength two and three. Results on improvements for arrays of general strength are contained in Section 2.4, while Section 2.5 pays special attention to orthogonal arrays in which all the factors are at two levels.

2.1 Introduction

An important problem in the study of orthogonal arrays is to determine the minimal number of runs N in any $OA(N,k,s,t)$, for given values of k,s and t. We denote this minimal value by $F(k,s,t)$.

A related problem, which approaches the existence question for orthogonal arrays in a slightly different way, may be formulated as follows. As we observed in Chapter 1, by deleting factors from an $OA(N,k,s,t)$ we may obtain an $OA(N,k',s,t)$ for any k' with $t \leq k' \leq k$. So for fixed values of N,s and t the problem of determining all values of k for which an $OA(N,k,s,t)$ exists can be solved if we know the maximal number of factors k in any $OA(N,k,s,t)$. We denote this maximal value by $f(N,s,t)$.

A moment's thought shows that $F(k,s,t)$ and $f(N,s,t)$ are related in the following way:

$$F(k,s,t) = \min\{N : f(N,s,t) \geq k\} , \tag{2.1}$$
$$f(N,s,t) \leq \max\{k : F(k,s,t) \leq N\} . \tag{2.2}$$

The values of $f(N,s,t)$ therefore completely determine those of $F(k,s,t)$, although the converse is not true. The values of $F(k,s,t)$ only provide upper bounds on the values of $f(N,s,t)$, and so determining $f(N,s,t)$ is a more difficult problem than determining $F(k,s,t)$.

The present chapter is primarily concerned with obtaining upper bounds on $f(N,s,t)$, while in Chapter 4 we give lower bounds on $F(k,s,t)$. Of course the orthogonal arrays constructed throughout the book imply lower bounds on $f(N,s,t)$ and upper bounds on $F(k,s,t)$. By the end of the book we will have determined the exact values of these two functions in a few special cases (see the summary in Chapter 12). However, the complete determination of $F(k,s,t)$ and $f(N,s,t)$ is a problem beyond the present capabilities of science.

For the remainder of this chapter we study the function $f(N,s,t)$. If N is not a multiple of s^t, that is if $N \not\equiv 0 \pmod{s^t}$, we take $f(N,s,t)$ to be 0. Also $f(N,s,0)$ and $f(\lambda s,s,1)$ are infinite. We are therefore mostly interested in the values of $f(N,s,t)$ for $t \geq 2$ and $N = \lambda s^t$ for some $\lambda \geq 1$. If $N \equiv 0 \pmod{s^t}$, we can immediately obtain the inequality $f(N,s,t) \geq t+1$. To see this we simply construct an $OA(N,t+1,s,t)$ for $t \geq 2$. Start with an $N \times t$ array which has each of the s^t possible t-tuples N/s^t times as a row. Considering the levels $0,1,\ldots,s-1$ as the residue classes modulo s, add the entries in each row and let the negative of this sum be the level for a new factor. Problem 2.1 invites the reader to verify that this results in an $OA(N,t+1,s,t)$. Since this array has the property that the levels in every run add to zero, it is called a *zero-sum array.* The array in Table 1.3 is an example of a zero-sum array.

In general, however, we can expect that more than $t+1$ factors can be accommodated. The remainder of this chapter will provide upper bounds for $f(N,s,t)$. Some of these will actually give the exact value for $f(N,s,t)$, but in most cases we won't be able to address this aspect until later chapters, when various methods of constructing orthogonal arrays are considered.

2.2 Rao's Inequalities

One of the first upper bounds on the maximal number of factors in an orthogonal array was obtained by Rao (1947). This is a remarkably general result since it applies to any values of N, s and t. (From now on we restrict our attention to the only interesting case, $N \equiv 0 \pmod{s^t}$.)

In this chapter we give a purely algebraic and rather non-intuitive proof of Rao's result; an interpretation of the right-hand side in (2.3) (in terms of orthogonally estimable functions) will be given in Chapter 11. The bounds for the number of factors are given implicitly and, in general, no explicit form is known. For $t = 2$ the result was already known from the work of Plackett and Burman (1946).

Theorem 2.1. *The parameters of an $OA(N, k, s, t)$ satisfy the following inequalities:*

$$N \geq \sum_{i=0}^{u} \binom{k}{i} (s-1)^i , \quad \text{if } t = 2u , \tag{2.3}$$

$$N \geq \sum_{i=0}^{u} \binom{k}{i} (s-1)^i + \binom{k-1}{u}(s-1)^{u+1} , \quad \text{if } t = 2u+1 , \tag{2.4}$$

for $u \geq 0$.

Proof: (May be skipped at the first reading.) Let $A = (a_{ij})$, $a_{ij} \in S$, $i = 1, \ldots, N$, $j = 1, \ldots, k$, be an $OA(N, k, s, t)$. We consider $S = \{0, 1, \ldots, s-1\}$ as a subset of the real numbers. Let M denote the expression on the right-hand side of either (2.3) or (2.4), depending on whether t is even or odd. We will use A to construct an $N \times M$ matrix H of rank M. That will obviously imply $M \leq N$, and establish the theorem.

The matrix H that we construct will not only be of rank M, but will have the property that $H^T H$ is an $M \times M$ diagonal matrix. To begin, let B be an $(s-1) \times s$ matrix with pairwise orthogonal rows, all of which are orthogonal to 1_s^T, the $1 \times s$ vector with all entries equal to 1. It will be convenient to label the rows of B by $1, 2, \ldots, s-1$, and its columns by $0, 1, \ldots, s-1$. Then, if $b(i,j)$ denotes the entry of B in position (i,j), we have

$$\sum_{j=0}^{s-1} b(i,j) b(i',j) = 0, \quad i \neq i' ,$$

and

$$\sum_{j=0}^{s-1} b(i,j) = 0 .$$

Case 1: $t = 2u$. For each $m \in \{1, 2, \ldots, u\}$ and each ordered m-tuple $(i_1, i_2, \ldots, i_m)$, where $i_j \in \{1, 2, \ldots, s-1\}$, we define an $N \times \binom{k}{m}$ matrix $H(i_1, i_2, \ldots, i_m)$. We start by labeling its rows and columns. The columns are labeled by the $\binom{k}{m}$ m-subsets of $\{1, 2, \ldots, k\}$, and the rows by $1, 2, \ldots, N$. An entry of $H(i_1, i_2, \ldots, i_m)$ is denoted by $h_{j,(\ell_1,\ell_2,\ldots,\ell_m)}^{(i_1,i_2,\ldots,i_m)}$, where the subscripts refer to the corresponding row and column labels, respectively. To avoid any ambiguity, we assume that the ℓ_i's are ordered so that $\ell_1 < \ell_2 < \cdots < \ell_m$. The entries of $H(i_1, i_2, \ldots, i_m)$ are now defined by using the orthogonal array A and the matrix B as follows:

$$h_{j,(\ell_1,\ell_2,\ldots,\ell_m)}^{(i_1,i_2,\ldots,i_m)} = b(i_1, a_{j\ell_1})b(i_2, a_{j\ell_2})\cdots b(i_m, a_{j\ell_m}) .$$

Finally, the matrix H is given by

$$H = [1_N, H(1), \ldots, H(s-1), H(1,1), H(1,2), \ldots, H(s-1, s-1, \ldots, s-1)] .$$

After a column of ones its building blocks are the matrices $H(i_1, i_2, \ldots, i_m)$, with $m \in \{1, 2, \ldots, u\}$ and $i_j \in \{1, 2, \ldots, s-1\}$. Clearly H has $1 + \binom{k}{1}(s-1) + \cdots + \binom{k}{u}(s-1)^u = M$ columns, and is thus an $N \times M$ matrix.

Case 2: $t = 2u+1$. Start with the matrix H as constructed in case 1. Further, for each $(u+1)$-tuple $(i_1, i_2, \ldots, i_{u+1})$ based on $\{1, 2, \ldots, s-1\}$ define an $N \times \binom{k-1}{u}$ matrix $\overline{H}(i_1, i_2, \ldots, i_{u+1})$ by labeling its columns as $(1, \ell_2, \ldots, \ell_{u+1})$, $1 < \ell_2 < \cdots < \ell_{u+1} \le k$, its rows as $1, 2, \ldots, N$, and by defining its entries by

$$h_{j,(1,\ell_2,\ldots,\ell_{u+1})}^{(i_1,i_2,\ldots,i_{u+1})} = b(i_1, a_{j1})b(i_2, a_{j\ell_2})\cdots b(i_{u+1}, a_{j\ell_{u+1}}) .$$

We obtain an $N \times M$ matrix for case 2 by appending the columns of these $(s-1)^{u+1}$ matrices to the columns of the matrix H constructed in case 1.

All columns of these $N \times M$ matrices are nonzero (Problem 2.3). It remains to be shown that the columns are orthogonal. Clearly, for $m \le u+1$,

$$\begin{aligned}
\sum_{j=1}^{N} h_{j,(\ell_1,\ldots,\ell_m)}^{(i_1,\ldots,i_m)} &= \sum_{j=1}^{N} (b(i_1, a_{j\ell_1})\cdots b(i_m, a_{j\ell_m})) \\
&= N\left(\sum_{j_1=0}^{s-1}\cdots\sum_{j_m=0}^{s-1}(b(i_1,j_1)\cdots b(i_m,j_m))\right)\Big/ s^m \\
&= N\left(\sum_{j_1=0}^{s-1}\cdots\sum_{j_{m-1}=0}^{s-1}(b(i_1,j_1)\cdots b(i_{m-1},j_{m-1}))\left(\sum_{j_m=0}^{s-1} b(i_m,j_m)\right)\right)\Big/ s^m \\
&= 0 ,
\end{aligned}$$

where we use the fact that any m columns of the orthogonal array A, $m \le t$, form an orthogonal array of strength m and index N/s^m. This shows that the columns of $H(i_1, \ldots, i_m)$ and $\overline{H}(i_1, \ldots, i_{u+1})$ are orthogonal to 1_N.

The inner product of any other two columns is of the form

$$\sum_{j=1}^{N} (b(i_1, a_{j\ell_1})\cdots b(i_m, a_{j\ell_m})b(f_1, a_{jk_1})\cdots b(f_n, a_{jk_n})) .$$

The basic argument that this vanishes is the same as before. From the construction of H, columns $\ell_1, \ldots, \ell_m$ and $k_1, \ldots, k_n$ in A contain at most t distinct columns. If there are c columns that occur in both the sequence of ℓ's and k's, say $\ell_{p_1} = k_{r_1}, \ldots, \ell_{p_c} = k_{r_c}$, then, with $\ell_{p_{c+1}}, \ldots, \ell_{p_m}$ and $k_{r_{c+1}}, \ldots, k_{r_n}$ as the remaining columns, the inner product reduces to

$$C \prod_{\nu=1}^{c} \left(\sum_{j=0}^{s-1} b(i_{p_\nu}, j) b(f_{r_\nu}, j) \right) \prod_{\nu=c+1}^{m} \left(\sum_{j=0}^{s-1} b(i_{p_\nu}, j) \right) \prod_{\nu=c+1}^{n} \left(\sum_{j=0}^{s-1} b(f_{r_\nu}, j) \right)$$

where $C = N/s^{m+n-c}$.

The properties of the matrix B guarantee that this expression vanishes, unless $m = n = c$ and $i_1 = f_1, \ldots, i_c = f_c$. But this cannot occur since we are considering two distinct columns of H. ■

The following example illustrates the construction of the matrix H in the preceding proof.

Example 2.2. Let A be the $OA(9, 4, 3, 2)$ in Table 2.3. Select B to be

$$B = \begin{pmatrix} -1 & 0 & 1 \\ 1 & -2 & 1 \end{pmatrix}.$$

Table 2.3. An $OA(9, 4, 3, 2)$ (transposed).

0	0	0	1	1	1	2	2	2
0	1	2	0	1	2	0	1	2
0	1	2	2	0	1	1	2	0
0	1	2	1	2	0	2	0	1

The matrix H will be a 9×9 matrix. One of its components is the 9×4 matrix $H(1)$, the transpose of which is

$$\begin{pmatrix} -1 & -1 & -1 & 0 & 0 & 0 & 1 & 1 & 1 \\ -1 & 0 & 1 & -1 & 0 & 1 & -1 & 0 & 1 \\ -1 & 0 & 1 & 1 & -1 & 0 & 0 & 1 & -1 \\ -1 & 0 & 1 & 0 & 1 & -1 & 1 & -1 & 0 \end{pmatrix}.$$

Another component is the matrix $H(2)$ whose transpose is

$$\begin{pmatrix} 1 & 1 & 1 & -2 & -2 & -2 & 1 & 1 & 1 \\ 1 & -2 & 1 & 1 & -2 & 1 & 1 & -2 & 1 \\ 1 & -2 & 1 & 1 & 1 & -2 & -2 & 1 & 1 \\ 1 & -2 & 1 & -2 & 1 & 1 & 1 & 1 & -2 \end{pmatrix}.$$

All the columns of $H(1)$ and $H(2)$ are pairwise orthogonal and are orthogonal to 1_9. ■

It is worth remarking that the Rao bound for odd t, (2.4), can be deduced from the bound for even t, (2.3). Suppose A is an $OA(N, k, s, 2u+1)$. By Property 7 of Chapter 1 there exists an $OA(N/s, k-1, s, 2u)$. From (2.3) we obtain

$$N/s \geq \sum_{i=0}^{u} \binom{k-1}{i} (s-1)^i ,$$

and rewriting $s = (s-1)+1$ leads to

$$\begin{aligned} N &\geq \sum_{i=0}^{u} \binom{k-1}{i} (s-1)^{i+1} + \sum_{i=0}^{u} \binom{k-1}{i} (s-1)^i \\ &= 1 + \sum_{i=1}^{u} \binom{k}{i} (s-1)^i + \binom{k-1}{u} (s-1)^{u+1} , \end{aligned}$$

which is (2.4).

The bounds in Theorem 2.1 are sharp in the sense that for both odd and even strengths there are infinitely many orthogonal arrays for which equality holds. An example of an array for which equality holds in (2.4) is that in Table 1.8. As we saw in Section 2.1, an $OA(\lambda s^t, t+1, s, t)$ exists for any λ, s and t. Taking $\lambda = 1$ and $s = 2$ we find that $f(2^t, 2, t) \geq t+1$. It is however easy to show that these parameters give equality in Rao's bounds for any t (see Problem 2.4). We conclude that $f(2^t, 2, t) = t+1$ for all $t \geq 2$.

An interesting but difficult problem is the determination of all parameters for which equality occurs in Rao's inequalities. Orthogonal arrays that achieve these bounds are called *tight* (or sometimes *complete*, although that is an overworked term). Noda (1979) obtained a characterization of the parameters of tight orthogonal arrays of strength 4. We state his result without proof.

Theorem 2.4. *The parameters of a tight $OA(N, k, s, 4)$ are of one of the following three types:*

a. $N = 16$, $k = 5$ *and* $s = 2$,

b. $N = 243$, $k = 11$ *and* $s = 3$, *or*

c. $N = 9m^2(9m^2-1)/2$, $k = (9m^2+1)/5$ *and* $s = 6$, *where* $m \equiv 0 \pmod 3$, $m \equiv \pm 1 \pmod 5$ *and* $m \equiv 5 \pmod{16}$.

The existence of an array with the parameters of case *a* was demonstrated in Section 2.1. An array with the parameters of case *b* also exists, as will be seen in Chapters 3 and 4. It is not known if any arrays exist with the parameters of case *c*. The smallest example of an array in case *c* would be an $OA(7874496, 794, 6, 4)$, corresponding to $m = 21$.

Theorem 2.4 can easily be strengthened (see Problem 2.5) to obtain a list of possible parameters for tight orthogonal arrays of strength 5. It takes more effort to show that these parameters can be further reduced, as was done by Noda (1986). This author also determined a list of possible parameters for tight orthogonal arrays of strength 3. Kageyama (1988) studied tight 2-symbol orthogonal arrays. The proof of his Lemma 2.3 contains an error, however, which invalidates the proof of his Theorem 1.1. The correctness of this theorem is still in doubt. See also Mukerjee and Kageyama (1994).

For orthogonal arrays of strength 2 or 3 we can easily obtain explicit bounds for $f(N, s, t)$ from Theorem 2.1. We formulate these here in the form of two corollaries.

Corollary 2.5. *In an $OA(\lambda s^2, k, s, 2)$ the inequality*

$$k \leq (\lambda s^2 - 1)/(s - 1) \tag{2.5}$$

must hold.

Corollary 2.6. *In an $OA(\lambda s^3, k, s, 3)$ the inequality*

$$k \leq (\lambda s^2 - 1)/(s - 1) + 1 \tag{2.6}$$

must hold.

Corollary 2.6 can also be deduced directly from Corollary 2.5 by the argument given just after Example 2.2. We will also see that the bounds in the two corollaries can be improved if $\lambda - 1 \not\equiv 0 \pmod{s-1}$. This condition characterizes those cases in which the right-hand sides in (2.5) and (2.6) are not integral.

2.3 Improvements on Rao's Bounds for Strength Two and Three

Rao's inequalities apply to any $OA(N, k, s, t)$. When considering specific values for one or more of the parameters or when considering some functional relation between the parameters, it is quite possible that the upper bound for the number of factors in an orthogonal array can be sharpened. Various results of this type are discussed in this and the following sections.

This section presents the results of Bose and Bush (1952), who sharpened Rao's bounds for some parameters when $t = 2$ or 3. We start with a lemma that is useful for proving the nonexistence of orthogonal arrays with certain parameters.

Lemma 2.7. *For a fixed run $u = (u_1, \ldots, u_k)$ in an $OA(\lambda s^t, k, s, t)$, let $A_i(u)$, $0 \le i \le k$, denote the number of runs $v = (v_1, \ldots, v_k)$ in the array with exactly i factors whose levels differ from those of u, i.e. such that $u_a \ne v_a$ for exactly i values of the subscript a. Then the following equalities hold:*

$$\sum_{i=h}^{k} \binom{i}{h} A_{k-i}(u) = \lambda s^{t-h} \binom{k}{h}, \quad \textit{for} \;\; 0 \le h \le t \; ,$$

independently of the choice of u.

Proof: Possibly after renaming the levels for some of the factors, we may take the fixed run to be $u = (0, 0, \ldots, 0)$. For $h = 0$, the alleged equality states that $\sum_{i=0}^{k} A_i(u) = \lambda s^t$, which is true since both sides are equal to N. For $1 \le h \le t$ the validity of the equality follows by counting the different h-tuples of zeros in the runs. Since the entire array is an orthogonal array of strength h and index λs^{t-h}, it follows that for any $N \times h$ subarray there are λs^{t-h} runs with all h factors at level 0. Since there are $\binom{k}{h}$ possible choices for the $N \times h$ subarray, there are $\lambda s^{t-h} \binom{k}{h}$ different h-tuples of zeros in the $N \times k$ array. On the other hand, any run in the $N \times k$ array with i zeros, $i \ge h$, contains $\binom{i}{h}$ h-tuples of zeros. Since there are $A_{k-i}(u)$ runs with i zeros, it follows that the number of different h-tuples of zeros in the $N \times k$ array must also be equal to $\sum_{i=h}^{k} \binom{i}{h} A_{k-i}(u)$. The result now follows. ∎

A far-reaching generalization of this result will be given in Theorem 4.9.

To formulate the results of Bose and Bush (1952), the following notation will be used. Let $\lambda - 1 = a(s-1) + b$, where $0 \le b \le s - 2$. Let $\theta = ((1 + 4s(s-1-b))^{1/2} - (2s - 2b - 1))/2$. Observe that (2.5) is now equivalent to $k \le \lambda(s+1) + a$. If $b > 0$, this can be improved as follows.

Theorem 2.8. *In an $OA(\lambda s^2, k, s, 2)$, if $\lambda - 1 \not\equiv 0 \pmod{s-1}$ the inequality*

$$k \le \lambda(s+1) + a - \lfloor \theta \rfloor - 1$$

must hold. (Here $\lfloor \theta \rfloor$ denotes the largest integer not exceeding θ.)

Proof: Let us consider any run u, and define

$$\Phi(x) = (k-x)(k-x-1)(A_0(u) - 1) + \sum_{i=0}^{k-1} (i-x)(i-x-1) A_{k-i}(u) \; ,$$

where the $A_i(u)$ are as in Lemma 2.7. Since $i - x$ and $i - x - 1$ are consecutive integers, $\Phi(x) \ge 0$ for integral values of x. By applying Lemma 2.7 we may write this, for integral x, as

$$0 \le k(k-1)(\lambda - 1) - 2kx(\lambda s - 1) + x(x+1)(\lambda s^2 - 1) \; ,$$

or

$$\lambda D \geq k(k-1) - 2kx + x(x+1) \;, \tag{2.7}$$

where $D = k(k-1) - 2kxs + x(x+1)s^2$.

For integral x, the right-hand side of (2.7) has a minimum of 0 at $x = k-1$ and at $x = k$. It follows now easily from (2.7) that $D > 0$ for integral x. The validity of the strict inequality for the cases $x = k-1$ and $x = k$ can be established by using the fact that $s > 1$.

Thus we have shown

$$\lambda \geq (k(k-1) - 2kx + x(x+1))/D \;,$$

or, equivalently, that

$$(\lambda s^2 - 1)/(s-1) \geq k(1 + (k-1-xs)(xs+s+1-k)/D) \;. \tag{2.8}$$

Observe that $(\lambda s^2 - 1)/(s-1) = \lambda(s+1) + a + b/(s-1)$. Now assume that the orthogonal array has $k = \lambda s + \lambda + a - n$ factors for some nonnegative integer n. We now select a particular value for x, namely $x = \lambda + a$. Then

$$\begin{aligned} D &= k(k-1) - 2kxs + x(x+1)s^2 \\ &= k(s-1) - (k-xs)(xs+s-k) \\ &= k(s-1) - (\lambda + a - n - as)(as + s - \lambda - a + n) \\ &= k(s-1) - (b-n+1)(s+n-b-1) \\ &< k(s-1) \;, \end{aligned}$$

where the last inequality holds provided $0 \leq n \leq b$. In that case, using (2.8), we have

$$(\lambda s^2 - 1)/(s-1) > k + (k - 1 - \lambda s - as)(\lambda s + as + s + 1 - k)/(s-1) \;,$$

which after some simple algebra reduces to

$$(b-n)(b+1-n) - s(b-2n) > 0 \;.$$

This last inequality is violated if $0 \leq n \leq \lfloor \theta \rfloor$. Notice that with such an n we indeed have $n \leq b$, as required for the inequality $D < k(s-1)$. Consequently, n must be at least $\lfloor \theta \rfloor + 1$, and k can be at most $\lambda s + \lambda + a - \lfloor \theta \rfloor - 1$. ∎

We point out that inequality (2.8) can also be used for an alternative proof of the result in Corollary 2.5. Problem 2.6 invites the reader to show this.

Example 2.9. From Theorem 2.1 we obtain $f(18,3,2) \leq 8$. Theorem 2.8 improves this to $f(18,3,2) \leq 7$. An $OA(18,7,3,2)$ can indeed be constructed and is exhibited in Table 2.10. Details of the construction of this array will be found in Section 3.3. ∎

Table 2.10. An $OA(18, 7, 3, 2)$ (transposed).

0	1	2	0	1	2	0	1	2	0	1	2	0	1	2	0	1	2
0	1	2	0	1	2	1	2	0	2	0	1	1	2	0	2	0	1
0	1	2	1	2	0	0	1	2	2	0	1	2	0	1	1	2	0
0	1	2	2	0	1	2	0	1	0	1	2	1	2	0	1	2	0
0	1	2	1	2	0	2	0	1	1	2	0	0	1	2	2	0	1
0	1	2	2	0	1	1	2	0	1	2	0	2	0	1	0	1	2
0	0	0	0	0	0	1	1	1	1	1	1	2	2	2	2	2	2

By arguments similar to those used in the proof of Theorem 2.8, Bose and Bush (1952) improved Rao's bounds for various orthogonal arrays of strength three. The notation is that of Theorem 2.8.

Theorem 2.11. *In an* $OA(\lambda s^3, k, s, 3)$, *if* $\lambda - 1 \not\equiv 0 \pmod{s-1}$, *then*

$$k \leq \lambda(s+1) + a - \lfloor \theta \rfloor .$$

Alternatively, the result in Theorem 2.11 can be obtained as a corollary of Theorem 2.8 and the following lemma. This lemma will also be useful in other arguments.

Lemma 2.12. *The maximal numbers of factors in orthogonal arrays of strengths* $t-1$ *and* t, $t \geq 2$, *are related by*

$$f(\lambda s^t, s, t) \leq f(\lambda s^{t-1}, s, t-1) + 1 .$$

Proof: The result will follow if we can show that an $OA(\lambda s^{t-1}, k-1, s, t-1)$ can be constructed from an $OA(\lambda s^t, k, s, t)$. But this is precisely what was done in Property 7 of Chapter 1. ∎

The result in Lemma 2.12 may also be written as

$$f(sN, s, t+1) \leq f(N, s, t) + 1, \quad t \geq 1 .$$

When $s = 2$ and t is even, a stronger result holds — see Corollary 2.25.

Example 2.13. For the $OA(8, 4, 2, 3)$ in Table 1.8, if we restrict our attention to the four runs in which the first factor is at level 0, and then delete the first factor, we are left with an $OA(4, 3, 2, 2)$ as exhibited in Table 2.14. ∎

Table 2.14. An $OA(4, 3, 2, 2)$.

1	0	0
0	1	0
0	0	1
1	1	1

Theorem 2.11 now follows immediately from Theorem 2.8 and Lemma 2.12. Also, as claimed in Section 2.2, Lemma 2.12 and Corollary 2.5 immediately imply Corollary 2.6.

Some further considerations in the spirit of those in the proof of Theorem 2.8 enabled Bose and Bush (1952) to formulate the following improvement to Corollary 2.6 when $\lambda - 1 \equiv 0 \pmod{s-1}$.

Theorem 2.15. *In an* $OA(\lambda s^3, k, s, 3)$, *if* $\lambda - 1 = a(s-1)$ *and* $(s-1)^2(s-2) \not\equiv 0 \pmod{as+2}$, *then*

$$k \leq \lambda(s+1) + a - 1 \ .$$

Proof: Assume that an orthogonal array with these parameters exists, but with $k = \lambda(s+1) + a$. Let u be any run, and let $n_i = A_{k-i}(u)$, $i = 0, \ldots, k-1$, $n_k = A_0(u) - 1$. An application of Lemma 2.7 shows that

$$\sum_{i=0}^{k} i(i-1-as)(i-2-as)n_i = 0 \ .$$

Hence n_i must be 0 unless $i = 0$, $as+1$ or $as+2$. The equations in Lemma 2.7 for $h = 1$ and $h = 2$ now reduce to

$$(as+1)n_{as+1} + (as+2)n_{as+2} = k(\lambda s^2 - 1)$$

and

$$(as+1)asn_{as+1} + (as+2)(as+1)n_{as+2} = k(k-1)(\lambda s - 1) \ ,$$

respectively.

Since $k - 1 = \lambda(s+1) + a - 1 = s(as+1)$, the second of these equations reduces to

$$asn_{as+1} + (as+2)n_{as+2} = ks(\lambda s - 1) \ .$$

Solving for n_{as+2} from the two equations yields

$$\begin{aligned} n_{as+2} &= s(s-1)k\lambda/(as+2) \\ &= k(s-1)^2 - s(s-1)(s-2) + (s-1)^2(s-2)/(as+2) \ . \end{aligned}$$

Since n_{as+2} must be an integer, we reach a contradiction if $(s-1)^2(s-2) \not\equiv 0 \pmod{as+2}$. Hence the result. ∎

Improvements to Corollary 2.5 for an $OA(\lambda s^2, k, s, 2)$ with $\lambda - 1 \equiv 0 \pmod{s-1}$ are less obvious. The condition $(s-1)^2(s-2) \not\equiv 0 \pmod{as+2}$ is certainly no longer the discriminating condition. In Chapter 3, for example, we will see that an $OA(27, 13, 3, 2)$ does exist. The number of factors in this array, 13, reaches the upper bound in Corollary 2.5, although $(s-1)^2(s-2) \not\equiv 0 \pmod{as+2}$.

Example 2.16. Consider $f(54,3,3)$. Corollary 2.6 implies that $f(54,3,3) \leq 9$, and Theorem 2.11 improves this to $f(54,3,3) \leq 8$. However, an $OA(54,8,3,3)$ does not exist. Recently, Hedayat, Seiden and Stufken (1997) established the nonexistence of an $OA(54,6,3,3)$. An $OA(54,5,3,3)$ can be constructed by taking the following 54 runs:

— two runs with all factors at level 1,
— the ten distinct runs with three factors at level 0 and two at level 1,
— the runs obtained from the preceding 12 runs by replacing 1's with 2's,
— the thirty distinct runs with one factor at level 0, two at level 1 and two at level 2.

Other examples of an $OA(54,5,3,3)$ can be found in Fujii, Namikawa and Yamamoto (1987). Hedayat, Seiden and Stufken (1997) showed that there are precisely four nonisomorphic arrays with these parameters. The existence of these arrays and the nonexistence of an $OA(54,6,3,3)$ imply $f(54,3,3) = 5$. Hedayat, Stufken and Su (1997) used this result to show that $f(2\cdot 3^t,3,t) = t+1$ if $t \geq 4$. ∎

Example 2.17. By Theorem 2.15 we obtain

$$f(81,3,3) \leq 12 .$$

This improves the bound from Corollary 2.6. However, further improvement is possible. Seiden (1955a,b) showed that $f(81,3,3) = 10$. ∎

Thus we see that although the general bounds are sometimes attainable, they can often be strengthened by special investigation of particular cases.

2.4 Improvements on Rao's Bounds for Arrays of Index Unity

Lemma 2.12 provides us with a simple tool for obtaining upper bounds on the maximal number of factors in orthogonal arrays of general strength $t \geq 4$ from those in arrays of strength $t = 3$. By repeated application of Lemma 2.12 we obtain for $t \geq 4$

$$f(\lambda s^t, s, t) \leq f(\lambda s^3, s, 3) + t - 3 . \tag{2.9}$$

We should not expect that this easily obtained bound will be very sharp, especially for large values of t, since for many parameters the inequality in Lemma 2.12 is strict.

Indeed, it can be verified that for large t the bounds from Rao's inequalities tend to be better than (2.9), even if we use the improved bounds from Section 2.3 for $f(\lambda s^3, s, 3)$.

It is rather remarkable therefore that there is an exception to this observation, namely when the array is of index unity and $s > t$. Under these conditions Hedayat and Stufken (1989a) showed that if we use Corollary 2.6 to bound $f(\lambda s^3, s, 3)$, (2.9) gives at least as strong a result as is obtained from Rao's inequalities.

Theorem 2.18. *If $s > t$, then the bound*

$$f(s^t, s, t) \leq s + t - 1$$

is at least as good as the bound for $f(s^t, s, t)$ obtained from Theorem 2.1.

Proof: It suffices to show that Rao's inequalities are satisfied if $k = s + t - 1$.

Case 1: $t = 2u$. We have to show that

$$s^{2u} \geq 1 + \binom{s+2u-1}{1}(s-1) + \cdots + \binom{s+2u-1}{u}(s-1)^u .$$

First, we observe that

$$\binom{u}{i}(s+1)^i \geq \binom{s+2u-1}{i}, \quad i = 0, 1, \ldots, u . \tag{2.10}$$

This is obvious for $i = 0$. For $1 \leq i \leq u - 1$, we can write (2.10) as

$$\prod_{j=0}^{i-1}(u-j)(s+1) \geq \prod_{j=0}^{i-1}(s + 2u - j - 1) .$$

Since $j \leq u - 2$ and $s > 2u$ it is easily seen that

$$(u-j)(s+1) \geq s + 2u - j - 1 .$$

Finally, if $i = u$, (2.10) reduces to

$$(s+1)^u \geq \binom{s+2u-1}{u} . \tag{2.11}$$

This holds since

$$u!(s+1)^u = u(s+1)^2 \prod_{i=2}^{u-1}(i(s+1)) \geq u(s+1)^2 \prod_{i=2}^{u-1}(s+2u-i) \geq \prod_{i=1}^{u}(s+2u-i) .$$

The proof for Case 1 is now completed by observing that

$$\begin{aligned} s^{2u} &= \sum_{i=0}^{u} \binom{u}{i}(s^2-1)^i = \sum_{i=0}^{u} \binom{u}{i}(s+1)^i(s-1)^i \qquad (2.12) \\ &\geq \sum_{i=0}^{u} \binom{s+2u-1}{i}(s-1)^i . \end{aligned}$$

Case 2: $t = 2u+1$. We will again use (2.11), and, based on (2.10), the inequality

$$\binom{u}{i-1}(s+1)^{i-1} + \binom{u}{i}(s+1)^i \geq \binom{s+2u-1}{i-1} + \binom{s+2u-1}{i} = \binom{s+2u}{i} .$$

As in (2.12) we now obtain

$$\begin{aligned} s^{2u+1} &= s\sum_{i=0}^{u} \binom{u}{i}(s+1)^i(s-1)^i \\ &= \sum_{i=0}^{u} \binom{u}{i}(s+1)^i(s-1)^{i+1} + \sum_{i=0}^{u} \binom{u}{i}(s+1)^i(s-1)^i \\ &= 1 + \sum_{i=1}^{u} \left[\binom{u}{i-1}(s+1)^{i-1} + \binom{u}{i}(s+1)^i\right](s-1)^i \\ &\quad + (s+1)^u(s-1)^{u+1} \\ &\geq 1 + \sum_{i=1}^{u} \binom{s+2u}{i}(s-1)^i + \binom{s+2u-1}{u}(s-1)^{u+1} , \end{aligned}$$

which is what had to be shown. ∎

The claim in Theorem 2.18 is probably somewhat modest. Numerical evidence indicates that the bound in Theorem 2.18 is usually strictly better than the bound in Theorem 2.1 if $s > t$.

It follows immediately from (2.9) that if we can improve the bound for $f(s^3, s, 3)$, we obtain better bounds for $f(s^t, s, t)$ than those in Theorem 2.18. It should be observed that no such improvements can be obtained from Theorems 2.11 or 2.15.

However, Bush (1952b) was able to obtain the following improved bounds for $f(s^t, s, t)$.

Theorem 2.19. *In an $OA(s^t, k, s, t)$, the following inequalities hold:*

$$k \leq t+1 \quad \text{if} \quad s \leq t , \qquad (2.13)$$

$$k \le s+t-2 \quad \textit{if} \quad s > t \ge 3 \ \textit{and } s \textit{ is odd}, \tag{2.14}$$
$$k \le s+t-1 \quad \textit{in all other cases}. \tag{2.15}$$

Proof: That (2.15) holds was already observed as a consequence of (2.9). To see that (2.13) holds, we use an argument of Mukerjee (1979). Assume that an array exists with $k = t+2$. Without loss of generality we may assume that one run has all factors at level 0. When restricted to the first t factors, there are $(s-1)(t-1)$ other runs in which the first factor and $t-2$ of the remaining $t-1$ factors are at level 0. In addition, there are $s-1$ other runs in which all the factors $2, \ldots, t$ are at level 0. We will only need one of the latter $s-1$ runs for the argument. Thus there are $2+(s-1)(t-1)$ runs as exhibited in Table 2.20. A star $(*)$ in this table indicates that the level is not yet determined. If any star for factor $t+1$ or factor $t+2$ were 0, the corresponding run would have at least t coincidences with the run in which all factors are at level 0. But since the array is of strength t and index unity that is clearly not possible. Since any two of the runs in Table 2.20 have at least $t-2$ coincidences, it follows by the same argument that the stars for factors $t+1$ and $t+2$ should form different ordered pairs from $\{1, \ldots, s-1\} \times \{1, \ldots, s-1\}$. Since there are $(s-1)^2$ such pairs and $1+(s-1)(t-1)$ runs to which we have to assign them, we need $(s-1)^2 \ge 1+(s-1)(t-1)$. This gives a contradiction for $s \le t$, and establishes inequality (2.13).

Table 2.20. Some runs in a putative $OA(s^t, t+2, s, t)$.

			Factors			
1	2	3	$\cdots$	t	$t+1$	$t+2$
0	0	0	$\cdots$	0	0	0
0	1	0	$\cdots$	0	$*$	$*$
.	.	.	$\cdots$	.	.	.
0	$s-1$	0	$\cdots$	0	$*$	$*$
0	0	1	$\cdots$	0	$*$	$*$
.	.	.	$\cdots$	.	.	.
0	0	$s-1$	$\cdots$	0	$*$	$*$
.	.	.	$\cdots$	.	.	.
0	0	0	$\cdots$	1	$*$	$*$
.	.	.	$\cdots$	.	.	.
0	0	0	$\cdots$	$s-1$	$*$	$*$
$*$	0	0	$\cdots$	0	$*$	$*$

For (2.14), it suffices to show that $f(s^3, s, 3) \le s+1$ when $s > 3$, s odd. That this is sufficient again follows from inequality (2.9).

Suppose therefore that the inequality does not hold and an $OA(s^3, s+2, s, 3)$ exists for $s > 3$, s odd. With the $A_i(u)$'s as in Lemma 2.7, we obviously have

$A_{k-i}(u) = 0$ if $3 \leq i < k$ and $A_0(u) = 1$, since the array has strength 3 and index unity. Using this in the equalities in Lemma 2.7 we find $A_{k-2}(u) = (s+2)(s+1)(s-1)/2$, $A_{k-1}(u) = 0$ and $A_k(u) = s(s-1)^2/2$. Since $A_k(u) > 0$, we can without loss of generality assume that the array contains one run with all factors at level 0 and one run with all factors at level 1. There are $s-1$ other runs, as exhibited in Table 2.21, in which the first two factors are both at level 1.

Table 2.21. Some runs in a putative $OA(s^3, s+2, s, 3)$.

$$\begin{array}{cccccc} 0 & 0 & 0 & \cdots & 0 \\ 1 & 1 & 1 & \cdots & 1 \\ 1 & 1 & * & \cdots & * \\ \cdot & \cdot & \cdot & \cdots & \cdot \\ 1 & 1 & * & \cdots & * \end{array}$$

Since the array has strength 3 and index unity, for each of the last s factors exactly one of the stars will be equal to 0. Since s is odd, the total number of 0's in the runs with the first two factors at level 1 will be odd. Hence at least one of the runs containing the stars will have an odd number of coincidences with the run in which all factors are at level 0. But that contradicts the fact that (apart from $A_0(u)$) only $A_k(u)$ and $A_{k-2}(u)$ are positive. ∎

Inequality (2.13) and the construction of an $OA(s^t, t+1, s, t)$ in Section 2.1 lead immediately to the following corollary when $s \leq t$.

Corollary 2.22. *$f(s^t, s, t) = t + 1$ if $s \leq t$.*

It should no longer be a surprise that the bounds in (2.14) and (2.15) can be improved for some values of s and t. The case that has received most attention in the literature is that of $t = 2$. An $OA(s^2, k, s, 2)$ exists if and only if $k-2$ pairwise orthogonal Latin squares of order s exist. Those familiar with this topic undoubtedly know that numerous values for $f(s^2, s, 2)$ are still unknown. This special case is further discussed in Chapter 8. We also refer the interested reader to Dénes and Keedwell (1974, 1991).

These improvements for $t = 2$ can generally be used to obtain improvements for larger t. Taking Lemma 2.12 one step further than in equation (2.9) we obtain

$$f(\lambda s^t, s, t) \leq f(\lambda s^2, s, 2) + t - 2 \ .$$

Although useful, there is little reason to believe that this inequality is sharp, even when restricted to $\lambda = 1$.

Nevertheless, for $t \geq 3$ there have been very few attempts in the literature to improve (2.14) or (2.15). This may be a reflection of the difficulty of this problem.

Kounias and Petros (1975) were able to obtain the following improvements to Theorem 2.19. The proof is rather long and will not be given here.

Theorem 2.23. *In an $OA(s^t, k, s, t)$ the following inequalities hold:*

$$k \leq s \quad \text{if } t = 3,\ s \equiv 2 \pmod 4 \text{ and } s \geq 6\ ; \tag{2.16}$$

$$k \leq s + t - 3 \quad \text{if } 4 \leq t < s,\ s \text{ even and } s \not\equiv 0 \pmod{36}\ ; \tag{2.17}$$

$$k \leq 6 \quad \text{if } t = 4 \text{ and } s = 5\ . \tag{2.18}$$

Before the reader attempts to obtain further improvements, it should be pointed out that there are arrays which achieve equality in some of the bounds in Theorems 2.19 and 2.23, and so no further improvement is possible. These arrays, and the equations in which they achieve equality, are the following.

For s a prime power, an $OA(s^2, s+1, s, 2)$ exists (equality in (2.15)). (2.19)

For s an odd prime power, an $OA(s^3, s+1, s, 3)$ exists (equality in (2.14)). (2.20)

For $s = 2^n$, an $OA(s^3, s+2, s, 3)$ exists (equality in (2.15)). (2.21)

For $s = 2^n$, an $OA(s^4, s+1, s, 4)$ exists (equality in (2.17)). (2.22)

An $OA(5^4, 6, 5, 4)$ exists (equality in (2.18)). (2.23)

We will return to the construction of these arrays in Chapters 3 and 5.

2.5 Orthogonal Arrays with Two Levels

For statistical applications the most frequently used orthogonal arrays are probably those with all factors at two levels, which we will usually refer to as *2-level* or *binary* orthogonal arrays. These applications will be discussed in Chapter 11. Most of the preceding theorems can be applied to 2-level arrays simply by setting $s = 2$. There are also a number of papers in the literature that are specifically concerned with 2-level arrays, for example Seiden (1954), Seiden and Zemach (1966), Blum, Schatz and Seiden (1970), Gulati (1971b), Chakravarty and Dey (1976), Chopra (1976), Yamamoto, Kuriki and Sato (1984), and Hedayat and Stufken (1988). Furthermore, in view of the fact that any binary linear code produces a 2-level array, as we shall see in Chapter 4, any paper on binary codes is potentially relevant here (and there are a very large number of such papers — see MacWilliams and Sloane, 1977, Pless and Huffman, 1998).

In this section we discuss some specific properties of 2-level arrays. We have already defined the set-theoretic complement of an orthogonal array in Chapter 1. For 2-level arrays there is a second kind of complement, called the *binary complement*, obtained simply by interchanging the two symbols in the array. We shall suppose in this section that the two levels are 0 and 1. Then binary complementation, which will be indicated by a bar, replaces 0's by 1's and 1's by 0's.

We begin with a basic result which seems to have first appeared in the paper of Seiden and Zemach (1966) (the case $u = 1$ was already given by Seiden, 1954).

Theorem 2.24. *An* $OA(N, k, 2, 2u)$ *exists if and only if an* $OA(2N, k+1, 2, 2u+1)$ *exists.*

Proof: The proof of Lemma 2.12 shows that an $OA(N, k, 2, 2u)$ can be obtained from an $OA(2N, k+1, 2, 2u+1)$. For the converse, let A be an $OA(N, k, 2, 2u)$, of index λ, and let B be the array consisting of the runs of A followed by 0 together with the runs of $\overline{A}$ followed by 1. We will show that B is an $OA(2N, k+1, 2, 2u+1)$, also of index λ. Consider any $t = 2u+1$ columns of B. We must show that any binary t-tuple appears λ times as a row in this $2N \times t$ subarray. If the last column of B is one of the t columns of the subarray the result follows because A has strength $2u$. If $k > 2u$ we must also consider any t of the first k columns. For simplicity of notation we suppose these are the first t columns. For any binary t-tuple $v = v_1 \cdots v_t$, let $n(v)$ denote the number of rows of A that begin with v. Then the number of rows of B that begin with v is equal to $n(v) + n(\overline{v})$, and we must show that this is equal to λ for all v.

Since A has strength $t-1$, we know that if v' differs from v in exactly one place, then

$$n(v) + n(v') = \lambda \ .$$

Therefore, if v and v'' differ in exactly two places,

$$n(v) - n(v'') = n(v) + n(v') - (n(v') + n(v'')) = 0 \ ,$$

where v' differs from both v and v'' in exactly one place. Repeating this argument, we see that if v and w differ in evenly many places, $n(v) = n(w)$. But $\overline{v}$ and v' differ in $2u$ places, so $n(\overline{v}) = n(v')$, and $n(v) + n(\overline{v}) = n(v) + n(v') = \lambda$, as required. ■

Corollary 2.25.

$$\begin{aligned} f(2N, 2, 2u+1) &= f(N, 2, 2u) + 1 \ , \\ F(k+1, 2, 2u+1) &= 2F(k, 2, 2u) \ . \end{aligned}$$

We now study some values of $f(N, 2, t)$. From Corollaries 2.5 and 2.6 we readily obtain

$$f(4\lambda, 2, 2) \le 4\lambda - 1 \tag{2.24}$$

and

$$f(8\lambda, 2, 3) \leq 4\lambda \ . \tag{2.25}$$

As we will see in Chapter 7, these bounds can be attained if and only if a Hadamard matrix of order 4λ exists. It is believed that such Hadamard matrices always exist, and certainly they are known for every $\lambda \leq 106$ (see Chapter 7).

Thus, at least for small or moderate values of λ, the first challenge is encountered when considering the case $t = 4$. We know already from Corollary 2.22 that $f(2^4, 2, 4) = 5$. Table 2.26 shows the first few values of $f(16\lambda, 2, 4)$.

Table 2.26. Values of $f(16\lambda, 2, 4)$ for $1 \leq \lambda \leq 5$.

λ	$f(16\lambda, 2, 4)$
1	5
2	6
3	5
4	8
5	6

The values in Table 2.26 were established by Seiden and Zemach (1966). The proofs consist essentially of two parts. First, an array $OA(16\lambda, k, 2, 4)$ with the claimed maximal number of factors is exhibited. Then it is shown that an $OA(16\lambda, k+1, 2, 4)$, where k is again the claimed maximum, does not exist. The first four arrays are easily constructed; for the fifth see Problem 2.17. We omit the arguments for the upper bounds, but mention that Lemma 2.7 plays a crucial role.

An alternative technique for deriving these results was presented by Chopra (1976). It would however appear that his technique is no more powerful than Seiden and Zemach's. Some of the results in Table 2.26 can also be derived from Yamamoto, Kuriki and Sato (1984). Furthermore, by constructing an appropriate array, these authors show that $f(96, 2, 4) \geq 7$.

From Theorem 2.24 and Table 2.26 we immediately obtain the following values of $f(32\lambda, 2, 5)$.

Table 2.27. Values of $f(32\lambda, 2, 5)$ for $1 \leq \lambda \leq 5$.

λ	$f(32\lambda, 2, 5)$
1	6
2	7
3	6
4	9
5	7

The values were originally obtained by Gulati (1971b) by a different method.

From Section 2.1 we know already that $f(\lambda 2^t, 2, t) \geq t+1$. We also saw that if $\lambda = 1$ then equality holds, i.e. $f(2^t, 2, t) = t+1$. A natural question to ask is whether there are other values of λ for which equality holds.

We first observe that equality does not hold if λ is even.

Lemma 2.28. *If $\lambda = q2^n$, where q is odd and $n \geq 1$, then*

$$f(\lambda 2^t, 2, t) \geq t+n+1 \ .$$

Proof: From Section 2.1 we know that an $OA(q2^{t+n}, t+n+1, 2, t+n)$ exists. When considered as an array of strength t this is an $OA(\lambda 2^t, t+n+1, 2, t)$ of index $\lambda = q2^n$, which establishes the lemma. ■

With λ as in Lemma 2.28, it can be shown that any $OA(\lambda 2^t, t+1, 2, t)$ can be extended to an $OA(\lambda 2^t, t+n+1, 2, t)$. In general, it is not possible to increase the strength of this extended array. This can be done only if the initial $OA(\lambda 2^t, t+1, 2, t)$ already has strength $t+1$, in which case it can be extended to an $OA(q2^{t+n}, t+n+1, 2, t+n)$ (see Problems 2.13 and 2.14).

What can be said for odd values of λ? Blum, Schatz and Seiden (1970) proved the following result.

Theorem 2.29. *If λ is odd and $t \geq \lambda+1$, then*

$$f(\lambda 2^t, 2, t) = t+1 \ .$$

Proof: We must show that under these hypotheses an $OA(\lambda 2^t, t+2, 2, t)$ does not exist. We assume the existence of such an array and show that it leads to a contradiction.

For $I \subset \{1, 2, \ldots, t+2\}$ let $a(I)$ denote the number of times that a run has all factors $i \in I$ at level 1 and all $t+2-|I|$ remaining factors at level 0. For simplicity of notation we will write $a_0, a_1, \ldots, a_{t+2}$ for $a(\emptyset)$, $a(\{1\}), \ldots, a(\{t+2\})$, respectively. Note that the values of $a(I)$ are completely determined by $a_0, a_1, \ldots, a_{t+2}$. This follows from the observation that if $|I| \geq 2$ and $i_1, i_2 \in I$, $i_1 \neq i_2$, then

$$a(I) + a(I \setminus \{i_1\}) + a(I \setminus \{i_2\}) + a(I \setminus \{i_1, i_2\}) = \lambda \ ,$$

since the left-hand side is the number of runs with specified values for exactly t factors. Therefore

$$a(I) = \lambda - a(I \setminus \{i_1\}) - a(I \setminus \{i_2\}) - a(I \setminus \{i_1, i_2\}) \ .$$

Using this repeatedly and denoting the cardinality of I by m, we obtain

$$a(I) = -\tfrac{1}{2}(m-1)\lambda + (m-1)a_0 + \sum_{i\in I} a_i \ , \quad \text{if } m \text{ is odd} \qquad (2.26)$$

$$a(I) = \tfrac{1}{2}m\lambda - (m-1)a_0 - \sum_{i\in I} a_i \ , \quad \text{if } m \text{ is even} \ . \qquad (2.27)$$

Thus a necessary and sufficient condition for the existence of an $OA(\lambda 2^t, t+2, 2, t)$ is the existence of nonnegative integers $a_0, a_1, \ldots, a_{t+2}$ such that all $a(I)$'s obtained from (2.26) and (2.27) are also nonnegative.

If there are solutions to (2.26) and (2.27) with all $a(I)$'s positive, including $a_0, a_1, \ldots, a_{t+2}$, we may remove one copy of each of the 2^{t+2} possible runs, and obtain an $OA((\lambda-4)2^t, t+2, 2, t)$. Repeating this argument, we may assume that $a(I)=0$ for some I. By permuting 1's and 0's for the factors corresponding to the elements of this I, we may assume that $a_0 = 0$. Also, after a possible permutation of the factors, we can assume that $a_1 \geq a_2 \geq \cdots \geq a_{t+2}$. We can also assume that t is even. If it is not, we use the construction in the proof of Lemma 2.12 to obtain an $OA(\lambda 2^{t-1}, t+1, 2, t-1)$. Since λ is odd and $\lambda \leq t-1$, which is even, we see that $\lambda \leq t-2$. Thus the conditions of the theorem are satisfied for this even-strength array, and its nonexistence will imply that of the original array.

Now take $I = \{1,2\}$ in (2.27). Since $a_0 = 0$ and $a_1 \geq a_2$, we obtain $0 \leq \lambda - 2a_2$, and since λ is odd

$$a_2 \leq \tfrac{1}{2}(\lambda - 1) \ . \qquad (2.28)$$

Next, take $I = \{2, \ldots, t+2\}$. Since t is even we use (2.26) and obtain

$$0 \leq -\tfrac{1}{2}t\lambda + (t+1)a_2 \ . \qquad (2.29)$$

It is easy to see that inequalities (2.28) and (2.29) yield a contradiction if $\lambda \leq t-1$. ■

Yamamoto, Kuriki and Sato (1984) also studied equations (2.26) and (2.27). Similar relations between the $a(I)$'s can be obtained for any orthogonal array $OA(\lambda 2^t, k, 2, t)$. Instead of $a_0, a_1, \ldots, a_{t+2}$, an appropriate choice of $\binom{k}{0} + \binom{k}{1} + \cdots + \binom{k}{k-t-1}$ of the $a(I)$'s now determines the others. Yamamoto, Kuriki and Sato (1984) used this technique to construct an $OA(96, 7, 2, 4)$, as mentioned after Table 2.26.

We will return to 2-level orthogonal arrays in our discussions of the construction of orthogonal arrays, in particular in Chapters 4, 5 and 7.

2.6 Concluding Remarks

As mentioned at the beginning of this chapter, we may also obtain bounds on $f(N,s,t)$ by studying $F(k,s,t)$. We take this approach in Chapter 4, where we use an analytic result known as the linear programming bound to obtain results that are always at least as strong as the Rao bounds (and are usually stronger).

It is clear that, even combining all these bounds, we are still a long way from knowing the exact value of $f(N,s,t)$, especially when t is large or s is not a prime power.

Even for $t = 4$ and $s = 2$ there is a considerable gap between the best bounds and the exact values as determined by Seiden and Zemach (1966), as can be seen in Table 2.30, where we compare the exact values with the Rao bound and the linear programming bound.

Table 2.30. Comparison of Rao bound, linear programming bound, and actual values of $f(16\lambda, 2, 4)$.

λ	Rao bound	LP bound	$f(16\lambda, 2, 4)$
1	5	5	5
2	7	6	6
3	9	7	5
4	10	8	8
5	12	8	6

Starting in Chapter 3 we will give a number of constructions for orthogonal arrays. But there is almost always a gap between the largest number of factors produced by these constructions and the tightest bounds known (see the tables in Chapter 12). The basic problem in the subject remains: find better constructions and better bounds! To focus the reader's attention, let us state:

Research Problem 2.31. Extend Tables 2.26 and 2.27!

Research Problem 2.32. While there are considerable gaps in our knowledge of the functions $F(k,s,t)$ and $f(N,s,t)$ (see the tables in Chapter 12), this is especially true when s is not a prime power and, for the function $f(N,s,t)$, when $\lambda = N/s^t$ is not a multiple of s. Find better bounds for these two functions, with special attention to the special cases mentioned.

From the perspective of statistical applications (see Chapter 11), the function $F(k,s,t)$ is more natural than $f(N,s,t)$. Often we first decide how many factors k to use in the experiment, how many levels s to use for the factors (of course this may lead to a mixed array, see Chapter 9), and what model to use (which determines the strength t). The value of $F(k, s, t)$, if available, would then tell us the minimal number of runs of an orthogonal array that meets our needs. First deciding on the number of runs, and then asking about the maximal number of factors that can be used in an orthogonal array (which is the type of question that $f(N, s, t)$ can help us with), is a much less common order of events. From a theoretical perspective though, since $f(N, s, t)$ completely determines $F(k, s, t)$ (see (2.1)), learning more about $f(N, s, t)$ will help us to fill in gaps for $F(k, s, t)$.

For many years the Rao bounds (together with the improvements mentioned in this chapter) were the only lower bounds known for the number of runs, and there was a tendency to think that they may be close to the truth. In fact, the opposite might often be true when k is large.

Research Problem 2.33. For fixed values of s and t, and large k, how far are the Rao bounds from the truth? (See also Section 10.6.)

2.7 Notes on Chapter 2

In view of the many parallels between orthogonal arrays and codes that will be mentioned in Chapters 4 and 5, it is worth remarking here that the coding counterpart to Theorem 2.24 is a much simpler result — see MacWilliams and Sloane (1977), Theorem 2 of Chapter 2. The coding theory analogue of the Rao bound (2.3) is also a lot simpler — this is the *Hamming* or *sphere-packing* bound (*op. cit.*, Theorem 6 of Chapter 1). We will sketch an alternative proof of the Rao bounds using linear programming in Section 4.5. For yet another proof see Beder (1998).

2.8 Problems

2.1. Show that if N is a multiple of s^t then the $N \times (t+1)$ zero-sum array constructed in Section 2.1 is an $OA(N, t+1, s, t)$.

2.2. Give a short proof of Theorem 2.1 for the special case $s = 2$.

2.3. a. Show that all the columns of the $N \times M$ matrices in the proof of Theorem 2.1 are nonzero.

b. Justify the last two sentences in the proof of Theorem 2.1.

2.4. a. Show that equality occurs in Rao's inequalities if $\lambda = 1$, $s = 2$ and $k = t + 1$, for any $t \geq 2$.

b. Why does the result in Part a imply that $f(2^t, 2, t) \leq t + 1$?

2.5. Use the results in Theorem 2.4 to show that the parameters of a tight $OA(N, k, s, 5)$ must be one of the following: (i) $N = 32$, $k = 6$ and $s = 2$; (ii) $N = 729$, $k = 12$ and $s = 3$, or (iii) $N = 27m^2(9m^2 - 1)$, $k = 3(3m^2 + 2)/5$ and $s = 6$, where $m \equiv 0 \pmod 3$, $m \equiv \pm 1 \pmod 5$ and $m \equiv 5 \pmod{16}$.

[Noda (1986) showed that the arrays in (iii) do not exist, so the arrays in (i) and (ii) (which do exist) are the only possibilities].

2.6. Use inequality (2.8) with an appropriate choice of x to prove the result in Corollary 2.5.

2.7. Show that the array with repeated runs in Example 2.16 is indeed an $OA(54, 5, 3, 3)$.

2.8. Hedayat, Seiden and Stufken (1997) showed the nonexistence of an $OA(54, 6, 3, 3)$, so no $OA(54, 5, 3, 3)$ can be extended to an $OA(54, 6, 3, 3)$. Without using this result, show that an $OA(54, 5, 3, 3)$ with repeated runs cannot be extended to an $OA(54, 6, 3, 3)$. (Hint: If such an extension were possible, an $OA(54, 6, 3, 3)$ would exist in which two runs have at least five coincidences. Argue that such an $OA(54, 6, 3, 3)$ cannot exist).

2.9. Give an example of an $OA(54, 5, 3, 3)$ without repeated runs. Can such an array be either isomorphic or statistically equivalent to an $OA(54, 5, 3, 3)$ with repeated runs?

2.10. Collect all the answers from your class to the first part of Problem 2.9. Are any of these arrays isomorphic or statistically equivalent?

2.11. Study the claim in Section 2.4 that the bounds from Rao's inequalities are generally better than those from (2.9), where $f(\lambda s^3, s, 3)$ is replaced by its best upper bound from Section 2.3.

2.12. Provide numerical evidence to support the claim that the bound in Theorem 2.18 is often strictly better than the bound in Theorem 2.1.

2.13. Let D be the $OA(2^t, t+1, 2, t)$ obtained via the construction in Section 2.1. Let D^c be the set-theoretic complement of D, as defined in Chapter 1. By property 7 in Chapter 1, D^c is also an $OA(2^t, t+1, 2, t)$.

a. Show that every orthogonal array with these parameters is statistically equivalent to D or D^c.
b. Show that any $OA(\lambda 2^t, t+1, 2, t)$ is statistically equivalent to the juxtaposition of μ copies of D and $\lambda - \mu$ copies of D^c, for some integer μ with $0 \le \mu \le \lambda$.

(Hint: Show that if two runs in an $OA(\lambda 2^t, t+1, 2, t)$ have

(i) an even number of factors at a different level, then they are repeated equally often in the array, and

(ii) an odd number of factors at a different level, then they form together λ runs in the array.)

2.14. a. Show that an $OA(\lambda 2^t, t+1, 2, t)$ with $\lambda = q2^n$, $n \geq 1$, can always be extended to an $OA(\lambda 2^t, t+n+1, 2, t)$.

b. For the extension in part a, show that a necessary condition for increasing the strength of the array is that the initial $OA(\lambda 2^t, t+1, 2, t)$ already has strength $t+1$.

c. Show that if the necessary condition in part b is satisfied, then the initial array can be extended to an $OA(q2^{t+n}, t+n+1, 2, t+n)$.

2.15. a. Show that D and D^c, as defined in Problem 2.13, are isomorphic orthogonal arrays.

b. Show that any two orthogonal arrays $OA(2^t, t+1, 2, t)$ are isomorphic.

c. Show that there are $\lfloor(\lambda+2)/2\rfloor$ nonisomorphic orthogonal arrays $OA(\lambda 2^t, t+1, 2, t)$.

2.16. Write k as $4a-1-b$, where $a \geq 1$, $0 \leq b \leq 3$. Use (2.3) to show that $F(4a-1-b, 2, 2) \geq 4a$. Hence show that $F(4a-b, 2, 3) \geq 8a$.

2.17. a. Construct arrays with parameters $OA(32, 6, 2, 5)$ (and hence $OA(32, 6, 2, 4)$), $OA(48, 5, 2, 4)$ and $OA(64, 8, 2, 4)$ as mentioned in Table 2.26.

b. Show that an $OA(80, 6, 2, 4)$ can be constructed by taking the following runs:

— three runs with all factors at level 0,
— two copies of each of the six runs of type 0 1 1 1 1 1,
— the 15 runs of type 0 0 1 1 1 1,
— the 20 runs of type 0 0 0 1 1 1,
— two copies of each of the 15 runs of type 0 0 0 0 1 1 .

Chapter 3

Orthogonal Arrays and Galois Fields

A large number of techniques are known for constructing orthogonal arrays. This chapter, the first of several describing these techniques, discusses some constructions due to Bush (1952b), Addelman and Kempthorne (1961a), Rao (1946a, 1947, 1949) and Bose and Bush (1952), that have the common theme of using Galois fields and finite geometries. We also describe a number of basic properties of orthogonal arrays, including the important concept of linearity.

3.1 Introduction

Ideally one would like to classify the different methods for constructing orthogonal arrays by the parameters of the arrays they produce. Anyone interested in a particular orthogonal array would then immediately know how to construct it. Unfortunately this idea fails because many arrays can be constructed in several different ways. Instead we have chosen to classify the different constructions by the essential ideas that underlie them, leading to a division into six chapters of which this is the first. Chapter 12 is intended to serve as a guide and overview to help in determining which construction will produce an orthogonal array with a particular set of parameters.

The underlying themes of the present chapter are Galois fields and finite geometries. To assist readers who are not entirely comfortable with these concepts we have included a brief appendix at the end of the book, giving the most essential definitions and results.

Throughout this chapter our computations will be performed in a Galois field $GF(s)$. We will use α to denote a primitive element of this field. When proving theorems we will usually write the elements of $GF(s)$ as $\alpha_0 = 0$, $\alpha_1 = \alpha$, $\alpha_2 = \alpha^2, \ldots, \alpha_{s-1} = \alpha^{s-1} = 1$. In the examples however it is more natural to take the elements of $GF(3)$ to be $0, 1, 2$ and to denote the elements $\{0, 1, \alpha, \alpha^2\}$ of $GF(4)$ by $0, 1, 2, 3$.

We will also use the symbol $GF(s)^n$ to denote the set of all n-tuples with entries from $GF(s)$. The elements of $GF(s)^n$ will sometimes be referred to as points or vectors, thinking of $GF(s)^n$ as a vector space over $GF(s)$.

The proofs of many of the theorems in this book will be in two parts, the first giving a construction, and the second a verification of its correctness.

3.2 Bush's Construction

Bush (1952b) studied the construction of orthogonal arrays of index unity. Such arrays are of special interest as they contain the smallest number of runs for a given number of levels and a given strength. Not only does this make them mathematically interesting, but they also reduce time and cost in statistical experiments. It should be mentioned however that there is generally a price to be paid when using the minimal number of runs. For example, the consequences of the failure of a run may be more severe, and also such arrays usually do not allow as many factors as arrays with a larger index.

Bush's main results are contained in Theorems 3.1 and 3.2.

Theorem 3.1. *If $s \geq 2$ is a prime power then an $OA(s^t, s+1, s, t)$ of index unity exists whenever $s \geq t - 1 \geq 0$.*

Proof:[1]
CONSTRUCTION: We first construct an $s^t \times s$ array whose columns are labeled with the elements of $GF(s)$ and whose rows are labeled by the s^t polynomials over $GF(s)$ of degree at most $t-1$. Let those polynomials (in the variable X, say) be denoted by $\phi_1, \ldots, \phi_{s^t}$. Then the entry in this array in the column labeled α_i and the row labeled ϕ_j is defined to be $\phi_j(\alpha_i)$, i.e. the value of the polynomial ϕ_j at the point α_i.

We add one additional factor to this array, taking the level of this factor in the row labeled ϕ_j to be the coefficient of X^{t-1} in ϕ_j. (It may be helpful to readers who are familiar with algebraic geometry to point out that the level of this additional factor is the value of ϕ_j at infinity, and then the column labels can be taken to be $GF(s) \cup \{\infty\}$.)

[1]Theorems 3.1 and 3.2 also follow from coding theory, using Reed-Solomon and extended Reed-Solomon code – see Section 5.5.

VERIFICATION: The crucial idea underlying this construction is that a polynomial of degree d over a field cannot have more than d zeros without being identically zero.

Any t-tuple should appear once as a row in any $s^t \times t$ subarray. To verify this we first consider the case when all t factors in the subarray are among the first s factors, say corresponding to the columns labeled $z_1, \ldots, z_t$. It suffices to verify that no two rows in the subarray are identical. Suppose on the contrary that

$$\phi_j(z_i) = \phi_{j'}(z_i), \quad \text{for} \quad i = 1, \ldots, t \ ,$$

where $j \neq j'$, or equivalently

$$\phi(z_i) = 0, \quad \text{for} \quad i = 1, \ldots, t \ ,$$

when $\phi = \phi_j - \phi_{j'}$. Since ϕ is also a polynomial of degree at most $t-1$, this implies (by Lemma A.6) that ϕ is identically zero, and so $j = j'$, a contradiction.

Secondly, suppose that the last factor is one of the t factors of the subarray. Let the other $t-1$ factors correspond to columns labeled $z_1, \ldots, z_{t-1}$. Suppose that

$$\phi_j(z_i) = \phi_{j'}(z_i), \quad \text{for} \quad i = 1, \ldots, t-1 \ ,$$

and that the coefficients of X^{t-1} in ϕ_j and $\phi_{j'}$ are equal. If $\phi = \phi_j - \phi_{j'}$, this means that

$$\phi(z_i) = 0, \quad \text{for} \quad i = 1, \ldots, t-1 \ ,$$

where ϕ is now of degree at most $t-2$. Again the conclusion is that $j = j'$, a contradiction. Thus in either case all rows in the subarray are distinct. ∎

Recall from Corollary 2.22 that $f(s^t, s, t) = t + 1$ if $s \leq t$. For $s > t$ we only had upper bounds on $f(s^t, s, t)$ from Theorems 2.19 and 2.23. Theorem 3.1 now gives us a lower bound on the number of factors in such arrays, provided that s is a prime power. Most of the time however the upper and lower bounds do not agree, so we cannot conclude that these arrays have the maximal number of factors (although this may in fact be the case — see the discussion below). Before analyzing this further, let us consider Theorem 3.2, which states that the result of Theorem 3.1 can be improved for some values of s and t.

Theorem 3.2. *If $s = 2^m$, $m \geq 1$, and $t = 3$ then there exists an $OA(s^3, s+2, s, t)$.*

Proof: For $m = 1$ we exhibited such an array in Table 1.3. Thus we may restrict our attention to $m \geq 2$.

CONSTRUCTION: We use the construction in Theorem 3.1 to obtain an $OA(2^{3m}, 2^m + 1, 2^m, 3)$, and adjoin another factor for which the level in row

ϕ_j is given by the coefficient of X in ϕ_j. (Note that now the ϕ_j have degree at most 2.)

VERIFICATION: If the three factors in a $2^{3m} \times 3$ subarray are all from the first $2^m + 1$ factors, the verification has already been given in Theorem 3.1. If the new factor is part of a $2^{3m} \times 3$ subarray involving only one of the first 2^m factors, say the one corresponding to the column z_1, arguments as in the proof of Theorem 3.1 lead to $\phi(z_1) = 0$, where ϕ is a polynomial of degree zero. Obviously ϕ is identically zero. If two factors are chosen from the first 2^m factors, corresponding to columns z_1 and z_2, say, we obtain $\phi(z_1) = \phi(z_2) = 0$, where ϕ is a polynomial of degree at most 2 in which the coefficient of X is 0. Using the fact that $z_1^2 = z_2^2$ in $GF(2^m)$ if and only if $z_1 = z_2$, we again conclude that ϕ is identically zero. Hence the rows in any $2^{3m} \times 3$ subarray are all distinct, and the array does indeed have strength 3. ■

The orthogonal arrays constructed in Theorems 3.1 and 3.2 have two special properties — they are *simple* and *linear.*

Definition 3.3. An orthogonal array is *simple* if the runs are distinct.

Definition 3.4. Let s be a prime power. An orthogonal array $OA(N, k, s, t)$ with levels from $GF(s)$ is said to be *linear* if it is simple and if, when considered as k-tuples from $GF(s)$, its N runs form a vector space over $GF(s)$ (i.e. satisfy the condition that if R_1 and R_2 are any two runs of the array then every k-tuple $c_1R_1 + c_2R_2$ is also a run, for any choice of c_1, $c_2 \in GF(s)$).

An orthogonal array that is isomorphic to a linear array need not itself be linear. However, for statistical applications such arrays are just as important as linear ones. Linearity *is* preserved under statistical equivalence.

If an orthogonal array is linear then it follows from elementary linear algebra that $N = s^n$ for some integer $n \geq 0$. The number n, the dimension of the vector space formed by the runs, is called the *dimension* of the array.

Linear orthogonal arrays have two advantages over orthogonal arrays that do not have this property.

(1) They have a very succinct description, for it is enough to give a basis for the vector space formed by the rows. This basis is usually given in the form of an $n \times k$ matrix called a *generator matrix*, whose rows are the basis. Thus if $u_1, \ldots, u_n$ are the rows of the generator matrix, then the set of all linear combinations

$$c_1u_1 + \cdots + c_nu_n \,, \tag{3.1}$$

where $c_1, \ldots, c_n \in GF(s)$, comprises the runs of the array. This is a very compact way to specify the array.

(2) If an orthogonal array is linear, it is possible to regard the collection of runs as the *codewords* in a linear *error-correcting code.* This is much more than just a change in view-point, for it enables one to use the techniques of coding theory to analyze the properties of the orthogonal array. We will return to this topic in Chapter 4.

Example 3.5. The orthogonal array $OA(8,4,2,3)$ given in Table 1.3 is linear, and has generator matrix

$$\begin{bmatrix} 1 & 0 & 0 & 1 \\ 0 & 1 & 0 & 1 \\ 0 & 0 & 1 & 1 \end{bmatrix} . \tag{3.2}$$

Indeed, the reader can easily verify that the set of all linear combinations of the rows of this matrix, with coefficients that are either 0 or 1 (and evaluated over $GF(2)$) are precisely the eight runs 0000, 1001, 0101, 1100, ... of the orthogonal array in Table 1.3. ■

Example 3.6. It follows immediately from the proofs of Theorems 3.1 and 3.2 that the orthogonal arrays constructed there are linear. The crucial point is that the set of polynomial functions of degree at most $t-1$ is a linear space.

In fact we will see in Section 5.5 that the runs in the orthogonal arrays constructed in Theorems 3.1 and 3.2 are precisely the same as the codewords in extended Reed-Solomon codes. These codes were discovered only many years after Bush's paper appeared. The original reference for these codes is Reed and Solomon (1960), while the extended codes are discussed by Wolf (1969), Tanaka and Nishida (1970) and Gross (1973). The recently developed algebraic-geometry codes that will also be mentioned in Chapter 5 are a profound generalization of these codes and of Bush's constructions. ■

For linear orthogonal arrays of index unity, such as those constructed in Theorems 3.1 and 3.2, there is a "duality" theorem, as follows (the reason for this terminology will appear in Section 4.3).

Theorem 3.7. *If s is a prime power and a linear array $OA(s^t, k, s, t)$ exists, then there also exists a linear array $OA(s^{k-t}, k, s, k-t)$.*

We postpone the proof to Chapter 5 — see Theorem 5.6. Theorems 3.2 and 3.7 imply the following result.

Corollary 3.8. *If $s = 2^m$, $m \geq 1$, then there exists an $OA(s^{s-1}, s+2, s, s-1)$.*

Comparing the number of factors in the arrays in Theorems 3.1 and 3.2 with the upper bounds in Chapter 2, we obtain the following result. This establishes the validity of the claims in equations (2.19)–(2.23).

Corollary 3.9. *If s is a prime power then*

(i) $f(s^2, s, 2) = s + 1$,

(ii) $f(s^3, s, 3) = s + 1$, *if s is odd,*

(iii) $f(2^{3m}, 2^m, 3) = 2^m + 2$, *where* $m \geq 1$,

(iv) $f(2^{4m}, 2^m, 4) = 2^m + 1$, *where* $m \geq 2$, *and*

(v) $f(5^4, 5, 4) = 6$.

The result in Corollary 3.9(i) corresponds to the existence of a complete set of pairwise orthogonal Latin squares of order s, as we shall see in Chapter 8. The construction in Theorem 3.1 can also be carried out if $s \leq t$. In this case, however, the arrays are not so interesting, since the simple construction in Section 2.1 already provides us with the maximal number of factors, $t + 1$, for any s, not just for prime powers. The construction of Theorem 3.1 is thus most interesting for $s > t \geq 3$.

At least two challenging problems remain unanswered for orthogonal arrays of index unity.

Research Problem 3.10. *Number of levels not a prime power.* The preceding constructions apply only if s is a prime power. Even for the case $t = 2$, very little is known about constructing OA's of index unity when s is not a prime power (see also Chapter 8). Can better constructions be found for such values of s, or can tighter upper bounds for $f(s^t, s, t)$ be obtained?

Second, even when s is a prime power, for most values of s and t there are still gaps between the best upper bounds known for $f(s^t, s, t)$ and the number of factors given by the constructions. There is evidence for believing that the exact values of $f(s^t, s, t)$ are as follows.

Conjecture 3.11. *If s is a prime power then*

$$f(s^t, s, t) = \begin{cases} s + 1, & \text{if } 2 \leq t \leq s, \\ t + 1, & \text{if } t \geq s, \end{cases} \tag{3.3}$$

except that

$$f(s^3, s, 3) = s + 2, \quad \text{if } s = 2^m, \tag{3.4}$$
$$f(s^{s-1}, s, s - 1) = s + 2, \quad \text{if } s = 2^m. \tag{3.5}$$

The second assertion of (3.3) is true, by Corollary 2.22, and (3.4) follows from Corollary 3.9. Also $f(s^{s-1}, s, s-1) \geq s+2$ if $s = 2^m$ by Corollary 3.8. Conjecture 3.11 is an analogue for orthogonal arrays of a long-standing conjecture in coding theory concerning maximal distance separable codes, as stated in MacWilliams and Sloane (1977), Chapter 11, Section 7, especially Figure 11.2. Conjecture 3.11 is implicitly stated in Research Problem (11.1f) on page 329 of that reference. We will say more about this conjecture in Section 5.6, but let us highlight its importance by stating:

Research Problem 3.12. *The index unity conjecture.* Establish the truth of Conjecture 3.11.

The following example illustrates both the construction in Theorem 3.2 and the linearity of the arrays constructed there.

Example 3.13. The first interesting case of Theorem 3.2 occurs when $s = 4$ and $t = 3$, which leads to an $OA(64, 6, 4, 3)$. As usual we denote the elements of $GF(4)$ by $0, 1, 2, 3$.

The 64 rows of the array are labeled by polynomials $\phi(X) = \beta_0 + \beta_1 X + \beta_2 X^2$, where $\beta_0, \beta_1, \beta_2 \in GF(4)$. The columns are labeled $0, 1, 2, 3, \beta_2, \beta_1$, as explained in the construction part of the proof of Theorem 3.2. Some selected runs of the array are shown in Table 3.14.

The linearity of the array follows since the polynomials of degree at most 2 form a linear space over $GF(4)$. For example, the sum of the rows labeled 1 0 0 and 0 1 0 is the row labeled 1 1 0. Multiplying row 0 1 0 by the field element 2 yields the row labeled 0 2 0, and so on. Instead of listing all 64 rows, we simply give a generator matrix, for which we use the rows labeled 1 0 0, 0 1 0 and 0 0 1:

$$\begin{bmatrix} 1 & 1 & 1 & 1 & 0 & 0 \\ 0 & 1 & 2 & 3 & 0 & 1 \\ 0 & 1 & 3 & 2 & 1 & 0 \end{bmatrix} . \tag{3.6}$$

The full array is then obtained by taking all linear combinations with coefficients from $GF(4)$ of these three rows. The full array can be found in the electronic library of orthogonal arrays described in Chapter 12. Note the substantial saving in space achieved by use of the generator matrix: three rows are enough to specify all 64. ∎

Table 3.14. Some runs of the array $OA(64, 6, 4, 3)$.

β_0	β_1	β_2	0	1	2	3	β_2	β_1
0	0	0	0	0	0	0	0	0
1	0	0	1	1	1	1	0	0
2	0	0	2	2	2	2	0	0
3	0	0	3	3	3	3	0	0
0	1	0	0	1	2	3	0	1
1	1	0	1	0	3	2	0	1
2	1	0	2	3	0	1	0	1
.	.	.	.	.	.	.	.	.
0	2	0	0	2	3	1	0	2
.	.	.	.	.	.	.	.	.
0	0	1	0	1	3	2	1	0
1	0	1	1	0	2	3	1	0
.	.	.	.	.	.	.	.	.

Remark 3.15. *Orthogonal arrays over rings; additive arrays.* By definition 3.4, the runs of a linear orthogonal array form a vector space over a field. It is sometimes useful to weaken this definition and to consider linear arrays with levels taken from a ring (see the Appendix for the definition of a ring). For example, the levels might be taken from $\mathbb{Z}_4$, the ring of integers mod 4. Linearity over a ring means that the set of runs is closed under vector addition and subtraction, and under multiplication by elements of the ring. Again one can define a generator matrix for an array. We will meet examples of orthogonal arrays over $\mathbb{Z}_4$ in Section 5.10.

A second generalization allows orthogonal arrays with levels from a field, requiring only that the set of runs is closed under vector addition and subtraction, but not necessarily under multiplication by field elements. These are called *additive* orthogonal arrays. A linear array is additive, but the converse need not hold. Again there is an obvious notion of generator matrix. An additive orthogonal array over $GF(4)$ will appear in Section 5.13.

3.3 Addelman and Kempthorne's Construction

The main result of this section is due to Addelman and Kempthorne (1961a). Again this is a method that can be applied only when the number of levels is a prime power. There is some resemblance to the method in the previous section in that the entries are also computed by evaluating functions over $GF(s)$. However, the number of runs is now not a power of s, implying that the arrays are no longer linear. The labeling of rows and columns is also completely different and is more complicated than in the previous construction.

Theorem 3.16. *If s is an odd prime power then an orthogonal array $OA(2s^n, 2(s^n-1)/(s-1)-1, s, 2)$ exists for all $n \geq 2$.*

Proof: Addelman and Kempthorne (1961a) gave a detailed description of their construction only for the case $n = 2$. Mukhopadhyay (1981) observed that the arrays for $n \geq 3$ can be obtained recursively from those for $n = 2$. Here we present the construction in the spirit of Addelman and Kempthorne (1961a), but for arbitrary $n \geq 2$. An alternative construction for these arrays will be given in Chapter 6.

Before we begin the construction it may be helpful to give an overview. In the proof of Theorem 3.1 the rows of the array were labeled with polynomials in one variable of bounded degree, the columns were labeled with the points of $GF(s) \cup \{\infty\}$, and the typical entry in the array was obtained by evaluating the polynomial corresponding to the row at the point corresponding to the column.

In the present construction the array is formed by juxtaposing two arrays F and G. Now the columns of F and G are labeled with certain cleverly chosen polynomials in n variables $X_1, \ldots, X_n$ which are linear in $X_2, \ldots, X_n$ but quadratic in X_1. The s^n rows of F and G are labeled with all points from $GF(s)^n$, and the entries are obtained by evaluating the polynomial corresponding to the column at the point corresponding to the row.

We now proceed with the formal proof.

CONSTRUCTION: Let $r = (s^{n-1}-1)/(s-1)$, and let $\ell_1, \ell_2, \ldots, \ell_r$ be the $(n-1)$-tuples in $PG(n-2, s)$ — in other words a maximal set of nonzero $(n-1)$-tuples no one of which is a scalar multiple of any other. If $n = 2$, for example, $r = 1$ and $\ell_1 = (1)$. Furthermore, let $X_1, \ldots, X_n$ be n indeterminates, and let $X = (X_2, \ldots, X_n)$. For each i, $1 \leq i \leq r$, we define the following $4s$ functions of X_1 and X, where b_j, c_j, d_j and d, for $1 \leq j \leq s-1$, are to be specified later.

$$
\begin{aligned}
f_1^{(i)} &= X\ell_i^T, & g_1^{(i)} &= X\ell_i^T, \\
f_2^{(i)} &= X_1 + \alpha_1 X\ell_i^T, & g_2^{(i)} &= X_1 + \alpha_1 X\ell_i^T + b_1, \\
&\cdots & &\cdots \\
f_s^{(i)} &= X_1 + \alpha_{s-1} X\ell_i^T, & g_s^{(i)} &= X_1 + \alpha_{s-1} X\ell_i^T + b_{s-1}, \\
f_{s+1}^{(i)} &= X_1^2 + X\ell_i^T, & g_{s+1}^{(i)} &= dX_1^2 + X\ell_i^T, \\
f_{s+2}^{(i)} &= X_1^2 + \alpha_1 X_1 + X\ell_i^T, & g_{s+2}^{(i)} &= dX_1^2 + d_1 X_1 + X\ell_i^T + c_1, \\
&\cdots & &\cdots \\
f_{2s}^{(i)} &= X_1^2 + \alpha_{s-1} X_1 + X\ell_i^T, & g_{2s}^{(i)} &= dX_1^2 + d_{s-1} X_1 + X\ell_i^T + c_{s-1}\,.
\end{aligned}
$$

We also define $f^{(0)} = g^{(0)} = X_1$.

We now construct two $s^n \times (2rs+1)$ arrays F and G as follows. The columns of F are to be labeled by

$$f^{(0)}, f_1^{(1)}, \ldots, f_{2s}^{(1)}, f_1^{(2)}, \ldots, f_1^{(r)}, \ldots, f_{2s}^{(r)} ,$$

and its rows by all possible n-tuples with entries from $GF(s)$. The entries are obtained by evaluating the polynomial specified by the column label at the n-tuple specified by the row label.

The array G is constructed in a similar way, but the column labels are now

$$g^{(0)}, g_1^{(1)}, \ldots, g_{2s}^{(1)}, g_1^{(2)}, \ldots, g_1^{(r)}, \ldots, g_{2s}^{(r)} .$$

The juxtaposition of the two arrays

$$\begin{bmatrix} F \\ G \end{bmatrix}$$

then has the dimensions of the desired orthogonal array. However, whether it has the required orthogonality property depends on the choice of b_j, c_j, d_j and d. We will use the following values:

$$b_j = (\alpha_1 - 1)/(4\alpha_{j+1}),\ c_j = \alpha_{2j}(\alpha_1 - 1)/4,\ d_j = \alpha_{j+1},\ d = \alpha_1 . \tag{3.7}$$

Here 1 is the multiplicative unit element of $GF(s)$, $4 = 1+1+1+1$ (remember s is odd!), and $\alpha_1 = \alpha$ is a primitive element of $GF(s)$.

VERIFICATION: We have to verify that any $2s^n \times 2$ subarray contains each of the s^2 possible pairs from $GF(s)$ in precisely $2s^{n-2}$ of its rows. There are a number of cases to be considered, depending on which two factors occur in the subarray. In most cases the arguments are straightforward. We will limit our verification to just one case and leave the others to the reader.

Consider the case where the two factors correspond to columns labeled $f_1^{(i)}$ and $f_{s+j}^{(i')}$ in F (and $g_1^{(i)}$ and $g_{s+j}^{(i')}$ in G), where $1 \le i, i' \le r$ and $2 \le j \le s$. With z_1, $z_2 \in GF(s)$, the number of times that (z_1, z_2) appears as a row in this subarray is equal to the number of pairs (X_1, X) such that

$$X\ell_i^T = z_1,\ \ X_1^2 + \alpha_{j-1}X_1 + X\ell_{i'}^T = z_2 , \tag{3.8}$$

plus the number of pairs (X_1, X) such that

$$X\ell_i^T = z_1,\ \ dX_1^2 + d_{j-1}X_1 + X\ell_{i'}^T + c_{j-1} = z_2 . \tag{3.9}$$

Suppose first that $i \ne i'$. Then for each value of X_1 we have two independent equations in X in (3.8), which leads to s^{n-3} solutions. Since there are s choices for X_1, there are s^{n-2} solutions to (3.8). A similar argument applies to (3.9), so the total number of solutions is $2s^{n-2}$, as required.

If $i = i'$, (3.8) will not have a solution unless

$$X_1^2 + \alpha_{j-1}X_1 = z_2 - z_1 . \tag{3.10}$$

For every X_1 satisfying (3.10) there are s^{n-2} solutions for X in (3.8). Similarly, we find from (3.7) and (3.9) that X_1 must satisfy

$$\alpha_1 X_1^2 + \alpha_j X_1 + \alpha_{2j-2}(\alpha_1 - 1)/4 = z_2 - z_1 . \tag{3.11}$$

Thus it suffices to show that there are precisely two values of X_1 that satisfy (3.10) or (3.11). It is easily seen that (3.10) has 0, 1 or 2 solutions depending on whether $\alpha_{2j-2} + 4(z_2 - z_1)$ is respectively a nonresidue, zero or a quadratic residue in $GF(s)$. Similarly, (3.11) has 0, 1 or 2 solutions depending on whether $\alpha_1(\alpha_{2j-2} + 4(z_2 - z_1))$ is respectively a nonresidue, zero or a residue. Since $\alpha_1 = \alpha$ is a nonresidue, it follows that the total number of solutions is always 2, as required. This completes the verification for this case. ∎

Addelman and Kempthorne (1961a) also suggest that their construction can be used when $s = 2^m$. As they point out, this requires a different choice for b_j, c_j, d_j and d. The proof in Addelman and Kempthorne (1961a) contains some inaccuracies and unclear statements, and is incomplete. They take $d = 1$ and $d_j = \alpha_j$. It can then be shown that the resulting array has the required orthogonality property if b_j and c_j satisfy the following conditions:

(i) for $1 \le j \le s-1$, b_j cannot be written as $\alpha_j z^2 + z$, $z \in GF(s)$,

(ii) for $1 \le j \le s-1$, c_j cannot be written as $z^2 + \alpha_j z$, $z \in GF(s)$, and

(iii) for $1 \le j, j' \le s-1$ and $j + j' \not\equiv 0 \pmod{s-1}$, $c_{j'} + b_j/\alpha_j$ cannot be written as $z^2 + (\alpha_{j'} + \alpha_{s-j-1})z$, $z \in GF(s)$.

To complete the proof, an argument would be needed to show that such b_j's and c_j's always exist. We do not know if this is always possible. We will not pursue this here, since for $s = 2^m$ a method to be discussed in Example 6.31 will provide us with a simpler construction for these orthogonal arrays.

Theorems 2.8 and 3.16 have the following corollary.

Corollary 3.17. *If s is an odd prime power then*

$$f(2s^n, s, 2) = 2(s^n - 1)/(s-1) - 1 .$$

Proof: We apply Theorem 2.8 with $\lambda = 2s^{n-2}$. Then $a = 2(s^{n-2} - 1)/(s-1)$, $b = 1$ and

$$\theta = \sqrt{(s-1)^2 - 3/4} - s + 3/2 ,$$

which is easily seen to be between 0 and 1/2 for all s. Thus $\lfloor \theta \rfloor = 0$, implying that

$$\begin{aligned} k &\leq \lambda(s+1)+a-1 \\ &= 2\frac{(s^n-1)}{s-1}-1 . \end{aligned}$$

■

In Chapter 6 we will see that the result of Corollary 3.17 is also valid if $s = 2^m$, where $m \geq 2$. If $s = 2$ we cannot apply Theorem 2.8 and in that case we will see in Chapter 6 that the upper bound from Theorem 2.1 can be achieved.

Example 3.18. As an illustration of the construction in Theorem 3.16 we consider the case $n = s = 3$. This will lead to an $OA(54, 25, 3, 2)$. We take $GF(3) = \{0, 1, 2\}$ with the usual addition and multiplication modulo 3. Thus $\alpha_0 = 0$, $\alpha_1 = 2$ and $\alpha_2 = 1$. The values for b_j, c_j, d_j and d are as follows:

$$b_1 = 1, b_2 = 2, c_1 = 1, c_2 = 1, d_1 = 1, d_2 = 2, d = 2 .$$

Furthermore, $r = 4$ and we can take

$$\ell_1 = (0,1),\ \ell_2 = (1,0),\ \ell_3 = (1,1),\ \ell_4 = (1,2) .$$

The functions $f^{(0)}$, $f_1^{(1)}$, $f_2^{(1)}$, $f_3^{(1)}$, $f_4^{(1)}, \ldots, f_6^{(4)}$ are

$$\begin{aligned} &X_1, X_3, X_1+2X_3, X_1+X_3, X_1^2+X_3, X_1^2+2X_1+X_3, \\ &X_1^2+X_1+X_3, X_2, X_1+2X_2, X_1+X_2, X_1^2+X_2, X_1^2+2X_1+X_2, \\ &X_1^2+X_1+X_2, X_2+X_3, X_1+2X_2+2X_3, X_1+X_2+X_3, \\ &X_1^2+X_2+X_3, X_1^2+2X_1+X_2+X_3, X_1^2+X_1+X_2+X_3, X_2+2X_3, \\ &X_1+2X_2+X_3, X_1+X_2+2X_3, X_1^2+X_2+2X_3, X_1^2+2X_1+X_2+2X_3 \\ &\text{and} \quad X_1^2+X_1+X_2+2X_3 . \end{aligned}$$

The corresponding g-functions are

$$\begin{aligned} &X_1, X_3, X_1+2X_3+1, X_1+X_3+2, 2X_1^2+X_3, 2X_1^2+X_1+X_3+1, \\ &2X_1^2+2X_1+X_3+1, X_2, X_1+2X_2+1, X_1+X_2+2, 2X_1^2+X_2, \\ &2X_1^2+X_1+X_2+1, 2X_1^2+2X_1+X_2+1, X_2+X_3, \\ &X_1+2X_2+2X_3+1, X_1+X_2+X_3+2, 2X_1^2+X_2+X_3, \\ &2X_1^2+X_1+X_2+X_3+1, 2X_1^2+2X_1+X_2+X_3+1, X_2+2X_3, \\ &X_1+2X_2+X_3+1, X_1+X_2+2X_3+2, 2X_1^2+X_2+2X_3, \\ &2X_1^2+X_1+X_2+2X_3+1 \quad \text{and} \quad 2X_1^2+2X_1+X_2+2X_3+1. \end{aligned}$$

The orthogonal array $OA(54, 25, 3, 2)$ obtained by using these functions is shown in Table 3.31 at the end of this chapter. ■

Research Problem 3.19. While we will see a simpler construction of the Addelman-Kempthorne arrays in Theorem 6.40, a more transparent construction of these arrays would be helpful for understanding these arrays, and perhaps generalizing the construction to other parameters. Find a simpler direct way to construct these arrays.

3.4 The Rao-Hamming Construction

Before his 1947 and 1949 papers introduced what are now called orthogonal arrays, in an earlier paper Rao (1946a) had introduced a special case of this concept called a hypercube[2] of strength t. In particular, the construction of a certain family of hypercubes of strength 2 given in Rao (1946) corresponds to orthogonal arrays with parameters $OA(s^n, (s^n-1)/(s-1), s, 2)$, where s is a prime power. Such arrays are the subject of this section.

Rao (1947, 1949) later gave an alternative and simpler construction for arrays with those parameters. The second construction produces linear arrays. Essentially the same construction is used in one of the best-known families of error-correcting codes, the Hamming (1950) codes. (The precise statement is that the runs in the orthogonal array are the codewords in the dual Hamming code. We shall say more about this in Section 5.3.) For this reason we call this the *Rao-Hamming construction* . Of course, the matrix used in Construction 3 below is a very natural one to consider, and similar constructions were used to obtain fractional factorial designs by Finney (1945) and Kempthorne (1947). However, since Rao was the first to apply it to orthogonal arrays, and since the term Hamming code is firmly established in coding theory, it seems appropriate to call this the Rao-Hamming construction.

In the following proof we present three versions of this construction, of which the first two are due to Rao and the third to Hamming.

Theorem 3.20. *If s is a prime power then an $OA(s^n, (s^n-1)/(s-1), s, 2)$ exists whenever $n \geq 2$.*

Proof:
CONSTRUCTION 1. The $(s^{n+1}-1)/(s-1)$ points of the projective geometry $PG(n,s)$ can be divided into (a) $(s^n-1)/(s-1)$ points on an $(n-1)$-flat (or hyperplane) at infinity, and (b) the remaining s^n finite points. The number of $(n-2)$-flats in the hyperplane at infinity is $(s^n-1)/(s-1)$. Each of these $(n-2)$-flats is also contained in s other hyperplanes (excluding the hyperplane at infinity). These s hyperplanes form a parallel pencil, and partition the s^n finite points into s sets of s^{n-1} points each. We label these s hyperplanes in some arbitrary way by $0, 1, \ldots, s-1$, and do this for each of the $(s^n-1)/(s-1)$ pencils.

We now construct an $s^n \times (s^n-1)/(s-1)$ array whose rows are labeled with the s^n finite points and whose columns are labeled with the pencils of parallel hyperplanes. Each entry in the array is given by the label of the hyperplane in the pencil corresponding to the column label that contains the point corresponding to the row label.

[2] A *hypercube* $[k, s, r, t]$ of strength t, as in Raghavarao (1971), Chapter 2, is the same as an orthogonal array $OA(s^r, k, s, t)$.

CONSTRUCTION 2. A simpler way to obtain an orthogonal array with these parameters is the following. This construction, unlike the first, always produces linear arrays. Form an $s^n \times n$ array whose rows are all possible n-tuples from $GF(s)$. Let $C_1, \ldots, C_n$ denote the columns of this array. The columns of the full orthogonal array then consist of all columns of the form

$$z_1 C_1 + \cdots + z_n C_n = [C_1, \ldots, C_n] z ,$$

where $z = (z_1, \ldots, z_n)^T$ is an n-tuple from $GF(s)$, not all the z_i are zero, and the first nonzero z_i is 1. There are $(s^n - 1)/(s - 1)$ such columns, as required.

CONSTRUCTION 3. Alternatively, form an $n \times (s^n - 1)/(s - 1)$ matrix whose columns are all nonzero n-tuples $(z_1, \ldots, z_n)^T$ from $GF(s)$ in which the first nonzero z_i is 1. The orthogonal array is then formed by taking all linear combinations of the rows of this generator matrix, as described in Section 3.2.

VERIFICATION FOR CONSTRUCTION 1: Two different factors correspond to two different pencils of parallel hyperplanes. Any hyperplane in one pencil has exactly s^{n-2} finite points in common with any hyperplane in the other pencil. Therefore any pair of labels (z_1, z_2), with $z_1, z_2 \in \{0, 1, \ldots, s-1\}$, appears exactly s^{n-2} times in any $s^n \times 2$ subarray.

VERIFICATION FOR CONSTRUCTION 2: If $z^{(1)} = (z_1^{(1)}, \ldots, z_n^{(1)})^T$ and $z^{(2)} = (z_1^{(2)}, \ldots, z_n^{(2)})^T$ are the n-tuples corresponding to two distinct columns in the array, then

$$Z = [z^{(1)} \; z^{(2)}]$$

is a matrix of rank 2 over $GF(s)$. Hence for every β_1 and β_2 in $GF(s)$ there are s^{n-2} row vectors y such that

$$yZ = [\beta_1 \; \beta_2] .$$

Consequently, there are s^{n-2} rows in the $s^n \times 2$ array

$$[C_1 \cdots C_n] Z$$

which are identical to (β_1, β_2). This shows that the array we have constructed has strength 2.

VERIFICATION FOR CONSTRUCTION 3: It is straightforward to show that the array produced by Construction 3 is isomorphic to that produced by Construction 2 — see Problem 3.3. Alternatively, the result can be deduced from Theorem 3.29 below: indeed, by construction, any two columns of the array are linearly independent. ∎

All three constructions are essentially the same: any orthogonal array with parameters $OA(s^n, (s^n - 1)/(s - 1), s, 2)$ (s a prime power, $n \geq 2$) which is isomorphic to a linear array is isomorphic to the array produced by Construction 3. We shall say that all such orthogonal arrays are of *Rao-Hamming type.*

Combining Theorem 3.20 and Corollary 2.5 results in the following corollary.

Corollary 3.21. *If s is a prime power and $n \geq 2$, then*

$$f(s^n, s, 2) = (s^n - 1)/(s - 1) .$$

Example 3.22. Taking $s = 2$ and $n = 3$ in Construction 3 we obtain an $OA(8, 7, 2, 2)$ with generator matrix

$$\begin{bmatrix} 1 & 0 & 0 & 1 & 1 & 0 & 1 \\ 0 & 1 & 0 & 1 & 0 & 1 & 1 \\ 0 & 0 & 1 & 0 & 1 & 1 & 1 \end{bmatrix} . \tag{3.12}$$

The full array is

$$\begin{bmatrix} 0 & 0 & 0 & 0 & 0 & 0 & 0 \\ 1 & 0 & 0 & 1 & 1 & 0 & 1 \\ 0 & 1 & 0 & 1 & 0 & 1 & 1 \\ 0 & 0 & 1 & 0 & 1 & 1 & 1 \\ 1 & 1 & 0 & 0 & 1 & 1 & 0 \\ 1 & 0 & 1 & 1 & 0 & 1 & 0 \\ 0 & 1 & 1 & 1 & 1 & 0 & 0 \\ 1 & 1 & 1 & 0 & 0 & 0 & 1 \end{bmatrix} . \tag{3.13}$$

Alternatively we may use Construction 2 to obtain an isomorphic array. We first form the 8×3 array shown on the left of (3.14) and then complete it to the full array by adjoining four more columns.

$$\begin{array}{|ccc|cccc|} \hline 0 & 0 & 0 & 0 & 0 & 0 & 0 \\ 1 & 0 & 0 & 1 & 1 & 0 & 1 \\ 0 & 1 & 0 & 1 & 0 & 1 & 1 \\ 1 & 1 & 0 & 0 & 1 & 1 & 0 \\ 0 & 0 & 1 & 0 & 1 & 1 & 1 \\ 1 & 0 & 1 & 1 & 0 & 1 & 0 \\ 0 & 1 & 1 & 1 & 1 & 0 & 0 \\ 1 & 1 & 1 & 0 & 0 & 0 & 1 \\ \hline \end{array} \tag{3.14}$$

■

Example 3.23. We will use Constructions 1 and 3 to obtain an $OA(27, 13, 3, 2)$. Using Construction 1, let (z_0, z_1, z_2, z_3) denote a point in $PG(3, 3)$. Here $z_i \in GF(3)$, $i = 0, 1, 2, 3$, not all z_i's are zero, and the first nonzero z_i is 1. Furthermore, let $GF(3) = \{0, 1, 2\}$ with the usual addition and multiplication modulo 3. We will often use -1 instead of 2 to make the description more symmetrical. Let $z_0 = 0$ be the hyperplane at infinity. The $(n-2)$-flats at infinity are then

$$\begin{aligned}
(z_0 &= 0,\ z_3 = 0)\ , \\
(z_0 &= 0,\ z_2 = 0)\ , \\
(z_0 &= 0,\ z_2 \pm z_3 = 0)\ , \\
(z_0 &= 0,\ z_1 = 0)\ , \\
(z_0 &= 0,\ z_1 \pm z_3 = 0)\ , \\
(z_0 &= 0,\ z_1 \pm z_2 = 0)\ , \\
(z_0 &= 0,\ z_1 + z_2 \pm z_3 = 0)\ , \\
(z_0 &= 0,\ z_1 - z_2 \pm z_3 = 0)\ ,
\end{aligned}$$

where the symbol $\pm$ indicates that the expression is to be taken with both choices for the sign. The three hyperplanes that intersect the first of these $(n-2)$-flats at infinity are

$$z_3 = 0, \quad z_3 + z_0 = 0, \quad z_3 - z_0 = 0\ .$$

This is the first pencil of parallel hyperplanes, and we label these hyperplanes $0, 1$ and 2 respectively. Consequently the entry corresponding to the column labeled $(z_0 = 0,\ z_3 = 0)$ and the row labeled $(1, 0, 2, 1)$, say, will be 2, since the point $(1, 0, 2, 1)$ is in the hyperplane $z_3 - z_0 = 0$.

The $OA(27, 13, 3, 2)$ obtained in this way is shown in Table 3.25. For simplicity of notation we use $(\beta_1, \beta_2, \beta_3)$ as a column label to denote the $(n-2)$-flat $(z_0 = 0,\ \beta_1 z_1 + \beta_2 z_2 + \beta_3 z_3 = 0)$, and (x_1, x_2, x_3) as a row label to denote the finite point $(1, x_1, x_2, x_3)$. We have carefully chosen the labeling of the pencils so that the resulting array is linear.

Construction 3 is considerably simpler. The generator matrix can be written down immediately (see Table 3.24), and the full array is shown in Table 3.26.

Table 3.24. Generator matrix for linear $OA(27, 13, 3, 2)$ obtained from Construction 3 in Theorem 3.20.

	C_1	C_2	C_3										
u_1	1	0	0	1	1	1	1	1	1	1	1	0	0
u_2	0	1	0	1	2	0	0	1	1	2	2	1	1
u_3	0	0	1	0	0	1	2	1	2	1	2	1	2

Table 3.25. An $OA(27, 13, 3, 2)$ obtained by the first construction in Theorem 3.17 (transposed).

β_1	β_2	β_3		
			x_1	0 0 0 0 0 0 0 0 0 1 1 1 1 1 1 1 1 1 2 2 2 2 2 2 2 2 2
			x_2	0 0 0 1 1 1 2 2 2 0 0 0 1 1 1 2 2 2 0 0 0 1 1 1 2 2 2
			x_3	0 1 2 0 1 2 0 1 2 0 1 2 0 1 2 0 1 2 0 1 2 0 1 2 0 1 2
0	0	1		0 2 1 0 2 1 0 2 1 0 2 1 0 2 1 0 2 1 0 2 1 0 2 1 0 2 1
0	1	0		0 0 0 2 2 2 1 1 1 0 0 0 2 2 2 1 1 1 0 0 0 2 2 2 1 1 1
0	1	1		0 2 1 2 1 0 1 0 2 0 2 1 2 1 0 1 0 2 0 2 1 2 1 0 1 0 2
0	1	2		0 1 2 2 0 1 1 2 0 0 1 2 2 0 1 1 2 0 0 1 2 2 0 1 1 2 0
1	0	0		0 0 0 0 0 0 0 0 0 2 2 2 2 2 2 2 2 2 1 1 1 1 1 1 1 1 1
1	0	1		0 2 1 0 2 1 0 2 1 2 1 0 2 1 0 2 1 0 1 0 2 1 0 2 1 0 2
1	0	2		0 1 2 0 1 2 0 1 2 2 0 1 2 0 1 2 0 1 1 0 2 1 0 2 1 0 2
1	1	0		0 0 0 2 2 2 1 1 1 2 2 2 1 1 1 0 0 0 1 1 1 0 0 0 2 2 2
1	1	1		0 2 1 2 1 0 1 0 2 2 1 0 1 0 2 0 2 1 1 0 2 0 2 1 2 1 0
1	1	2		0 1 2 2 0 1 1 2 0 2 0 1 1 2 0 0 1 2 1 2 0 0 1 2 2 0 1
1	2	0		0 0 0 1 1 1 2 2 2 2 2 2 0 0 0 1 1 1 1 1 1 2 2 2 0 0 0
1	2	1		0 2 1 1 0 2 2 1 0 2 1 0 0 2 1 1 0 2 1 0 2 2 1 0 0 2 1
1	2	2		0 1 2 1 2 0 2 0 1 2 0 1 0 1 2 1 2 0 1 2 0 2 0 1 0 1 2

If the rows of the generator matrix are labeled u_1, u_2, u_3, then the rows of the full array are the linear combinations $a_1u_1 + a_2u_2 + a_3u_3$, where a_1, a_2, $a_3 \in GF(3)$. The values of a_1, a_2, a_3 are plainly visible in the first three rows (since the array has been transposed) of Table 3.26. Again, illustrating Construction 2, if the first three columns of Table 3.24 and the first three rows of Table 3.26 are labeled C_1, C_2, C_3, then the rows of the full array are the linear combinations $z_1C_1 + z_2C_2 + z_3C_3$ when $(z_1, z_2, z_3)^T$ is visible in the first three columns of Table 3.26. It can be shown that the arrays in Tables 3.25 and 3.26 are in fact isomorphic. ∎

Table 3.26. The full array $OA(27, 13, 3, 2)$ (transposed) obtained from the generator matrix of Table 3.24.

C_1	0 1 0 0 2 0 0 1 1 1 1 0 0 2 2 2 2 0 0 1 1 1 1 2 2 2 2
C_2	0 0 1 0 0 2 0 1 0 2 0 1 1 2 0 1 0 2 2 1 1 2 2 2 2 1 1
C_3	0 0 0 1 0 0 2 0 1 0 2 1 2 0 2 0 1 2 1 1 2 1 2 2 1 2 1
	0 1 1 0 2 2 0 2 1 0 1 1 1 1 2 0 2 2 2 2 2 0 0 1 1 0 0
	0 1 2 0 2 1 0 0 1 2 1 2 2 0 2 1 2 1 1 0 0 2 2 0 0 1 1
	0 1 0 1 2 0 2 1 2 1 0 1 2 2 1 2 0 2 1 2 0 2 0 1 0 1 0
	0 1 0 2 2 0 1 1 0 1 2 2 1 2 0 2 1 1 2 0 2 0 2 0 1 0 1
	0 1 1 1 2 2 2 2 2 0 0 2 0 1 1 0 0 1 0 0 1 1 2 0 2 2 1
	0 1 1 2 2 2 1 2 0 0 2 0 2 1 0 0 1 0 1 1 0 2 1 2 0 1 2
	0 1 2 1 2 1 2 0 2 2 0 0 1 0 1 1 0 0 2 1 2 0 1 2 1 0 2
	0 1 2 2 2 1 1 0 0 2 2 1 0 0 0 1 1 2 0 2 1 1 0 1 2 2 0
	0 0 1 1 0 2 2 1 1 2 2 2 0 2 2 1 1 1 0 2 0 0 1 1 0 0 2
	0 0 1 2 0 2 1 1 2 2 1 0 2 2 1 1 2 0 1 0 2 1 0 0 1 2 0

To conclude this section, we note that the factors corresponding to rows C_1, C_2, C_3 and $C_1 + C_2 + C_3$ in Table 3.26 form an $OA(27, 4, 3, 3)$. This is the maximal number of factors in such an orthogonal array, by (2.13). In Problem 3.5 the reader is invited to identify other 4×27 subarrays in Table 3.26 that form orthogonal arrays of strength 3.

3.5 Conditions for a Matrix to be an Orthogonal Array

In this section we present some fundamental properties of linear orthogonal arrays with entries from a Galois field $GF(s)$ and of some more general orthogonal arrays. We also give some conditions on a matrix which guarantees that it forms an orthogonal array. Since these results involve Galois fields and the concept of linearity, it is appropriate that they appear in this chapter.

The first result is an immediate consequence of the definition of an orthogonal array.

Theorem 3.27. *Let A be an orthogonal array $OA(N, k, s, t)$ with entries from $GF(s)$. Then any t columns of A are linearly independent over $GF(s)$.*

Proof: We remind the reader that $N \times 1$ vectors $v_1, \ldots, v_t$ with components from a ring R are said to be linearly independent over R if the relation

$$c_1 v_1 + \cdots + c_t v_t = 0, \quad c_1, \ldots, c_t \in R\,, \tag{3.15}$$

implies that $c_1 = \cdots = c_t = 0$. An equivalent condition is that the matrix $[v_1 \cdots v_t]$ has rank t over R.

Now let $v_1, \ldots, v_t$ be any t columns of A, and suppose (3.15) holds. There is a run i with the first entry equal to 1 and others 0. Then (3.15) implies $c_1 = 0$. Similarly $c_2 = \cdots = c_t = 0$. ■

Exactly the same argument establishes the following result.

Theorem 3.28. *If A is an $OA(N, k, s, t)$ with entries from a ring R that contains 1, then any t columns of A are linearly independent over R.*

There is a converse to Theorem 3.27 for linear arrays.

Theorem 3.29. *Let A be an $N \times k$ matrix whose rows form a linear subspace of $GF(s)^k$. If any t columns of A are linearly independent over $GF(s)$, then A is an orthogonal array $OA(N, k, s, t)$.*

Proof: Suppose $N = s^n$, and let G be an $n \times k$ generator matrix for A, so that the rows of A consist of all k-tuples ξG, where $\xi = (\xi_1, \ldots, \xi_n)$, $\xi_i \in GF(s)$.

Choose t columns of A, and let G_1 be the corresponding $n \times t$ submatrix of G. Clearly the columns of G_1 are linearly independent. The number of times that a t-tuple z appears as a row in these t columns of A is equal to the number of ξ such that

$$\xi G_1 = z \ . \tag{3.16}$$

Since G_1 has rank t, this number is s^{n-t}, for all z. Therefore A is an orthogonal array of strength t. ∎

Once the requirement of linearity is dropped, however, the converse to Theorem 3.27 is certainly *not* true. Indeed, arrays with the property that any t columns are linearly independent exist with N much less than k. These are called arrays with "Property P_t"—see Section 11.9.

The final theorem gives a necessary and sufficient condition for a matrix to be an orthogonal array, which does not assume linearity nor even that the entries belong to a field or ring.

Theorem 3.30. *An $N \times k$ matrix A with entries from $\{0, 1, \ldots, s-1\}$ is an $OA(N, k, s, t)$ if and only if*

$$\sum_{u=\text{row of A}} \zeta^{uv^T} = 0 \tag{3.17}$$

for all k-tuples v over $\{0, 1, \ldots, s-1\}$ with w nonzero entries, for all w in the range $1 \le w \le t$, where $\zeta = e^{2\pi i/s}$, and uv^T is evaluated modulo s.

Note that the special case when $s = 2$ is particularly simple: the condition is that an $N \times k$ matrix A with $0, 1$ entries is an orthogonal array $OA(N, k, 2, t)$ if and only if

$$\sum_{u=\text{row of A}} (-1)^{uv^T} = 0 \tag{3.18}$$

for all $0, 1$ vectors v containing w 1's, for all w in the range $1 \le w \le t$, where the sum is over all rows u of A.

Proof: If A is an orthogonal array then it is easy to see that (3.18) and (3.17) hold. We begin the proof of the converse statements with an example that will illustrate the method. Suppose $s = t = 2$ and (3.18) holds. We will show that in the first two columns of A all pairs $00, 01, 10$ and 11 occur equally often. Let $n_{00}, \ldots, n_{11}$ be the number of occurrences of these pairs. Since the total number of runs is N, and by taking v in (3.18) to be respectively $010\ldots0$, $100\ldots0$ and $110\ldots0$, we obtain the four equations

$$\begin{aligned} n_{00} + n_{01} + n_{10} + n_{11} &= N \ , \\ n_{00} - n_{01} + n_{10} - n_{11} &= 0 \ , \\ n_{00} + n_{01} - n_{10} - n_{11} &= 0 \ , \\ n_{00} - n_{01} - n_{10} + n_{11} &= 0 \ . \end{aligned}$$

Plainly $n_{00} = n_{01} = n_{10} = n_{11} = N/4$ is a solution. Since the coefficient matrix

$$\begin{bmatrix} +1 & +1 & +1 & +1 \\ +1 & -1 & +1 & -1 \\ +1 & +1 & -1 & -1 \\ +1 & -1 & -1 & +1 \end{bmatrix}$$

is invertible (for any of four reasons: direct verification, or because it is a Vandermonde matrix, or because it is a Hadamard matrix of Sylvester type, see Chapter 7, or because it is the character table of the elementary abelian group of type 2^2 — all of which are invertible matrices!), the solution is unique.

The general proof of the converse statements uses the same argument. Let $n(i_1, i_2, \ldots, i_t)$ denote the number of occurrences of the t-tuple $(i_1, \ldots, i_t)$ in the t columns under consideration, where each i_r is in the range $0 \le i_r \le s-1$. By choosing the vector v to have all possible s^t different values in these t coordinates, and to be zero elsewhere, we obtain s^t equations for the s^t unknowns $n(i_1, \ldots, i_t)$. If v is identically zero the right-hand side of the equation is N, otherwise it is 0. Certainly setting all $n(i_1, \ldots, i_t)$ equal to N/s^t is a solution. The coefficient matrix is the character table of an elementary abelian group of type s^t, which (by the orthogonality of characters) is an invertible matrix. Therefore the solution is unique, and the proof is complete. ∎

3.6 Concluding Remarks

In this chapter we have used Galois fields and finite geometries to construct orthogonal arrays. The main results are that if s is a prime power then we can find orthogonal arrays with parameters $OA(s^t, s+1, s, t)$ for $s \ge t \ge 2$ (Theorem 3.1), $OA(s^3, s+2, s, 3)$ for $s = 2^m$ (Theorem 3.2), $OA(s^{s-1}, s+2, s, s-1)$ for $s = 2^m$ (Corollary 3.8), $OA(2s^n, 2(s^n-1)/(s-1)-1, s, 2)$ for s odd and $n \ge 2$ (Theorem 3.16), and $OA(s^n, (s^n-1)/(s-1), s, 2)$ for $n \ge 2$ (Theorem 3.20).

We also introduced the notion of a linear orthogonal array and showed that this leads to simpler ways to present the arrays.

In Chapters 4 and 5 we will continue to use Galois fields and will investigate the connections between orthogonal arrays and error-correcting codes. Most of the arrays we obtain will be linear. In particular, we will give an alternative way of looking at the constructions of Theorems 3.1 and 3.20.

3.7 Problems

3.1. To complete the proof of Theorem 3.16, check that the $2s^n \times 2$ subarray has the desired property in the following cases:

(i) The two factors correspond to the column labels $f_1^{(i)}(g_1^{(i)})$ and $f_j^{(i')}(g_j^{(i')})$, $1 \le i,\ i' \le r$, $2 \le j \le s$.

(ii) The two factors correspond to the column labels $f_1^{(i)}(g_1^{(i)})$ and $f_{s+1}^{(i')}(g_{s+1}^{(i')})$, $1 \le i,\ i' \le r$.

(iii) The two factors correspond to the column labels $f_j^{(i)}(g_j^{(i)})$ and $f_{s+j'}^{(i')}(g_{s+j'}^{(i')})$, $1 \le i,\ i' \le r$, $2 \le j,\ j' \le s$.

3.2. a. Following the proof of Theorem 3.16 we discussed the use of this construction in the case $s = 2^m$. If we choose d_j and d as in that discussion, show that the conditions on b_j and c_j stated there are indeed necessary.

b. For $m = 2$, find values for b_j and c_j, $1 \le j \le 3$, that satisfy the conditions discussed in part a. Use these values to construct an $OA(32, 9, 4, 2)$.

3.3. Show that Constructions 2 and 3 in Theorem 3.20 produce isomorphic arrays.

3.4. For each of the following parameters, consider whether an orthogonal array can be constructed by a method given in this chapter. If so, identify the method(s); if not, can we conclude from Chapter 2 that there is no orthogonal array with such parameters?

(i) $N = 8$, $k = 7$, $s = 2$, $t = 2$,

(ii) $N = 18$, $k = 6$, $s = 3$, $t = 2$,

(iii) $N = 27$, $k = 5$, $s = 3$, $t = 3$,

(iv) $N = 64$, $k = 10$, $s = 4$, $t = 2$.

3.5. For the orthogonal array in Table 3.26 we were able to identify four factors that formed an $OA(27, 4, 3, 3)$.

a. Give another selection of four factors from the orthogonal array in this table that has this property.

b. If we select five factors from the array in Table 3.26, we know that the resulting orthogonal array cannot have strength three. Thus for any 27×5 subarray of the orthogonal array in Table 3.26 there are three factors for which not all level combinations appear. There are $\binom{5}{3} = 10$ ways to select three of the five factors. Can we select the five factors in such a way that for only one of these ten possible selections of three factors the requirement for an orthogonal array of strength three fails to hold? Also, what would you consider to be a poor choice of five factors?

3.6. Examine the orthogonal arrays in Tables 1.3, 1.8 and Problem 1.4, and determine which are linear.

3.7. Use Theorem 3.20 to construct a linear $OA(9, 4, 3, 2)$.

3.8. Show that if a linear orthogonal array for k factors contains a run with weight k, then it is isomorphic to one containing the runs 00...0, 11...1, ... , $s-1, s-1, ..., s-1$.

3.9. Show by choosing a different labeling for the pencils in Example 3.23 that Construction 1 in Theorem 3.20 does not always produce a linear array.

3.10. The following is an alternative test for the orthogonality of a four-level array (compare Theorem 3.30). Let A be an $N \times k$ array with entries from $GF(4)$. We may write the elements of $GF(4)$ in the form $a = b + c\alpha$, $b, c \in GF(2)$, where α is a primitive element of $GF(4)$. Define an inner product on $GF(4)$ by $\langle a, a' \rangle = bb' + cc'$ for $a, b \in GF(4)$, and extend it to vectors $u = (a_1, \ldots, a_k)$, $u' = (a'_1, \ldots, a'_k)$ in $GF(4)^k$ by

$$\langle u, u' \rangle = \langle (a_1, \ldots, a_k), (a'_1, \ldots, a'_k) \rangle = \langle a_1, a'_1 \rangle + \cdots + \langle a_k, a'_k \rangle .$$

Show that A is an orthogonal array of strength t if and only if

$$\sum_{u=\text{row of A}} (-1)^{\langle u, u' \rangle} = 0$$

summed over all rows u of A, for all nonzero $u' \in GF(4)^k$ having at most t nonzero entries.

Table 3.31. An $OA(54, 25, 3, 2)$ (transposed). (a) The matrix F, containing the first 27 runs.

X_1	0	0	0	0	0	0	0	0	0	1	1	1	1	1	1	1	1	1	2	2	2	2	2	2	2	2	2
X_2	0	0	0	1	1	1	2	2	2	0	0	0	1	1	1	2	2	2	0	0	0	1	1	1	2	2	2
X_3	0	1	2	0	1	2	0	1	2	0	1	2	0	1	2	0	1	2	0	1	2	0	1	2	0	1	2
X_1	0	0	0	0	0	0	0	0	0	1	1	1	1	1	1	1	1	1	2	2	2	2	2	2	2	2	2
X_3	0	1	2	0	1	2	0	1	2	0	1	2	0	1	2	0	1	2	0	1	2	0	1	2	0	1	2
X_1+2X_3	0	2	1	0	2	1	0	2	1	1	0	2	1	0	2	1	0	2	2	1	0	2	1	0	2	1	0
X_1+X_3	0	1	2	0	1	2	0	1	2	1	2	0	1	2	0	1	2	0	2	0	1	2	0	1	2	0	1
$X_1^2+X_3$	0	1	2	0	1	2	0	1	2	1	2	0	1	2	0	1	2	0	1	2	0	1	2	0	1	2	0
$X_1^2+2X_1+X_3$	0	1	2	0	1	2	0	1	2	0	1	2	0	1	2	0	1	2	2	0	1	2	0	1	2	0	1
$X_1+X_1+X_3$	0	1	2	0	1	2	0	1	2	2	0	1	2	0	1	2	0	1	0	1	2	0	1	2	0	1	2
X_2	0	0	0	1	1	1	2	2	2	0	0	0	1	1	1	2	2	2	0	0	0	1	1	1	2	2	2
X_1+2X_2	0	0	0	2	2	2	1	1	1	1	1	1	0	0	0	2	2	2	2	2	2	1	1	1	0	0	0
X_1+X_2	0	0	0	1	1	1	2	2	2	1	1	1	2	2	2	0	0	0	2	2	2	0	0	0	1	1	1
$X_1^2+X_2$	0	0	0	1	1	1	2	2	2	1	1	1	2	2	2	0	0	0	1	1	1	2	2	2	0	0	0
$X_1^2+2X_1+X_2$	0	0	0	1	1	1	2	2	2	0	0	0	1	1	1	2	2	2	2	2	2	0	0	0	1	1	1
$X_1+X_1+X_2$	0	0	0	1	1	1	2	2	2	2	2	2	0	0	0	1	1	1	0	0	0	1	1	1	2	2	2
X_2+X_3	0	1	2	1	2	0	2	0	1	0	1	2	1	2	0	2	0	1	0	1	2	1	2	0	2	0	1
$X_1+2X_2+2X_3$	0	2	1	2	1	0	1	0	2	1	0	2	0	2	1	2	1	0	2	1	0	1	0	2	0	2	1
$X_1+X_2+X_3$	0	1	2	1	2	0	2	0	1	1	2	0	2	0	1	0	1	2	2	0	1	0	1	2	1	2	0
$X_1^2+X_2+X_3$	0	1	2	1	2	0	2	0	1	1	2	0	2	0	1	0	1	2	1	2	0	2	0	1	0	1	2
$X_1^2+2X_1+X_2+X_3$	0	1	2	1	2	0	2	0	1	0	1	2	1	2	0	2	0	1	2	0	1	0	1	2	1	2	0
$X_1^2+X_1+X_2+X_3$	0	1	2	1	2	0	2	0	1	2	0	1	0	1	2	1	2	0	0	1	2	1	2	0	2	0	1
X_2+2X_3	0	2	1	1	0	2	2	1	0	0	2	1	1	0	2	2	1	0	0	2	1	1	0	2	2	1	0
$X_1+2X_2+X_3$	0	1	2	2	0	1	1	2	0	1	2	0	0	1	2	2	0	1	2	0	1	1	2	0	0	1	2
$X_1+X_2+2X_3$	0	2	1	1	0	2	2	1	0	1	0	2	2	1	0	0	2	1	2	1	0	0	2	1	1	0	2
$X_1+X_2+2X_3$	0	2	1	1	0	2	2	1	0	1	0	2	2	1	0	0	2	1	1	0	2	2	1	0	0	2	1
$X_1^2+2X_1+X_2+2X_3$	0	2	1	1	0	2	2	1	0	0	2	1	1	0	2	2	1	0	2	1	0	0	2	1	1	0	2
$X_1+X_1+X_2+2X_3$	0	2	1	1	0	2	2	1	0	2	1	0	0	2	1	1	0	2	0	2	1	1	0	2	2	1	0

Table 3.31 (cont.) An $OA(54, 25, 3, 2)$ (transposed). (b) The matrix G, containing the second 27 runs.

X_1	0	0	0	0	0	0	0	0	0	1	1	1	1	1	1	1	1	1	2	2	2	2	2	2	2	2	2
X_2	0	0	0	1	1	1	2	2	2	0	0	0	1	1	1	2	2	2	0	0	0	1	1	1	2	2	2
X_3	0	1	2	0	1	2	0	1	2	0	1	2	0	1	2	0	1	2	0	1	2	0	1	2	0	1	2
X_1	0	0	0	0	0	0	0	0	0	1	1	1	1	1	1	1	1	1	2	2	2	2	2	2	2	2	2
X_3	0	1	2	0	1	2	0	1	2	0	1	2	0	1	2	0	1	2	0	1	2	0	1	2	0	1	2
$X_1 + 2X_3 + 1$	1	0	2	1	0	2	1	0	2	2	1	0	2	1	0	2	1	0	0	2	1	0	2	1	0	2	1
$X_1 + 2X_3 + 1$	1	0	2	1	0	2	1	0	2	2	1	0	2	1	0	2	1	0	0	2	1	0	2	1	0	2	1
$X_1 + X_3 + 2$	2	0	1	2	0	1	2	0	1	0	1	2	0	1	2	0	1	2	1	2	0	1	2	0	1	2	0
$2X_1 + X_3$	0	1	2	0	1	2	0	1	2	2	0	1	2	0	1	2	0	1	2	0	1	2	0	1	2	0	1
$2X_1^2 + X_1 + X_3 + 1$	1	2	0	1	2	0	1	2	0	1	2	0	1	2	0	1	2	0	2	0	1	2	0	1	2	0	1
$2X_1^2 + 2X_1 + X_3 + 1$	1	2	0	1	2	0	1	2	0	2	0	1	2	0	1	2	0	1	1	2	0	1	2	0	1	2	0
X_2	0	0	0	1	1	1	2	2	2	0	0	0	1	1	1	2	2	2	0	0	0	1	1	1	2	2	2
$X_1 + 2X_2 + 1$	1	1	1	0	0	0	2	2	2	2	2	2	1	1	1	0	0	0	0	0	0	2	2	2	1	1	1
$X_1 + X_2 + 2$	2	2	2	0	0	0	1	1	1	0	0	0	1	1	1	2	2	2	1	1	1	2	2	2	0	0	0
$2X_1^2 + X_2$	0	0	0	1	1	1	2	2	2	2	2	2	0	0	0	1	1	1	2	2	2	0	0	0	1	1	1
$2X_1^2 + X_1 + X_2 + 1$	1	1	1	2	2	2	0	0	0	1	1	1	2	2	2	0	0	0	2	2	2	0	0	0	1	1	1
$2X_1 + 2X_1 + X_2 + 1$	1	1	1	2	2	2	0	0	0	2	2	2	0	0	0	1	1	1	1	1	1	2	2	2	0	0	0
$X_2 + X_3$	0	1	2	1	2	0	2	0	1	0	1	2	1	2	0	2	0	1	0	1	2	1	2	0	2	0	1
$X_1 + 2X_2 + X_3 + 1$	1	0	2	0	2	1	2	1	0	2	1	0	1	0	2	0	2	1	0	2	1	2	1	0	1	0	2
$X_1 + X_2 + X_3 + 2$	2	0	1	0	1	2	1	2	0	0	1	2	1	2	0	2	0	1	1	2	0	2	0	1	0	1	2
$2X_1 + X_2 + X_3$	0	1	2	1	2	0	2	0	1	2	0	1	0	1	2	1	2	0	2	0	1	0	1	2	1	2	0
$2X_1^2 + X_1 + X_2 + X_3 + 1$	1	2	0	2	0	1	0	1	2	1	2	0	2	0	1	0	1	2	2	0	1	0	1	2	1	2	0
$2X_1^2 + 2X_1 + X_2 + X_3 + 1$	1	2	0	2	0	1	0	1	2	2	0	1	0	1	2	1	2	0	1	2	0	2	0	1	0	1	2
$X_2 + 2X_3$	0	2	1	1	0	2	2	1	0	0	2	1	1	0	2	2	1	0	0	2	1	1	0	2	2	1	0
$X_1 + 2X_2 + X_3 + 1$	1	2	0	0	1	2	2	0	1	2	0	1	1	2	0	0	1	2	0	1	2	2	0	1	1	2	0
$X_1 + X_2 + 2X_3 + 2$	2	1	0	0	2	1	1	0	2	0	2	1	1	0	2	2	1	0	1	0	2	2	1	0	0	2	1
$2X_1 + X_2 + 2X_3$	0	2	1	1	0	2	2	1	0	2	1	0	0	2	1	1	0	2	2	1	0	0	2	1	1	0	2
$2X_1^2 + X_1 + X_2 + 2X_3 + 1$	1	0	2	2	1	0	0	2	1	1	0	2	2	1	0	0	2	1	2	1	0	0	2	1	1	0	2
$2X_1^2 + 2X_1 + X_2 + 2X_3 + 1$	1	0	2	2	1	0	0	2	1	2	1	0	0	2	1	1	0	2	1	0	2	2	1	0	0	2	1

Chapter 4

Orthogonal Arrays and Error-Correcting Codes

In this chapter we introduce error-correcting codes and discuss their connections with orthogonal arrays. The two subjects are very closely related, since we can use the codewords in an error-correcting code as the runs of an orthogonal array, or conversely we can regard the runs of an orthogonal array as forming a code.

Sections 4.1 and 4.2 give a self-contained introduction to coding theory. The main theoretical results relating codes and orthogonal arrays, due to Delsarte (1973), are presented in Section 4.4. However, the connections between linear codes and linear orthogonal arrays (which are due to Kempthorne, 1947, and Bose, 1961) are somewhat simpler, and are given earlier, in Section 4.3. These results are a special case of those in Section 4.4. In Section 4.5 we give Delsarte's (1973) linear programming bound as it applies to orthogonal arrays. This bound is always at least as strong as the Rao bounds of Chapter 2, and is usually much stronger. The principal results in this chapter are Theorems 4.6, 4.9 and 4.15.

4.1 An Introduction to Error-Correcting Codes

Error-correcting codes are designed to correct errors in the transmission of data over noisy communication channels. We fix a set of symbols S of size s called the *alphabet*. Then S^k denotes the set of all s^k vectors of length k, and an *error-correcting code*, or simply a *code*, is any collection C (with repetitions allowed) of vectors in S^k. The vectors in C are called *codewords*.

The *Hamming weight* $w(u)$ of a vector $u = (u_1, \ldots, u_k) \in S^k$ is defined to be the number of nonzero components u_i. The *Hamming distance* $\text{dist}(u, v)$ between two vectors $u, v \in S^k$ is defined to be the number of positions where they differ, or in other words

$$\text{dist}(u, v) = w(u - v) .$$

We define the *minimal distance* d of a code C to be the minimal distance between *distinct* codewords:

$$d = \min_{\substack{u,v \in C \\ u \neq v}} \text{dist}(u, v) . \tag{4.1}$$

If the code is empty, d is undefined, and if there is only one distinct codeword then d is defined to be $k + 1$, by convention.

It is not usual in coding theory to allow repeated vectors, but it is appropriate to do so here since orthogonal arrays may contain repeated runs. Note that even if there are repeated codewords, d is still the minimal distance between *distinct* codewords.

If C contains N codewords then we say that it is a code of *length* k, *size* N and minimal distance d over an alphabet of size s, or simply a $(k, N, d)_s$ code.

For data transmission we are only interested in codes without repeated codewords: we call these *simple* codes. Let $e = \lfloor (d-1)/2 \rfloor$. Then a simple code with minimal distance d can correct any number of from 1 to e errors when used on a noisy channel, and so is called an *e-error-correcting* code.

Example 4.1. In the code $\{00000, 11111\}$ the two codewords are at Hamming distance 5 apart, and so the code can correct 2 errors: this is a $(5, 2, 5)_2$ code. Indeed, suppose the codeword 00000 is selected and two errors occur in transmission, so that 01001 (say) is received. The decoder operates by looking for the closest codeword to the received vector, if there is one, or declares that an uncorrectable error has occurred if not. In the present example the decoder would observe that 01001 is closer to 00000 in Hamming distance than to 11111, and so would decide that 00000 was transmitted. In this way the two errors would be corrected. It is even easier if only one error occurs. But if three errors occur, and 01101 (say) is received, the receiver will mistakenly assume that 11111 was sent. Thus zero, one or two errors can be corrected by this code, but not three. (See Problem 4.1.) This is a *double-error-correcting* code. ■

Example 4.2. The code consisting of the single vector $00\ldots0$ of length k has $d = k + 1$, by convention. The reason for this convention will become clear when we state Theorem 4.6. ■

Example 4.3. The Rao-Hamming orthogonal array $OA(8, 7, 2, 2)$ given in Example 3.22, after some rearranging of rows and columns (to make it cyclic —

see Section 5.2), can be written as:

$$\begin{bmatrix} 0 & 0 & 0 & 0 & 0 & 0 & 0 \\ 1 & 1 & 1 & 0 & 1 & 0 & 0 \\ 0 & 1 & 1 & 1 & 0 & 1 & 0 \\ 0 & 0 & 1 & 1 & 1 & 0 & 1 \\ 1 & 0 & 0 & 1 & 1 & 1 & 0 \\ 0 & 1 & 0 & 0 & 1 & 1 & 1 \\ 1 & 0 & 1 & 0 & 0 & 1 & 1 \\ 1 & 1 & 0 & 1 & 0 & 0 & 1 \end{bmatrix} . \tag{4.2}$$

The runs of this array form a $(7, 8, 4)_2$ code. This is a single-error correcting code since $\lfloor (d-1)/2 \rfloor = 1$. It is a member of the class of codes called Hamming codes—see Section 5.3. ■

One of the central problems in coding theory is to find the maximal value of N for a simple code (with distinct codewords) having specified s, k and d. In practice, of course, one also wants codes that are easy to encode and decode.

The theory we shall present in this chapter applies to codes (and orthogonal arrays) with symbols from any abelian group. For simplicity, however, we shall usually only discuss codes over a Galois field $GF(s)$ of order s. Codes over the fields of orders 2, 3 and 4 have been the most widely studied, and are called *binary*, *ternary* and *quaternary* codes respectively. We shall sometimes use these same adjectives when talking about orthogonal arrays based on $GF(2)$, $GF(3)$ or $GF(4)$, especially when they are constructed from codes.

4.2 Linear Codes

As defined in Section 4.1, a code need have no mathematical structure whatsoever. In practice, in order to make encoding and decoding feasible, one uses codes that have a considerable amount of mathematical structure.

A very useful property is that of linearity, a notion we have already introduced for orthogonal arrays in Section 3.2. The definitions of linearity, dimension and generator matrix given in that section apply equally well to codes. Thus a code C of length k is said to be *linear* if the codewords are distinct and C is a vector subspace of S^k. This implies that C has size $N = s^n$ for some nonnegative integer n, $0 \le n \le k$, called the *dimension* of the code[1]. For a linear code the minimal distance d is equal to the minimal weight of any nonzero codeword:

$$d = \min_{u \in C, u \neq 0} w(u) . \tag{4.3}$$

[1]It is traditional in coding theory to use the symbol n for the length of a code and k for its dimension. Here the roles of n and k have been reversed, so any coding theorist reading this chapter should remember a simple rule: "interchange n and k".

The reader is asked to verify this in Problem 4.2.

A linear code may be concisely specified by giving an $n \times k$ *generator matrix* G, whose rows form a basis for the code. The code then consists of all vectors $u = xG$, where x runs through S^n. Example 4.1 described a linear code with generator matrix

$$G = [1\ 1\ 1\ 1\ 1]\ ,$$

and the code in Example 4.3 is linear with generator matrix

$$G = \begin{bmatrix} 1 & 1 & 1 & 0 & 1 & 0 & 0 \\ 0 & 1 & 1 & 1 & 0 & 1 & 0 \\ 0 & 0 & 1 & 1 & 1 & 0 & 1 \end{bmatrix} . \qquad (4.4)$$

By reordering (if necessary) the coordinates of a linear code, it is always possible to find a generator matrix of the form

$$G = [I_n\ A]\ , \qquad (4.5)$$

where A is an $n \times (k-n)$ matrix with entries from S.

An alternative way to specify a linear code is to give a *parity check matrix* P. This is a $(k-n) \times k$ matrix with entries from S whose rows span the orthogonal space to the code. The code then consists of all vectors $u \in S^k$ such that $Pu^T = 0$. A parity check matrix for the code in Example 4.1 is

$$P = \begin{bmatrix} 1 & 1 & 0 & 0 & 0 \\ 1 & 0 & 1 & 0 & 0 \\ 1 & 0 & 0 & 1 & 0 \\ 1 & 0 & 0 & 0 & 1 \end{bmatrix} , \qquad (4.6)$$

and a parity check matrix for the code in Example 4.3 is

$$P = \begin{bmatrix} 1 & 0 & 1 & 1 & 0 & 0 & 0 \\ 0 & 1 & 0 & 1 & 1 & 0 & 0 \\ 0 & 0 & 1 & 0 & 1 & 1 & 0 \\ 0 & 0 & 0 & 1 & 0 & 1 & 1 \end{bmatrix} . \qquad (4.7)$$

If the generator matrix is given by (4.5), a corresponding parity check matrix is

$$P = [-A^T\ I_{k-n}]\ . \qquad (4.8)$$

Associated with any linear code C is another linear code called its *dual*, and denoted by $C^\perp$. This consists of all vectors $v \in S^k$ such that[2]

$$uv^T = 0 \quad \text{for all} \quad u \in C\ . \qquad (4.9)$$

[2] If the symbols are from a field with a natural *conjugation* operation (such as $GF(4)$, where conjugation sends x to x^2), it is better to replace (4.9) with "$u\bar{v}^T = 0$ for all $u \in C$."

Duality is very important, because it is the concept that ties together codes and orthogonal arrays (see Theorem 4.9). The reader will easily verify the properties stated in the following theorem (see Problem 4.4).

Theorem 4.4. *Let C be a $(k, s^n, d)_s$ linear code. Then (a) the dual code $C^\perp$ has length k, dimension $k-n$, and minimal distance $d^\perp$ for some number $d^\perp$, i.e. $C^\perp$ is a $(k, s^{k-n}, d^\perp)_s$ code; (b) a generator matrix for C is a parity check matrix for $C^\perp$, a parity check matrix for C is a generator matrix for $C^\perp$; and (c) $(C^\perp)^\perp = C$.*

The number $d^\perp$ is called the *dual distance* of C. In Section 4.5 we shall see how to define the dual distance even for a nonlinear code.

For example, the dual to the code in Example 4.3 has generator matrix (4.7). This is a $(7, 16, 3)_2$ code.

A code which is its own dual is called *self-dual.* Many of the most important of all codes are self-dual: the Golay codes mentioned in Section 5.7, for instance.

4.3 Linear Codes and Linear Orthogonal Arrays

We shall associate to any orthogonal array $OA(N, k, s, t)$, linear or not, the code formed by its runs. This is a $(k, N, d)_s$ code for some d. Conversely, to any $(k, N, d)_s$ code we associate the $N \times k$ array whose rows are the codewords. This is an orthogonal array $OA(N, k, s, t)$ for some t.

Theorem 4.5. *The orthogonal array associated with a code is linear if and only if the code is linear.*

Proof: This follows immediately from the definitions of linearity given in Sections 3.2 and 4.2. ∎

We can now define a *parity check matrix* for an orthogonal array to be any parity check matrix for the associated code, and the *dual orthogonal array* to be the orthogonal array corresponding to the dual of the code associated with the original array.

It follows from (4.9) that the rows of the dual array are all the k-tuples v such that $uv^T = 0$ for all the runs u in the original array. Thus the pair of arrays associated to a linear code and its dual are respectively a linear array and *its* dual.

The main theorem of this section, first stated explicitly by Bose (1961), specifies how the strength of a linear orthogonal array is determined by the associated code. The first half of the theorem had essentially been given by Kempthorne (1947). Although Theorem 4.6 is a special case of a much more profound result due to Delsarte (1973) that will be presented in Section 4.5, it seems worthwhile giving a separate proof, especially as the result follows easily from the theorems of Section 3.5.

Theorem 4.6. *If C is a $(k, N, d)_s$ linear code over $GF(s)$ with dual distance $d^\perp$ then the codewords of C form the rows of an $OA(N, k, s, d^\perp - 1)$ with entries from $GF(s)$. Conversely, the rows of a linear $OA(N, k, s, t)$ over $GF(s)$ form a $(k, N, d)_s$ linear code over $GF(s)$ with dual distance $d^\perp \geq t+1$. If the orthogonal array has strength t but not $t+1$, $d^\perp$ is precisely $t+1$.*

Proof: Suppose C is a code with the properties stated in the first part of the theorem, and let A be the array formed by the codewords. Any $d^\perp - 1$ columns of A must be linearly independent over $GF(s)$, or else there would be a codeword of weight less than $d^\perp$ in the dual code, contradicting the hypothesis that $d^\perp$ is the minimal nonzero weight in the dual. By Theorem 3.29, A is an $OA(N, k, s, d^\perp - 1)$.

Conversely, let C be the code associated with a linear $OA(N, k, s, t)$. By Theorem 3.27, any t columns of the array are linearly independent; and so there cannot be a codeword of weight t or less in $C^\perp$. If the array does not have strength $t+1$, some $t+1$ columns are dependent, and so there is a codeword of weight $t+1$ in the dual code, hence $d^\perp = t+1$. ■

Note that there is no direct relationship between d and t in the theorem, only between $d^\perp$ and t.

To illustrate the theorem, consider first the $(5, 2, 5)_2$ code C in Example 4.1, whose dual we have seen is a $(5, 16, 2)_2$ code. From Theorem 4.6, the codewords of C form a (not very interesting) $OA(2, 5, 2, 1)$, while the codewords of $C^\perp$ form an $OA(16, 5, 2, 4)$. The latter is an instance of a zero-sum array, as defined in Section 2.1. (Problem 4.5 gives the connection between zero-sum arrays and codes in the general case.)

Second, consider the $(7, 8, 4)_2$ code C in Example 4.3, whose dual is a $(7, 16, 3)_2$ code. From Theorem 4.6, the codewords of C form an $OA(8, 7, 2, 2)$, which is what we started from in (4.2), while the codewords of $C^\perp$ form an $OA(16, 7, 2, 3)$. If we had started from the $(7, 16, 3)_2$ code, we would have found the $OA(8, 7, 2, 2)$ by taking the set of all codewords of the dual code. This is an instance of the Rao-Hamming construction already described in Section 3.4. We give the general coding-theory formulation of this construction in Section 5.3.

Third, the $(5, 32, 1)_2$ code consisting of all 5-tuples from $GF(2)$ has as dual the zero code $\{00000\}$, which by convention has minimal distance 6. Indeed, the associated orthogonal array is an $OA(32, 5, 2, 5)$, with strength $5 = 6 - 1$.

This is the reason we adopted that convention: without it, orthogonal arrays that contained all possible runs would have to be handled separately from the others.

Two codes are said to be *isomorphic* if one can be obtained from the other by permuting the coordinate portions and then permuting the symbols in each coordinate position. If C and C' are codes with associated orthogonal arrays A and A', then it follows immediately from the definitions that C is isomorphic to C' if and only if A is isomorphic to A'.

Two linear codes are *isomorphic as linear codes* if one can be obtained from the other by permuting the coordinate positions and multiplying each coordinate position by a nonzero element of the field. If two linear orthogonal arrays are statistically equivalent then the associated codes are isomorphic as linear codes.

In Chapter 5 we will see many further examples of orthogonal arrays that can be constructed from codes.

We can weaken the notion of linearity to allow linear codes over rings and additive codes over fields, just as we did for orthogonal arrays in Remark 3.15. Linear codes over the ring $\mathbb{Z}_4$ will appear in Section 5.10.

4.4 Weight Enumerators and Delsarte's Theorem

In this section we present a powerful generalization of Theorem 4.6 that applies even to nonlinear orthogonal arrays. This is important because it leads to the linear programming bound for orthogonal arrays. To state the theorem we must define the dual distance of a nonlinear code, and with that goal in mind we now introduce the weight distribution and weight enumerator of a code.

For a $(k, N, d)_s$ code C we define the *weight distribution with respect to a codeword* $u \in C$ to be the $(k+1)$-tuple of nonnegative integers $(A_0(u), \ldots, A_k(u))$, where $A_i(u)$ is the number of codewords at Hamming distance i from u. We have already encountered the $A_i(u)$ in the context of orthogonal arrays in Lemma 2.7.

The *weight distribution* of the code is the $(k+1)$-tuple $(A_0, \ldots, A_k)$, where

$$A_i = \frac{1}{N} \sum_{u \in C} A_i(u) , \quad 0 \le i \le k .$$

The A_i are nonnegative rational numbers. The minimal distance of the code is the largest positive integer d such that

$$A_1 = \cdots = A_{d-1} = 0 .$$

Finally, the *weight enumerator*[3] of C is

$$W_C(x,y) = \sum_{i=0}^{k} A_i x^{k-i} y^i \ .$$

This is a homogeneous polynomial whose degree is equal to the length of the code

For a linear code, $A_i(u)$ is independent of u, and $A_i = A_i(u)$ for all $u \in C$. In any case we have $\sum_{i=0}^{k} A_i = N$ and

$$\begin{aligned} A_0 &\geq 1 \ , \\ A_1 &= A_2 = \cdots = A_{d-1} = 0 \ , \\ A_i &\geq 0 \ , \quad \text{for } 0 \leq i \leq k \ . \end{aligned} \tag{4.10}$$

The code is simple if and only if $A_0 = 1$.

If the code is linear there is a beautiful formula for the weight enumerator of the dual code, due to F. J. MacWilliams.

Theorem 4.7. *For a $(k, s^n, d)_s$ linear code C,*

$$W_{C^\perp}(x,y) = \frac{1}{N} W_C(x + (s-1)y, x - y) \ . \tag{4.11}$$

To illustrate this theorem, we see that the code in Example 4.1 has weight enumerator

$$W_C(x,y) = x^5 + y^5 \ , \tag{4.12}$$

and its dual (the zero-sum code with generator matrix (4.6)) has weight enumerator

$$W_{C^\perp}(x,y) = x^5 + 10x^3y^2 + 5xy^4 \tag{4.13}$$

(since there are $\binom{5}{2} = 10$ codewords with two 1's, and $\binom{5}{4} = 5$ codewords with four 1's). The reader will readily verify that (4.12) and (4.13) satisfy (4.11). Again, the code in Example 4.3 has weight enumerator

$$W_C(x,y) = x^7 + 7x^3y^4 \ .$$

The reader is invited to verify that the dual code $C^\perp$, with generator matrix (4.7), has weight enumerator

$$W_{C^\perp}(x,y) = x^7 + 7x^4y^3 + 7x^3y^4 + y^7$$

[3] There are considerable advantages in writing the weight enumerator as a homogeneous polynomial in two variables. Logically, one variable would be enough. We could, for example, set $x = 1$.

and that (4.11) holds.

Since we shall not make any direct use of Theorem 4.7, other than to motivate the definition of "dual distance", we omit the proof — see MacWilliams and Sloane (1977, Chapter 5).

Still considering linear codes, let $(A_0^\perp, \ldots, A_k^\perp)$ be the weight distribution of the dual code $C^\perp$. Equation (4.11) states that

$$\sum_{i=0}^{k} A_i^\perp x^{k-i} y^i = \frac{1}{N} \sum_{j=0}^{k} A_j (x + (s-1)y)^{k-j} (x-y)^j .$$

By expanding the right-hand side we find that

$$A_i^\perp = \frac{1}{N} \sum_{j=0}^{k} A_j P_i(j) , \quad 0 \le i \le k , \tag{4.14}$$

where

$$P_i(z) = \sum_{r=0}^{i} (-1)^r (s-1)^{i-r} \binom{z}{r} \binom{k-z}{i-r} , \quad 0 \le i \le k , \tag{4.15}$$

and

$$\binom{z}{r} = \frac{z(z-1)\cdots(z-r+1)}{r!} .$$

The polynomials defined in (4.15) are *Krawtchouk polynomials*, one of the standard families of orthogonal polynomials (Szegö, 1967; Delsarte, 1973; MacWilliams and Sloane, 1977, Chapter 5).

Since the numbers $(A_0^\perp, \ldots, A_k^\perp)$ are the weight distribution of the dual code, they satisfy $\sum_{i=0}^{k} A_i^\perp = s^k/N$ and

$$\begin{aligned} A_0^\perp &= 1 , \\ A_1^\perp &= A_2^\perp = \cdots = A_{d^\perp - 1}^\perp = 0 , \\ A_i^\perp &\ge 0 , \quad \text{for} \quad 0 \le i \le k . \end{aligned} \tag{4.16}$$

We come now to the heart of the matter. *For a nonlinear code we still define the dual weight distribution by (4.11) and (4.14)*, calling the numbers $(A_0^\perp, \ldots, A_k^\perp)$ the *MacWilliams transform* of the weight distribution. Then it is still true that $A_i^\perp \ge 0$ for all i (see Theorem 4.11 below), and we define the *dual distance* to be the largest positive integer $d^\perp$ such that

$$A_1^\perp = \cdots = A_{d^\perp - 1}^\perp = 0 .$$

Thus if $A_1^\perp = \cdots = A_k^\perp = 0$, we define $d^\perp$ to be $k+1$. It also follows (from (4.14)) that $A_0^\perp = 1$.

Example 4.8. Consider the $OA(12, 11, 2, 2)$ obtained by deleting the first column of the matrix in Table 7.15. Although this Hadamard array is nonlinear, $A_i(u)$ is still independent of u, and the weight enumerator of the code C formed by the runs is

$$W_C(x, y) = x^{11} + 11x^5y^6 .$$

The dual weight distribution is found by calculating

$$\begin{aligned} & \frac{1}{12} W(x + y, x - y) \\ = & \; x^{11} + \frac{55}{3}x^8y^3 + \frac{110}{3}x^7y^4 + \frac{88}{3}x^6y^5 \\ & + \frac{88}{3}x^5y^6 + \frac{110}{3}x^4y^7 + \frac{55}{3}x^3y^8 + y^{11} , \end{aligned}$$

and is

$$A_0^\perp = 1, \quad A_1^\perp = A_2^\perp = 0, \quad A_3^\perp = \frac{55}{3}, \dots ,$$

so that $d^\perp = 3$. ■

We can now state the main theorem of this section, due to Delsarte (1973).

Theorem 4.9. *If C is a $(k, N, d)_s$ code with dual distance $d^\perp$ then the corresponding orthogonal array is an $OA(N, k, s, d^\perp - 1)$. Conversely, the code corresponding to an $OA(N, k, s, t)$ is a $(k, N, d)_s$ code with dual distance $d^\perp \geq t + 1$. If the orthogonal array has strength t but not $t + 1$, $d^\perp$ is precisely $t + 1$.*

We shall prove Theorem 4.9 only in the binary case. The general case is a straightforward extension of this argument but requires the use of group characters (see Delsarte, 1973; MacWilliams and Sloane, 1977, Chapter 5).

Before giving the proof we state a simple property of Krawtchouk polynomials in the case $s = 2$.

Theorem 4.10. *If $v \in GF(2)^k$ has weight j then*

$$\sum_{w(u)=i} (-1)^{uv^T} = P_i(j) , \tag{4.17}$$

the sum being taken over all $u \in GF(2)^k$ of weight i.

Proof: There are $\binom{j}{r}\binom{k-j}{i-r}$ ways to choose u so that it has precisely r ones in common with v, so the left-hand side of (4.17) is

$$\sum_{r=0}^{i} (-1)^r \binom{j}{r} \binom{k-j}{i-r} ,$$

which is $P_i(j)$. ■

The next result will be used both in the proof of Theorem 4.9 and in Section 4.5.

Theorem 4.11. (Delsarte, 1973). *For any* $(k, N, d)_s$ *code* C *over* $GF(s)$, *the dual weight distribution satisfies*

$$A_i^{\perp} \geq 0 \,, \quad 0 \leq i \leq k \,. \tag{4.18}$$

The theorem is actually true for codes over abelian groups (i.e. for all codes), although we shall prove it only for codes over $GF(2)$.

Proof: By definition,

$$A_j = \frac{1}{N} \sum_{\substack{x \in C \quad y \in C \\ \operatorname{dist}(x,y) = j}} 1 \,.$$

Then

$$\begin{aligned} A_i^{\perp} &= \frac{1}{N} \sum_{j=0}^{k} A_j P_i(j) \\ &= \frac{1}{N^2} \sum_{j=0}^{k} \sum_{x \in C} \sum_{y \in C} \sum_{w(v)=i} (-1)^{(x-y)v^T} \,, \end{aligned}$$

by Theorem 4.10, where $\operatorname{dist}(x, y) = j$,

$$\begin{aligned} &= \sum_{w(v)=i} \left(\frac{1}{N} \sum_{x \in C} (-1)^{xv^T} \right)^2 \\ &\geq 0 \,. \end{aligned} \tag{4.19}$$

■

Proof of Theorem 4.9. If C has dual distance $d^{\perp}$, then $A_1^{\perp} = \cdots = A_{d^{\perp}-1}^{\perp} = 0$. From (4.19), this implies

$$\sum_{x \in C} (-1)^{xv^T} = 0$$

for all v with $1 \leq w(v) \leq d^{\perp} - 1$. Therefore, by Theorem 3.30, the matrix of codewords of C forms an orthogonal array of strength $d^{\perp} - 1$. Conversely, if the orthogonal array has strength t, from Theorem 3.30 and (4.19) we have $A_1^{\perp} = \cdots = A_{d^{\perp}-1}^{\perp} = 0$, and so $d^{\perp} \geq t + 1$. ■

Example 4.12. Let C be the nonlinear code defined in Example 4.8 by the runs of the $OA(12, 11, 2, 2)$. As we saw, this code has dual distance $d^{\perp} = 3$, and indeed the array does have strength 2, in agreement with the theorem. ■

Example 4.13. Let C be the nonlinear (and not simple) code {000, 000, 011, 011, 101, 101, 110, 110}. This has weight enumerator

$$W_C(x, y) = 2(x^3 + 3xy^2) ,$$

and from (4.11) we find

$$W_{C^{\perp}}(x, y) = x^3 + y^3 ,$$

so that $d^{\perp} = 3$. The codewords of C do indeed form an orthogonal array of strength 2. ■

Example 4.14. The nonlinear code $\{00, 00, 01, 01, 10, 10, 11, 11\}$ has

$$\begin{aligned} W_C(x,y) &= 2(x+y)^2 , \\ W_{C^{\perp}}(x,y) &= x^2 , \end{aligned}$$

so $d^{\perp} = 3$, and again the array has strength 2. ■

Of course the linear codes and arrays given at the end of Section 4.3 are also illustrations of Theorem 4.9.

4.5 The Linear Programming Bound

Theorem 4.11 is not as innocent as it appears. The inequalities (4.18) are strong linear constraints on the weight distribution of a code.

Suppose a $(k, N, d)_s$ code C exists with dual distance $d^{\perp}$, or equivalently that an orthogonal array exists with parameters $OA(N, k, s, d^{\perp} - 1)$. Then the weight distribution $(A_0, \ldots, A_k)$ of the code satisfies the equalities and inequalities given in (4.10) and (4.16), and the total number of codewords (or runs in the array) is given by

$$N = A_0 + A_1 + \cdots + A_k .$$

(We ignore the conditions $A_1 = A_2 = \cdots = A_{d-1} = 0$, since for orthogonal arrays it is only the dual distance that is important.) If we now use linear programming (see for example Chvatal, 1983; Schrijver, 1986; Sierksma, 1996; Simonnard, 1966) to find the smallest possible value of $A_0 + A_1 + \cdots + A_k$ that can be attained without violating these equalities and inequalities, we will certainly have a lower bound on N. Of course this bound may be very weak.

Nevertheless we have established (at least for $s = 2$) the following result, also due to Delsarte (1973), which is the third main theorem of this chapter.

Theorem 4.15. The linear programming bound. *Let $N_{LP}(k, d^\perp)$ be the solution to the following linear programming problem: choose real numbers $A_0, A_1, \ldots, A_k$ so as to*

$$\textit{minimize} \quad A_0 + A_1 + \cdots + A_k \tag{4.20}$$

subject to the constraints

$$\begin{aligned} A_0 &\geq 1, \quad A_i \geq 0, \quad 1 \leq i \leq k \,, \\ B_0 &= 1, \quad B_i \geq 0, \quad 1 \leq i \leq k \,, \\ B_1 &= \cdots = B_t = 0 \,, \end{aligned} \tag{4.21}$$

where

$$B_i = \sum_{j=0}^{k} A_j P_i(j), \quad 0 \leq i \leq k \,, \tag{4.22}$$

and $t = d^\perp - 1$. Then the size of any orthogonal array $OA(N, k, s, t)$ satisfies

$$N \geq N_{LP}(k, d^\perp) \,. \tag{4.23}$$

Note that $B_i = NA_i^\perp$ (compare Eq. (4.14)). We refer the reader to Delsarte (1973) for the proof in the general case. In the notation of Section 2.1, Equation (4.23) says that

$$F(k, s, t) \geq N_{LP}(k, d^\perp) \,. \tag{4.24}$$

There is an analogous result for codes, also due to Delsarte (1973). Since we can always repeat a codeword without changing the minimal distance (see Section 4.1), it only makes sense to obtain bounds for simple codes. The bound for codes states that if $M_{LP}(k, d)$ is the solution to the following linear programming problem: choose real numbers A_0, $A_1, \ldots, A_k$ so as to

$$\text{maximize } A_0 + A_1 + \cdots + A_k \tag{4.25}$$

subject to the constraints

$$\begin{aligned} A_0 &\geq 1, \; A_i \geq 0, \; 1 \leq i \leq k, \\ B_0 &= 1, \; B_i \geq 0, \; 1 \leq i \leq k, \\ B_1 &= \cdots = B_{d-1} = 0, \end{aligned} \tag{4.26}$$

where

$$B_i = \sum_{j=0}^{k} A_j P_i(j), \quad 0 \leq i \leq k \,, \tag{4.27}$$

then the size N of any simple $(k, N, d)_s$ code satisfies

$$N \leq M_{LP}(k, d) \,. \tag{4.28}$$

Again $B_i = NA_i^\perp$. This is proved in exactly the same way as Theorem 4.15 — see Chapter 17, Theorem 18 of MacWilliams and Sloane (1977).

These two bounds (Theorem 4.15 and (4.25) & (4.26)) have the same form, except that the roles of A_i and B_i have been interchanged. So a solution to one problem can be converted into a solution to the other. Let C be a set of vectors with minimal distance d and dual distance $d^\perp$. Then we have

$$N_{LP}(k, d^\perp) \le |C| \le \frac{s^k}{N_{LP}(k, d)} \tag{4.29}$$

and

$$\frac{s^k}{M_{LP}(k, d^\perp)} \le |C| \le M_{LP}(k, d) \ , \tag{4.30}$$

where the right-hand side bounds assume that C is a simple code. We will return to the discussion of such pairs of bounds later in this section.

To illustrate Theorem 4.15, in the case $k = 16$, $s = 2$ and $t = 5$ we obtain $N_{LP}(16, 6) = 256$. The optimal solution found by the computer has

$$A_0 = A_{16} = 1, \quad A_6 = A_{10} = 112, \quad A_8 = 30 \ , \tag{4.31}$$

with the other $A_i = 0$, and furthermore the values of $A_0^\perp, \ldots, A_{16}^\perp$ in the optimal solution coincide with $A_0, \ldots, A_{16}$. In fact there is a code (and hence an orthogonal array) with this weight distribution, the Nordstrom-Robinson code, which we will describe in Section 5.10. The code and the array are nonlinear. Remarkably, even though the code is nonlinear, its dual weight distribution does indeed coincide with its weight distribution. (In this particular case the Rao bound gives the same answer as the linear programming bound, but does not give the weight distribution (4.31) of the optimal code.)

The above example is quite exceptional. Usually the values of N_{LP} and A_i are not integers. When they are, there is always the exciting possibility of the existence of an exceptional orthogonal array.

The following is an example where although the A_i are integers, no array exists. For the parameters $k = 22$, $s = 3$, $t = 2$ the Rao bound of Theorem 2.1 and the linear programming bound coincide, giving $N \ge 45$. If an array $OA(45, 22, 3, 2)$ that meets the bound were to exist, then the runs would form a $(22, 45, 15)_3$ code, which by the solution to the linear program would necessarily have the weight distribution

$$A_0 = 1, \quad A_{15} = 44 \ ,$$

with the other $A_i = 0$. Furthermore, such a code would also meet the Plotkin bound for ternary codes (Mackenzie and Seberry, 1984; Plotkin, 1960). Unfortunately neither the code nor the orthogonal array exist. The proof will be given in Section 4.7.

Research Problem 4.16. The following are some examples where the linear programming bound N_{LP} is an integer, the weight distributions A_i and $A_i^\perp$ look especially interesting, but it is not known if the arrays exist: $OA(192, 17, 2, 4)$, $OA(320, 10, 2, 6)$, $OA(768, 12, 2, 6)$, $OA(2^{14}, 20, 2, 8)$, $OA(6144, 14, 2, 10)$, $OA(135, 23, 3, 3)$, $OA(486, 15, 3, 4)$, $OA(891, 20, 3, 4)$, $OA(1458, 16, 3, 5)$. Do any of these exist?

Theorem 4.15 has the drawback that one usually needs a computer to apply it, or to verify bounds that someone else has obtained from it. Furthermore, since the coefficients in (4.14) are large and alternate in sign, one must always worry about the reliability of the computer's answers.

The following corollary to Theorem 4.15 removes some of these drawbacks. In particular, bounds obtained from it can be verified with much less effort than those obtained from Theorem 4.15. (Instead of running the simplex algorithm, one need only check that certain numbers have the correct sign.) We state this result for both orthogonal arrays and codes, since the coding version is slightly simpler and will be useful later in this section (so strictly speaking this is not a corollary).

Theorem 4.17. The dual linear programming bound. *(i) Orthogonal arrays. Suppose we can find a polynomial $f(x)$ of the form*

$$f(x) = 1 + \sum_{i=1}^{k} f_i P_i(x) \tag{4.32}$$

such that the following conditions hold:

$$\begin{aligned} f_i &\le 0 \quad \text{for} \quad i = t+1, t+2, \ldots, k, \\ f(j) &\ge 0 \quad \text{for} \quad j = 0, \ldots, k \,. \end{aligned} \tag{4.33}$$

Then the number of runs in any $OA(N, k, s, t)$ satisfies

$$N \ge f(0) \,. \tag{4.34}$$

(ii) Codes. Suppose we can find a polynomial $f(x)$ of the form (4.32) such that the following conditions hold:

$$\begin{aligned} f_i &\ge 0 \quad \text{for} \quad i = 1, \ldots, k, \\ f(j) &\le 0 \quad \text{for} \quad j = d, d+1, \ldots, k \,. \end{aligned} \tag{4.35}$$

Then the number of distinct codewords in any $(k, N, d)_s$ code satisfies

$$N \le f(0) \,. \tag{4.36}$$

Proof: The dual linear program to the linear program defined by (4.20) and (4.21) (see any of Chvatal, 1983, p. 56; Sierksma, 1996, p. 114; Simonnard, 1966, p. 92 for the definition of the dual program) is to choose real numbers $f_1, \ldots, f_k$ so as to

$$\text{maximize } 1 + \sum_{i=1}^{k} f_i P_i(0) \tag{4.37}$$

subject to the constraints

$$\begin{aligned} & f_i \leq 0 \quad \text{for} \quad i = t+1, \ldots, k, \\ & \sum_{i=1}^{k} f_i P_i(j) \geq -1 \quad \text{for} \quad j = 0, \ldots, k \, . \end{aligned} \tag{4.38}$$

It follows from the duality theorem of linear programming (see the above references) that any feasible solution to the dual problem gives a lower bound on the optimal solution to the primal problem. So if $f_1, \ldots, f_k$ satisfy (4.38), then

$$N_{LP} \geq 1 + \sum_{i=1}^{k} f_i P_i(0) \, .$$

With $f(x)$ as in (4.32), the first assertion of the theorem follows.

The proof for the coding theory case is analogous — see Chapter 17, Theorem 20 of MacWilliams and Sloane (1977). ∎

Example 4.18. We give just one example to illustrate the application of Theorem 4.17. Many others can be found in Sloane and Young (1999). For $s = t = 3$, and $k = 3a$, where a is a positive integer, consider the polynomial

$$f(x) = 1 + \sum_{i=1}^{3} f_i P_i(x) \, ,$$

where

$$f_1 = \frac{2a-1}{a(3a-2)}, \quad f_2 = \frac{1}{3a-2}, \quad f_3 = \frac{1}{a(3a-2)} \, .$$

When expanded, this is

$$f(x) = -\frac{9}{2a(3a-2)}(x - 2a + 1)(x - 2a)(x - 3a) \, , \tag{4.39}$$

so

$$f(0) = \frac{27a(2a-1)}{3a-2} \, . \tag{4.40}$$

The first condition in (4.33) is trivially satisfied, since $f_4 = f_5 = \cdots = 0$, and the second condition is easily checked by drawing a sketch of the function $f(x)$.

We may conclude that the number of runs in an $OA(N, 3a, 3, 3)$ is at least (4.40). If that expression is not a multiple of 27, we may of course round up to the next multiple of 27. The most interesting case is when k is a multiple of 9, say $k = 3a = 9b$, for then (4.40) rounds up to

$$54b + 27 \; . \tag{4.41}$$

For comparison, the Rao bound in this case is $54b - 3$, which rounds up to $54b$. So the linear programming bound is better by (at least) 27 in this case. Similar polynomials can be found for other values of k (see Sloane and Young, 1999), but in fact for $s = t = 3$ the linear programming bound is better than the Rao bound only when k is a multiple of 9. ■

Table 4.19. Comparison of Rao and linear programming bounds for binary orthogonal arrays of strength 4 with k factors. The entries give the numerical bound (in parentheses) and the next multiple of 16. The last line gives the size of the smallest array known. An asterisk indicates this is the smallest array possible.

k	4	5	6	7	8	9
N_{Rao}	16(11)	16	32(22)	32(29)	48(37)	48(46)
N_{LP}	16	16	32(26.7)	48(42.7)	64	96(85.3)
Known	16*	16*	32*	64*	64*	128

k	10	11	12	13	14	15
N_{Rao}	64(56)	80(67)	80(79)	96(92)	112(106)	128(121)
N_{LP}	96(85.3)	96(85.3)	112(102)	128	128	128
Known	128	128	128	128*	128*	128 *

k	16	17	18	19	20	21
N_{Rao}	144(137)	160(154)	176(172)	192(191)	224(211)	240(232)
N_{LP}	160(154)	192	208(203)	224(220)	240(236)	272(271)
Known	256	256	256	256	512	512

k	22	23	24	25
N_{Rao}	256(254)	288(277)	304(301)	336(326)
N_{LP}	320(305)	352(347)	352(348)	368(358)
Known	512	512	1024	1024

Delsarte (1973) showed that the general Rao bound (Theorem 2.1) is implied by Theorem 4.15, so the linear programming bound is never weaker than (2.3).

In fact the linear programming bound is usually much stronger. Let N_{Rao} denote the value given by the Rao lower bound. Table 4.19 compares N_{Rao} and

N_{LP} for binary orthogonal arrays of strength 4 and k factors, where $4 \le k \le 25$. We give both the numerical values of the bounds (in parentheses) and the value obtained by rounding up to the next multiple of 16. The last line gives the size of the smallest orthogonal array presently known. An asterisk indicates that this is the smallest possible size. It can be seen that N_{LP} is sometimes considerably larger than N_{Rao}, enough so to establish the optimality of the arrays with 8, 13 and 14 factors, even though this does not follow from the Rao bound.

In Chapter 12, Tables 12.1 to 12.3 give the best lower and upper bounds presently known on the size of orthogonal arrays with 2, 3 and 4 levels. It will be seen that, with a few exceptions, the best lower bounds come from linear programming. The exceptions are particular cases of small parameters (such as those described in Section 2.5) where special arguments give better bounds.

The values of the linear programming bound given in the tables in Chapter 12 were obtained as follows. The problem as formulated in Theorem 4.15 was written, almost unchanged, in the AMPL mathematical programming language (Fourer, Gay and Kernigham, 1993). AMPL was then instructed to call the CPLEX program (CPLEX, 1991) to calculate numerical values of N_{LP}. Although the linear programming bound for orthogonal arrays has been known since 1973, this may be the first time that it has been systematically evaluated.

It is worth commenting briefly on how the Rao bound can be deduced from the linear programming bound. The idea is to apply part (ii) of Theorem 4.17, to obtain a bound on codes known as the *Hamming* or *sphere-packing bound* (given on the right-hand side of (4.42) for a code of minimal distance d). We then invoke (4.30) to get the Rao bound (given on the left-hand side of (4.42) for an orthogonal array of strength t). In the case of binary codes of minimal distance $d = 2e + 1$, we define $f(x)$ in (4.32) by taking f_i to be

$$\left\{ P_e(i-1, k-1) / \sum_{j=0}^{e} \binom{k}{j} \right\}^2 ,$$

for $i = 1, \ldots, k$, where we write $P_i(x, k)$ for the polynomial defined in (4.15), to indicate the dependence on k. One can verify that $f(j) = 0$ for $j = d, d+1, \ldots, k$, so (4.35) is satisfied, and

$$f(0) = 2^k / \sum_{j=0}^{e} \binom{k}{j} .$$

A similar definition of $f(x)$ works in the general case (cf. Delsarte, 1973; Levenshtein, 1995), and we obtain the following pair of bounds:

$$B_k(t+1) \le |C| \le \frac{s^k}{B_k(d)} , \tag{4.42}$$

where

$$B_k(d) = \begin{cases} B_{k,e}, & \text{if} \quad d = 2e+1, \\ sB_{k-1,e}, & \text{if} \quad d = 2e+2, \end{cases} \tag{4.43}$$

and

$$B_{k,e} = \sum_{i=0}^{e} \binom{k}{i} (s-1)^i, \quad e = 0, 1, \ldots, k \ , \tag{4.44}$$

and we assume in the right-hand side of (4.42) (and (4.45) and (4.46) below) that the code is simple.

There are several other pairs of bounds of the form (4.30). One such pair is given in the next theorem, from which we will be able to deduce that orthogonal arrays of index unity are exactly the same as what are called MDS codes.

Theorem 4.20. The Singleton bounds *Let C be a set of vectors of length k with minimal distance d and strength t. Then*

$$s^t \leq |C| \leq s^{k-d+1} \ , \tag{4.45}$$

where the right-hand side bound assumes that C is a simple code.

Proof: The left-hand inequality is obvious. To prove the right-hand side, suppose C is a simple $(k, N, d)_s$ code. By deleting any particular coordinate from all the vectors we obtain a $(k-1, N, d-1)_s$ code, and, iterating, a $(k-d+1, N, 1)_s$ code. (The true minimal distances may of course be larger.) Since we have deleted fewer than d coordinates, the vectors remain distinct, and so $N \leq s^{k-d+1}$. ∎

If equality holds on the left-hand side of (4.45), C is of course an $OA(s^t, k, s, t)$ of index unity. The bound on the right-hand side is the Singleton (1964) bound of coding theory, and it seems reasonable to refer to (4.45) as the Singleton bounds. A code for which equality holds on the right-hand side of (4.45) is called a *Maximal Distance Separable*, or *MDS*, code.

It is a remarkable fact, first proved by Delsarte (1973), that if equality holds on either side of (4.45) then it holds on both sides.

Theorem 4.21. *Let C be as in Theorem 4.20. If equality holds on either side of* (4.45) *it holds on both sides, and then C is simultaneously an $OA(s^t, k, s, t)$ of index unity and an MDS code with size s^t, minimal distance $d = k - t + 1$, dual distance $d^\perp = t + 1$, and $d + d^\perp = k + 2$.*

Proof: The following proof is an elaboration of the argument used in the proof of Theorem 4.20. If T is a subset of $\{1, 2, \ldots, k\}$ and u is a vector of length k, u_T will denote the restriction of u to T. Then $|C| = s^t$ if and only if for any t-set T, u_T attains every possibility exactly once as u ranges over C. This is equivalent to having $u_T \neq v_T$ for $u, v \in C$, $u \neq v$. But this says that the set of coordinates in which u differs from v intersects every t-set; the only way for this to happen is for u to differ from v in more than $k - t$ positions. Thus $|C| = s^t$ if and only if $d \geq k - t + 1$, i.e. $t \geq k - d + 1$.

So if $|C| = s^t$, then $|C| \geq s^{k-d+1}$ and $|C| = s^{k-d+1}$ by Theorem 4.20. Conversely, suppose $|C| = s^{k-d+1}$. The proof of Theorem 4.20 shows that if we delete any $d - 1$ coordinates, the shortened vectors are distinct, and so C has strength $t \geq k - d + 1$. By the above argument, $|C| = s^t$. ∎

We shall say more about MDS codes in Section 5.6.

Another pair of bounds is obtained from part (ii) of Theorem 4.17 by restricting $f(x)$ in (4.32) to have degree 1. Here we consider only the binary case. The best choice for $f(x)$ turns out to be

$$1 + \frac{k - 2x}{2d - k}$$

(see p. 543 of MacWilliams and Sloane, 1977). From (4.30) we obtain the following result: if $d > k/2$ and $t > k/2 - 1$ then

$$2^k \left(1 - \frac{k}{2t + 2}\right) \leq |C| \leq \frac{2d}{2d - k} \,. \tag{4.46}$$

The bound on the right-hand side is the *Plotkin bound* for codes (Plotkin, 1960), while the bound on the left was first found by Friedman (1992). This proof of the left-hand side bound is due to Bierbrauer, Gopalakrishnan and Stinson (1996).

Yet another pair of bounds comes from a remarkable result of Levenshtein, the main theorem of Levenshtein (1995), which gives the general analytic solution to the "continuous" versions of the linear programming bounds for codes and orthogonal arrays, where the conditions $f(j) \geq 0$ for $j = 0, \ldots, k$ (in (4.33)) or $f(j) \leq 0$ for $j = d, \ldots, k$ (in (4.35)) are replaced by $f(x) \geq 0$ for $0 \leq x \leq k$ or $f(x) \leq 0$ for $d \leq x \leq k$, respectively. These bounds are too complicated to state here (see Levenshtein, 1995). They also yield an analogue for orthogonal arrays of the McEliece-Rodemich-Rumsey-Welch (1977) bound for codes (cf. Section 10.6). See also Bierbrauer, Gopalakrishnan and Stinson (1998), Delsarte and Levenshtein (1998) and Levenshtein (1998, 1999).

We conclude this section with two other existence criteria for orthogonal arrays that arise from coding theory. The first is another result of Delsarte (1973), and gives a necessary condition for an orthogonal array to meet the Rao bound.

Theorem 4.22. *If an $OA(N, k, s, 2u)$ meets the Rao bound (2.1) with equality, then all zeros of the polynomial*

$$\sum_{r=0}^{u}(-1)^r(s-1)^{u-r}\binom{x-1}{r}\binom{k-x}{u-r} \tag{4.47}$$

belong to the set $\{1, 2, \ldots, k\}$.

This is an analogue of a similar theorem for codes due to Lloyd (1957), and the polynomial in (4.47) is usually called a *Lloyd polynomial.* It is a sum of Krawtchouk polynomials. Lloyd's theorem played an important role in the classification of codes that meet the Hamming bound (the coding-theory analogue of the Rao bound) — see Chapter 6 of MacWilliams and Sloane (1997).

We omit the proof of Theorem 4.22 and just give an example. If an array $OA(45, 22, 3, 2)$ were to exist (it does not — see Corollary 4.27), it would meet the Rao bound. Theorem 4.22 states that

$$\sum_{r=0}^{1}(-1)^r 2^{1-r}\binom{x-1}{2}\binom{22-x}{1-r} = 3(15-x)$$

would then have all its zeros in $\{1, 2, \ldots, 22\}$, which is indeed true (so unfortunately the theorem has not told us anything).

The second is a bound due to Griesmer (1960) that applies specifically to linear codes. When transformed into the context of orthogonal arrays this states the following.

Theorem 4.23. *Let s be a prime power. For positive integers m, t let*

$$k_0 = \sum_{i=0}^{m}\left\lceil \frac{t+1}{s^i} \right\rceil - 1 .$$

Then no linear $OA(s^{k_0-1-m}, k_0, s, t)$ exists.

For example, when $s = 4$, taking $t = 5$ and $m = 1$ we obtain

$$k_0 = \left\lceil \frac{6}{1} \right\rceil + \left\lceil \frac{6}{4} \right\rceil - 1 = 7 .$$

The theorem asserts (correctly, see Table 12.3) that no linear $OA(4^5, 7, 4, 5)$ exists. The Griesmer bound has been strengthened in various ways: see for example Baumert and McEliece (1973), Dodunekov and Manev (1985) and Brouwer (1993, 1998).

4.6 Concluding Remarks

The relationship between codes and orthogonal arrays can now be summarized in one sentence: *a good error-correcting code is a large set of vectors of given length whose distance apart is as large as possible; a good orthogonal array is a small set of vectors of given length whose dual distance is as large as possible.* This is the unifying principle behind everything in the chapter. The second assertion is the message of Theorem 4.9.

In view of this relationship it is not surprising that the two fields contain similar results. In fact there are a number of cases where there are parallel or analogous theorems, roughly following this dictionary:

orthogonal arrays	codes	
number of levels	alphabet size	
number of factors	length of code	(4.48)
number of runs	number of codewords	
strength	minimal distance of dual	

We have already seen several instances of this, in the pairs of bounds displayed in (4.42), (4.45) and (4.46), where the right-hand side gives a bound for codes, and the left-hand an analogous bound for orthogonal arrays. There are also parallel constructions in the two fields. For example, Rao's (1946a, 1947, 1949) construction of orthogonal arrays (described in Theorem 3.20) is essentially identical to the construction of what are universally called Hamming single-error-correcting codes, described by Golay (1949) and Hamming (1950) (see Section 5.3). We have therefore called this the Rao-Hamming construction. Again, Bush's (1952b) construction of orthogonal arrays of index unity (described in Theorem 3.1) is essentially identical to the Reed-Solomon (1960) construction for codes (described in Section 5.5).

4.7 Notes on Chapter 4

Error-correcting codes, like orthogonal arrays, began in the late 1940's. The first papers on the subject were by Shannon (1948, 1992), Golay (1949) and Hamming (1950), among others.

Codes were originally introduced in order to correct errors in the transmission of data over noisy communication channels. However they have also been used in a variety of other applications, such as increasing the storage capacity of compact disks, and by now coding theory has developed into a major area of study in its own right, with a rich mathematical theory.

Although coding theory and orthogonal arrays study analogous problems and have a number of parallel results, with a few notable exceptions until recently the workers in the two areas seem to have been largely unaware of each others' work and the subjects have proceeded almost independently of each other.

The main exceptions are R. C. Bose and his students I. M. Chakravarti, D. K. Ray-Chaudhuri and J. N. Srivastava, and the coding theorist P. Delsarte, all of whom have contributed to both fields. Nevertheless, in spite of the work of these authors, coding theory still has much to contribute to orthogonal arrays. These contributions are of three kinds. (i) Direct constructions of orthogonal arrays from codes, simply by taking the codewords as the runs of the array. This is the subject of Chapter 5. (ii) New general construction techniques for orthogonal arrays that are modifications of existing construction techniques for codes (for example, methods for combining two or more orthogonal arrays to form a new array). These constructions will be discussed in Chapter 10. (iii) Bounds on the minimal number of runs in an orthogonal array, obtained by forming a code from the runs of the array and then using bounds from coding theory. The most important of these bounds is Delsarte's (1973) linear programming bound, given in Theorem 4.15.

It is fair to say that there have been many more papers written about codes during the past fifty-five years than about orthogonal arrays (the bibliography of MacWilliams and Sloane, 1977, already lists 1478 items, and the subject has greatly expanded since then), so it is not surprising that coding theory has so much to contribute.

For more comprehensive information about codes than is given here we refer the reader to the books by McEliece (1977), van Lint (1998), Cameron and van Lint (1991), Tsfasman and Vlăduţ (1991) and Pless and Huffman (1998). Our main reference is MacWilliams and Sloane (1977).

Most of this chapter has borrowed from coding theory to benefit orthogonal arrays. It would be interesting to see if orthogonal arrays can reciprocate and help coding theory. For example:

Research Problem 4.24. *Addelman-Kempthorne codes?* Is there a construction of (necessarily nonlinear) codes that is an analogue of the Addelman-Kempthorne construction given in Theorem 3.16? No such construction is presently known.

Of course there are also a number of fundamental results in coding theory that as yet have no counterpart for orthogonal arrays:

Research Problem 4.25. *Johnson bounds.* Are there any analogues of the Johnson bounds from coding theory (MacWilliams and Sloane, 1977, Chap. 17) that apply to orthogonal arrays?

There are many Ph.D. dissertations still to be written on the subject of this chapter.

The equivalence of orthogonal arrays of index unity and MDS codes stated in Theorem 4.21 was first established by Delsarte (1973). This equivalence was also given in Theorem 12 of Chapter 11 of the first printing of MacWilliams and Sloane (1977). However, our proof there only applies to linear orthogonal arrays and linear codes, so in later editions we weakened the assertion so that the result was only claimed for the linear case. Theorem 4.21 remedies the situation and gives the result in full generality.

Theorem 4.21 can also be deduced from the linear programming bound (Delsarte, 1973). An elegant version of this proof due to Eric Rains is sketched in Problem 4.8; this is a modification of his proof of a generalized Singleton bound for quantum codes (Rains, 1999).

Covering radius. Let us mention yet another connection between codes and orthogonal arrays.

If several orthogonal arrays are available, all with the same parameters, one way to choose between them is to pick the one with the smallest covering radius.

The *covering radius* of an orthogonal array is the maximal Hamming distance of any potential run not in the array from the closest run actually in the array. There is an analogous definition for codes.

An orthogonal array with a small covering radius has the property that no potential treatment combination is too far from one that is actually used. Such arrays have been called *maximin distance designs* (Johnson, Moore and Ylvisaker, 1990; Duckworth, 1999). Several papers by Tietäväinen (1990, 1991), Laihonen (1998), Laihonen and Litsyn (1998, 1999), and others have investigated the relationship between the covering radius of a code and its dual distance. In our terms, these papers relate the strength of an orthogonal array to its covering radius.

We conclude this section by showing the nonexistence of a certain hypothetical orthogonal array mentioned in Section 4.5. The result was established independently by Vladimir Tonchev and Noam Elkies (personal communications). The following argument is due to Noam Elkies.

Theorem 4.26. *There is no* $(22, 45, 15)_3$ *code in which the Hamming distance between every pair of codewords is* 15.

Proof: Assuming that such a code were to exist, let one of the codewords be the zero vector, and form a 44×22 matrix from the other words. Replace 0 by

$1, 0$; 1 by $-\frac{1}{2}, \frac{\sqrt{3}}{2}$ and 2 by $-\frac{1}{2}, -\frac{\sqrt{3}}{2}$, obtaining a 44×44 matrix M. Evaluating the determinant of M in the naive way, we see that $\det M$ is a rational number (since for any diagonal the product of its entries is 0 or contains $\sqrt{3}$ to the power 22). On the other hand the diagonal entries of MM^T are 22 and the off-diagonal entries are $7 - 15/2 = -\frac{1}{2}$. Thus

$$MM^T = \frac{45}{2} I_{44} - \frac{1}{2} J_{44} \,, \quad \det MM^T = \frac{45^{43}}{2^{44}} \,,$$

so $\det M$ is irrational, a contradiction. ■

Corollary 4.27. *No $OA(45, 22, 3, 2)$ exists.*

Proof: As mentioned in Section 4.5, linear programming shows that if such an array were to exist, its rows would form a $(22, 45, 15)_3$ code with the property stated in Theorem 4.26. ■

4.8 Problems

4.1. a. Show that if the minimal distance between codewords is 5, then any error pattern of 0, 1 or 2 errors can be corrected.

b. More generally, show that a simple code with minimal distance d can correct $e = \lfloor (d-1) \rfloor / 2$ errors. [Hint: Use the triangle inequality to show that spheres of Hamming radius e around the codewords are disjoint. So a decoder which replaces a received word by the codeword that is closest to it in Hamming distance will correct up to e errors.]

4.2. Prove that the minimal distance of a linear code is equal to the minimal weight of any nonzero codeword (see (4.3)).

4.3. Let P be a $(k-n) \times k$ parity check matrix for a linear code C over $GF(s)$. Show that C has minimal distance d if and only if every $d-1$ columns of P are linearly independent over $GF(s)$ and some d columns are linearly dependent.

4.4. Prove Theorem 4.4.

4.5. a. Let C be the trivial $(k, s, k)_s$ *repetition code* with generator matrix consisting of the single vector $[1, 1, \ldots, 1]$ over the field $GF(s)$, which forms a (trivial) $OA(s, k, s, 1)$. Show that the dual code $C^\perp$ is a *zero-sum code* with parameters $(k, s^{k-1}, 2)_s$, whose codewords form a zero-sum orthogonal array $OA(s^{k-1}, k, s, k-1)$ as in Section 2.1.

b. Show that the same pair of orthogonal arrays exists even if s is not a prime power, provided we work over $\mathbb{Z}_s$, the ring of integers mod s. Show that these are still duals in the sense of (4.9), provided we interpret that equation in $\mathbb{Z}_s$.

4.6. Which of the codes defined by Examples 4.1, 4.3 and (5.1) are self-dual?

4.7. In Example 4.18 we used the dual linear programming bound to show that any $OA(N, 9b, 3, 3)$ must satisfy $N \geq 54b + 27$. Give an alternative proof using Theorem 2.11.

4.8. a. Make a change of variables in (4.11) and deduce the following more symmetrical version:

$$W_{C^\perp}\left(x + \frac{y}{\sqrt{s}}, \frac{y}{\sqrt{s}}\right) = \frac{(\sqrt{s})^k}{|C|} W_C\left(y + \frac{x}{\sqrt{s}}, \frac{x}{\sqrt{s}}\right) . \qquad (4.49)$$

b. Show that the left-hand side is equal to

$$\sum_{i=0}^{k} \sum_{h=0}^{k-i} \binom{k-i}{h-i} x^{k-h} \left(\frac{y}{\sqrt{s}}\right)^h A_i^\perp \qquad (4.50)$$

in which the coefficient of $x^m y^{k-m}$ is

$$\sum_{i=0}^{k-m} \binom{k-i}{k-m-i} (\sqrt{s})^{-(k-m)} A_i^\perp .$$

c. From a similar expansion of the right-hand side of (4.49) show that

$$\sum_{i=0}^{m} \binom{k-i}{m-i} A_i = \frac{|C|}{s^{k-m}} \sum_{i=0}^{k-m} \binom{k-i}{k-m-i} A_i^\perp , \qquad (4.51)$$

for $m = 0, 1, \ldots$.

d. Deduce Theorems 4.20 and 4.21 from (4.51). [Hint: If C has minimal distance d, take $m = d - 1$. If C has dual distance $d^\perp = t + 1$, take $m = k - d^\perp + 1$.]

Chapter 5

Construction of Orthogonal Arrays from Codes

In this chapter we present some of the most important families of codes and the orthogonal arrays that are derived from them.

The first two sections describe some general constructions, and then Sections 5.3 to 5.10 describe specific classes of codes. One of our goals is to give explicit constructions for many of the orthogonal arrays that are mentioned in the tables in Chapter 12. These arrays are given in Sections 5.11, 5.12 and 5.13, corresponding to $s = 2, 3$ and 4 respectively. Most of these arrays are not well known in the statistics literature.

An important fact from the previous chapter is that if C is a $(k, N, d)_s$ linear code over $GF(s)$, and if the dual code $C^\perp$ has minimal distance $d^\perp$, then the codewords of $C^\perp$ can be used as the rows of an $OA(N, k, s, d^\perp - 1)$.

While this chapter contains many fundamental and beautiful theorems and contributions from coding theory, a reader who wants to see detailed proofs and verifications would do well to keep a copy of MacWilliams and Sloane (1977) nearby.

5.1 Extending a Code by Adding More Coordinates

We begin by reminding the reader that, as we saw in Theorem 2.24, binary orthogonal arrays have a special property, not shared by arrays with greater numbers of levels: an $OA(N, k, 2, 2u)$ exists if and only if an $OA(2N, k + 1, 2, 2u + 1)$ exists.

There is a parallel theorem for codes — as we might expect, following the discussion in Section 4.6 — which states that a $(k, N, 2u+1)_2$ code exists if and only if a $(k+1, N, 2u+2)_2$ code exists — see MacWilliams and Sloane (1977), Chapter 2, Theorem 2.

Given a $(k, N, 2u+1)_2$ code, we obtain the $(k+1, N, 2u+2)_2$ code by appending a 0 to every codeword of even weight, and a 1 to every codeword of odd weight. This process is called *extending* a code, and the new code is called an *extended* code. This construction applies to both linear and nonlinear codes. If the original code was linear, so is the extended code.

For example, extending the $(7, 16, 3)_2$ code with generator matrix (4.7) yields an $(8, 16, 4)_2$ code with generator matrix

$$\begin{bmatrix} 1 & 0 & 1 & 1 & 0 & 0 & 0 & 1 \\ 0 & 1 & 0 & 1 & 1 & 0 & 0 & 1 \\ 0 & 0 & 1 & 0 & 1 & 1 & 0 & 1 \\ 0 & 0 & 0 & 1 & 0 & 1 & 1 & 1 \end{bmatrix} . \tag{5.1}$$

More generally, we say that a linear code has been extended whenever we modify it by adding columns to its generator matrix.

5.2 Cyclic Codes

A linear code over $GF(s)$ is said to be *cyclic* if whenever

$$(c_0, c_1, \ldots, c_{k-2}, c_{k-1}) \tag{5.2}$$

is a codeword, so also is

$$(c_1, c_2, \ldots, c_{k-1}, c_0) . \tag{5.3}$$

As with most of the notions defined in this chapter, we will apply the same terminology to orthogonal arrays. Thus an $OA(N, k, s, t)$ is *cyclic* if it is linear and if whenever (5.2) is a run, so is (5.3). For example, the $OA(8, 7, 2, 2)$ and the $(7, 8, 4)_2$ code shown in (4.2) are both cyclic.

Cyclic codes and arrays have an even more economical description than linear codes. For one can show that any cyclic code or array can be described by a single vector (cf. MacWilliams and Sloane, 1977, Chapter 7, Theorem 1). Suppose the code contains $N = s^n$ codewords, or the array contains $N = s^n$ runs. Then there is always a single *generating vector*

$$g = (g_0, g_1, \ldots, g_{k-1}) \tag{5.4}$$

such that the generator matrix consists of this vector and its first $n - 1$ cyclic shifts. In the case of the orthogonal array and code in (4.2), for example, a generating vector is

$$g = (1110100) , \tag{5.5}$$

as we can see in (4.4).

The generating vector for a $(k, s^n, d)_s$ cyclic code over $GF(s)$ has several important properties. It is convenient to represent codewords in a cyclic code by polynomials, so that (5.2) is represented by the polynomial

$$c_0 + c_1X + c_2X^2 + \cdots + c_{k-1}X^{k-1} .$$

A similar notation is used for the runs of a cyclic orthogonal array. Then the generating vector g (given in (5.4)) is represented by a polynomial

$$g(X) = g_0 + g_1X + \cdots + g_{k-1}X^{k-1} , \tag{5.6}$$

called a *generator polynomial* for the code (or orthogonal array). It is not difficult to prove the following result.

Theorem 5.1. *The generator polynomial $g(X)$ for a cyclic $(k, s^n, d)_s$ code or a cyclic $OA(s^n, k, s, t)$ can always be chosen to have the following properties: (a) the degree of $g(X)$ is $k - n$, (b) the leading coefficient, g_{k-n}, is 1, (c) $g(X)$ divides $X^k - 1$ over $GF(s)$, and (d) the set of all codewords or runs are represented by the polynomials $a(X)g(X)$, where $a(X)$ runs through all polynomials with coefficients from $GF(s)$ and degree not exceeding $n - 1$. There is a unique generator polynomial with these properties.*

The proof is quite elementary, but we shall not give it here. See MacWilliams and Sloane (1977) or Berlekamp (1968), where the reader will find much more about the beautiful algebraic theory of cyclic codes. These references only discuss codes, of course, and not orthogonal arrays, but the proofs are the same, provided we remember the dictionary given in (4.48).

If a code is cyclic, so is its dual.

Theorem 5.2. *If C is a cyclic code of length k over $GF(s)$, with generator polynomial $g(X)$, then the dual code $C^\perp$ is also cyclic and has generator polynomial*

$$\frac{X^k - 1}{\overleftarrow{g}(X)} , \tag{5.7}$$

where $\overleftarrow{g}(X) = X^{\deg(g)}g(X^{-1})$ is the reciprocal polynomial to $g(X)$.

The reader is asked to verify this in Problem 5.1.

For example, the $(7, 8, 4)_2$ cyclic code in Example 4.3 has generator polynomial

$$g(X) = 1 + X + X^2 + X^4 , \tag{5.8}$$

from (5.5), and the $(7, 16, 3)_2$ dual code has generator polynomial

$$\frac{X^7 - 1}{1 + X^2 + X^3 + X^4} = 1 + X^2 + X^3 ,$$

in agreement with the generator matrix given in (4.7).

Quasicyclic and constacyclic codes are straightforward generalizations of the notion of a cyclic code. A code of length $k = ab$ is *quasicyclic* if it is linear and has the property that whenever

$$(c_0, c_1, \ldots, c_{a-1}, c_a, c_{a+1}, \ldots, c_{2a}, \ldots)$$

is a codeword, so is the vector

$$(c_1, c_2, \ldots, c_0, c_{a+1}, c_{a+2}, \ldots, c_a, \ldots) ,$$

obtained by cycling the components in b blocks of size a. Another generalization, especially useful for codes over $GF(3)$ and $GF(4)$, is that of a *constacyclic* code, which is a linear code with the property that there is a constant $\beta \in GF(s)$ such that if $(u_0, u_1, \ldots, u_{k-1})$ is a codeword, then so is $(u_1, u_2, \ldots, u_{k-1}, \beta u_0)$.

It is convenient to describe cyclic, quasicyclic and constacyclic codes and arrays by the use of pointed brackets $\langle\ \rangle$. A vector containing one or more pointed brackets, possibly subscripted, will be used as shorthand notation for a set of vectors. To illustrate the use of pointed brackets, instead of the generating vector (5.5) we can write

$$\langle 1110100 \rangle , \tag{5.9}$$

which is an abbreviation for the seven vectors 1110100, 0111010, ..., 1101001 that are obtained by successive cyclic shifts of this generating vector. The code is in fact generated by the first three of these seven vectors (see (4.4)), so the last four vectors aren't needed. Of course they don't do any harm either: the linear span of all seven vectors is the same as the linear span of the first three vectors alone. What is important is that this provides a very condensed description of the code or orthogonal array.

Similarly, instead of the generator matrix (5.1) we can write

$$\langle 1011000 \rangle 1$$

which is an abbreviation for the seven vectors 10110001, 01011001, 00101101, ... 01100011. Again we only need some of the vectors to generate the code — in this case the first four vectors — but, again, the extra generators do no harm.

The description of a quasicyclic code may involve several sets of pointed brackets. An important code over $GF(4)$ is the $(6, 64, 4)_4$ code known as the *hexacode*, which has generator matrix

$$\begin{bmatrix} 100122 \\ 010212 \\ 001221 \end{bmatrix}, \tag{5.10}$$

where we write the elements of $GF(4)$ as $0, 1, 2, 3$. This is described by the single generator

$$\langle 100 \rangle \langle 122 \rangle , \tag{5.11}$$

which is a shorthand for the three vectors 100122, 010212, 001221. The corresponding orthogonal array, an $OA(64, 6, 4, 3)$, is isomorphic to that produced by Bush's construction in Theorem 3.2.

The formal meaning of the pointed brackets is the following. A vector containing pointed brackets, say

$$\langle x_1 \quad x_2 \quad x_3 \rangle \ x_4 \ x_5 \ \langle x_6 \quad x_7 \rangle \ x_8 ,$$

is an abbreviation for the set of vectors obtained by repeatedly applying the permutation defined by the locations of the pointed brackets, in this case

$$(1 \ \ 2 \ \ 3) \ \ 4 \ \ 5 \ \ (6 \ \ 7) \ \ 8 ,$$

to the underlying vector

$$x_1 \ x_2 \ x_3 \ x_4 \ x_5 \ x_6 \ x_7 \ x_8 .$$

(In this case we would obtain six vectors.)

We use a similar notation for constacyclic codes, writing the constant β as a subscript on the generator, as in

$$\langle g_0 g_1 \cdots g_{k-1} \rangle_\beta .$$

For example $\langle 1120 \rangle_{-1}$ is an abbreviation for the four ternary vectors 1120, 1202, 2022, 0221, which generate a $(4, 9, 3)_3$ constacyclic code (the last two generators are redundant).

5.3 The Rao-Hamming Construction Revisited

We now begin our descriptions of specific classes of codes.

A *Hamming code* (Hamming, 1950) is a single-error correcting code with the parameters $(k, s^{k-m}, 3)_s$, where $k = (s^m - 1)/(s - 1)$, $m \geq 2$, and s is any prime power. It is defined by a parity check matrix P, of size $m \times k$, whose columns consist of all distinct nonzero m-tuples from $GF(s)$ in which the top nonzero entry is a 1. That the minimal distance is 3 follows at once from Problem 4.3, since by construction no two columns of P are linearly dependent.

The dual code, which uses P as a generator matrix, is a $(k, s^m, s^{m-1})_s$ code — see Problem 5.4.

In view of Theorem 4.6, the codewords of this dual code form the rows of an orthogonal array with parameters

$$OA(s^m, k, s, 2) , \tag{5.12}$$

where $k = (s^m - 1)/(s - 1)$ and $m \geq 2$. This orthogonal array was constructed in exactly this way, using P as a generator matrix, in Construction 3 of Theorem 3.20.

The codewords of the Hamming code itself form an orthogonal array with parameters

$$OA(s^{k-m}, k, s, s^{m-1} - 1) . \tag{5.13}$$

We will refer to *linear* arrays with the parameters in (5.12) and (5.13) as orthogonal arrays of *Rao-Hamming type*.

When $s = 2$, the coordinate positions of a Hamming code can be arranged to make the code cyclic, and so a more concise description can be given for these codes and for orthogonal arrays of Rao-Hamming type. As generator polynomial for the Hamming code, and for the array (5.13), we can use any primitive irreducible polynomial $\pi(X)$ of degree m over $GF(2)$.

The dual code and the orthogonal array (5.12) are then also cyclic, with generator polynomial $(X^k - 1)/\overleftarrow{\pi}(X)$.

For example, when $m = 4$, we may take the primitive polynomial $\pi(X) = 1 + X + X^4$ as generator polynomial (see Table A.19 of the Appendix), obtaining a $(15, 2^{11}, 3)_2$ Hamming code and an $OA(2^{11}, 15, 2, 7)$. The dual code has generator polynomial

$$\frac{X^{15} - 1}{1 + X^3 + X^4} = 1 + X^3 + X^4 + X^6 + X^8 + X^9 + X^{10} + X^{11} . \tag{5.14}$$

This has parameters $(15, 16, 8)_2$ and the corresponding orthogonal array is an $OA(16, 15, 2, 2)$.

For $s = 2$, Hamming codes can be extended, as described in Section 5.1, to give a dual pair of codes with parameters

$$(2^m, 2^{2^m-1-m}, 4)_2 \quad \text{and} \quad (2^m, 2^{m+1}, 2^{m-1})_2 . \tag{5.15}$$

The corresponding orthogonal arrays have parameters

$$OA(2^{2^m-1-m}, 2^m, 2, 2^{m-1} - 1) \quad \text{and} \quad OA(2^{m+1}, 2^m, 2, 3) . \tag{5.16}$$

Taking $m = 2, 3, 4, 5$ for example we obtain arrays

$$\begin{array}{ll} OA(2,4,2,1), & OA(2^3,4,2,3), \\ OA(2^4,8,2,3), & OA(2^4,8,2,3), \\ OA(2^{11},16,2,7), & OA(2^5,16,2,3), \\ OA(2^{26},32,2,15), & OA(2^6,32,2,3). \end{array}$$

If $s > 2$ then the Hamming codes, their duals and the Rao-Hamming orthogonal arrays can all be made to be constacyclic.

The tables in Chapter 12 contain several examples of Rao-Hamming arrays. Table 12.2 for example contains both an $OA(3^3, 13, 3, 2)$ and its dual, an $OA(3^{10}, 13, 3, 8)$.

5.4 BCH Codes

BCH codes (the name honors Bose and Ray-Chaudhuri, 1960a, 1960b, and Hocquenghem, 1959) are a natural generalization of Hamming codes that correct any number of errors.

To construct a BCH code of minimal distance d and length k over $GF(s)$, where k and s are relatively prime, we choose a primitive k-th root of unity, ξ say, in some field $GF(s^m)$ containing $GF(s)$.

Let $M^{(i)}(X)$ denote the *minimal polynomial* of ξ^i, that is, the lowest degree polynomial with coefficients from $GF(s)$ and leading coefficient 1 that has ξ^i as a root. Choose an integer a in the range $0 \le a \le k-1$.

Then the cyclic code of length k with generator polynomial

$$g(X) = \text{l.c.m. } \{M^{(a)}(X), M^{(a+1)}(X), \ldots, M^{(a+d-2)}(X)\} \tag{5.17}$$

is a *BCH code* with minimal distance at least d and, by Theorem 5.1, of size $s^{k-\deg g}$. The actual minimal distance may sometimes be larger than d, but is never less.

The crucial point in the construction is that the generator polynomial must have $d-1$ consecutive powers of ξ as zeros. Usually one chooses $a = 0$ or 1, but sometimes other values give a higher minimal distance.

For binary codes it is easy to see that $M^{(2i)}(X) = M^{(i)}(X)$, for $0 \le i \le k-1$. This means that, in the case $a = 1$, the generator polynomial (5.17) can be replaced by

$$g(X) = \text{l.c.m. } \{M^{(1)}(X), M^{(3)}(X), \ldots, M^{(d-2)}(X)\}\ , \tag{5.18}$$

if d is odd, or by

$$g(X) = \text{l.c.m. } \{M^{(1)}(X), M^{(3)}(X), \ldots, M^{(d-1)}(X)\}\ , \tag{5.19}$$

if d is even.

Example 5.3. A binary BCH code with $d = 3$ and $k = 2^m - 1$ (and $a = 1$) is the same as a Hamming code. Indeed, ξ is now a primitive element of $GF(2^m)$, $M^{(1)}(X)$ is a primitive irreducible polynomial $\pi(X)$, and from (5.18) the generator polynomial is simply $\pi(X)$. ∎

Example 5.4. The binary double-error-correcting BCH code with $d = 5$ and $k = 2^m - 1$ (and $a = 1$) has generator polynomial

$$g(X) = M^{(1)}(X)M^{(3)}(X) , \tag{5.20}$$

and parameters

$$(2^m - 1, 2^{2^m-1-2m}, 5)_2, \quad \text{for } m \geq 4 . \tag{5.21}$$

It can be shown (see MacWilliams and Sloane, 1977, Chapter 18) that the dual to this code has parameters

$$(2^m - 1, 2^{2m}, 2^{m-1} - 2^{\lfloor m/2 \rfloor})_2 , \quad \text{for } m \geq 4 . \tag{5.22}$$

In view of Theorem 4.6 we have thus constructed orthogonal arrays

$$OA(2^{2^m-1-2m}, 2^m - 1, 2, 2^{m-1} - 2^{\lfloor m/2 \rfloor} - 1) , \tag{5.23}$$

and

$$OA(2^{2m}, 2^m - 1, 2, 4) , \tag{5.24}$$

for $m \geq 4$.

For example, take $m = 5$, so that $k = 2^5 - 1 = 31$, and let ξ be a primitive element of $GF(32)$. As shown in Table A.19 of the Appendix, we can take $M^{(1)}(X) = 1+X^2+X^5$. Then we find that $M^{(3)}(X) = 1+X^2+X^3+X^4+X^5$, and

$$g(X) = M^{(1)}(X)M^{(3)}(X) = 1 + X^3 + X^5 + X^6 + X^8 + X^9 + X^{10} \tag{5.25}$$

is the generator polynomial for a $(31, 2^{21}, 5)_2$ double-error-correcting BCH code. The codewords form an $OA(2^{21}, 31, 2, 11)$, and the codewords of the dual code form an $OA(2^{10}, 31, 2, 4)$. The latter has generator polynomial

$$\begin{aligned} \frac{X^{31} - 1}{\overleftarrow{g}(X)} &= \frac{X^{31} - 1}{1 + X + X^2 + X^4 + X^5 + X^7 + X^{10}} \\ &= 1 + X + X^3 + X^5 + X^7 + X^8 + X^9 + X^{10} + X^{13} \\ &\quad + X^{16} + X^{18} + X^{21} , \end{aligned}$$

by Theorem 5.2. ∎

Although BCH codes are good codes, there are sometimes even better ones. In particular, in most cases it is possible to find codes with twice as many codewords (and the same minimal distance) as those described in (5.21), and orthogonal arrays with half as many runs (and the same strength) as those described in (5.24). For odd m we do this by using duals of BCH codes of length $2^m + 1$ (see MacWilliams and Sloane, 1977, page 586), and for even m by using certain nonlinear codes and arrays to be described in Section 5.10.

5.5 Reed-Solomon Codes

There is an especially important subclass of BCH codes known as Reed-Solomon codes. They are important both because of their mathematical interest and because of their widespread use in compact disks and very high-speed data transmission.

In their simplest (cyclic) form, Reed-Solomon codes are the special case of BCH codes when ξ is actually in $GF(s)$, rather than in some extension field, and $k = s - 1$. Thus $M^{(i)}(X) = X - \xi^i$, and the generator polynomial

$$g(X) = (X - \xi^a)(X - \xi^{a+1}) \cdots (X - \xi^{a+d-2}) \tag{5.26}$$

produces a cyclic code with minimal distance d. Thus we have constructed cyclic Reed-Solomon codes with parameters $(s-1, s^n, s-n)_s$, for all $n = 1, \ldots, s-1$. It is known — see MacWilliams and Sloane (1977), Chapter 11 — that in every case the length can be increased by 2, yielding codes with parameters

$$(s+1, s^n, s-n+2)_s\ , \qquad 1 \leq n \leq s+1\ , \tag{5.27}$$

and that if $s = 2^m$ there are two cases when the length can be increased by 3, yielding extended Reed-Solomon codes with parameters

$$(s+2, s^3, s)_s \tag{5.28}$$

and

$$(s+2, s^{s-1}, 4)_s \tag{5.29}$$

where $s = 2^m$.

When $s = 5$ for example we may take $\xi = 2$ to be a primitive element of $GF(5)$. With $a = 1$, $d = 3$ we see that the generator polynomial for a $(4, 25, 3)_5$ Reed-Solomon code is

$$g(X) = (X-2)(X-4) = X^2 + 4X + 3\ .$$

A generator matrix for this code is

$$\begin{bmatrix} 3 & 4 & 1 & 0 \\ 0 & 3 & 4 & 1 \end{bmatrix} .$$

This can be extended to a $(6, 25, 5)_5$ code with generator matrix

$$\begin{bmatrix} 3 & 4 & 1 & 0 & 3 & 1 \\ 0 & 3 & 4 & 1 & 3 & 3 \end{bmatrix} .$$

The dual distances for the codes in (5.27), (5.28), (5.29) are respectively $n + 1$, 4 and 2^m (this will follow from Theorem 5.6), so the corresponding orthogonal arrays have parameters

$$OA(s^n, s+1, s, n), \tag{5.30}$$
$$OA(s^3, s+2, s, 3), \quad s = 2^m, \tag{5.31}$$
$$OA(s^{s-1}, s+2, s, s-1), \quad s = 2^m . \tag{5.32}$$

These are the same parameters as the arrays found by Bush (1952b) and described in Theorems 3.1, 3.2 and Corollary 3.8. In fact it can be shown that the orthogonal arrays obtained from Reed-Solomon codes are isomorphic to those constructed in Section 3.2.

5.6 MDS Codes and Orthogonal Arrays of Index Unity

As we saw in Section 4.5, the *Singleton* (1964) *bound* for simple codes, linear or nonlinear, states that in any $(k, N, d)_s$ code over $GF(s)$,

$$N \leq s^{k-d+1} . \tag{5.33}$$

A code achieving this bound is called a *Maximal Distance Separable* or *MDS* code. Theorem 4.21 showed that MDS codes are the same as orthogonal arrays of index unity.

All Reed-Solomon codes are MDS codes. It is remarkable that in every case where an MDS code (over a finite field) is presently known to exist, there is a linear code, in fact even a Reed-Solomon code, with the same parameters. The same thing is true for all known orthogonal arrays of index unity over a finite field. We conjecture, perhaps rashly, that this is always the case.

Research Problem 5.5. *The linearity conjecture for MDS codes and orthogonal arrays of index unity.* Prove (or disprove) the following conjecture. Let s be a prime power. Then for every MDS code or orthogonal array of index unity there is a linear code or array with the same parameters.

If an MDS code *is* linear we can say more about its properties. For an arbitrary linear $(k, s^n, d)_s$ code the Singleton bound (5.33) states that

$$d \leq k - n + 1 , \tag{5.34}$$

and the code is an MDS code if and only if equality holds in (5.34).

Theorem 5.6. *Let C be a linear $(k, s^t, k-t+1)_s$ MDS code and (equivalently) a linear $OA(s^t, k, s, t)$. Then the dual code $C^\perp$ is a linear $(k, s^{k-t}, t+1)_s$ MDS code and a linear $OA(s^{k-t}, k, s, k-t)$.*

Proof: This is an immediate consequence of Theorem 4.6. ∎

Corollary 5.7. *Let C be a linear $(k, s^t, d)_s$ code. Then the following statements are equivalent. (a) C is a linear MDS code, with $d = k - t + 1$. (b) C is a linear $OA(s^t, k, s, t)$. (c) Every set of t columns of a generator matrix for C are linearly independent. (d) Every set of $k - t$ columns of a parity check matrix for C are linearly independent.*

Proof: We already know that (a) and (b) are equivalent. For (d), note that C contains a codeword of weight w if and only if some w columns of the parity check matrix P are linearly dependent. So C has $d \geq k-t+1$ if and only if no $k-t$ columns of P are linearly dependent. Finally, (c) is equivalent to (d) by Theorem 5.6. ∎

Note that Theorem 5.6 also establishes Theorem 3.7.

For $t \geq 2$, let us denote by $m(t,s)$ the maximal length k of any linear $(k, s^t, k-t+1)_s$ MDS code. In view of Corollary 5.7, $m(t,s)$ gives the maximal number of factors k in any linear $OA(s^t, k, s, t)$, and so

$$m(t,s) \leq f(s^t, s, t) . \tag{5.35}$$

The conjecture in Research Problem 5.5 asserts that if s is a prime power and $t \geq 2$, then

$$m(t,s) = f(s^t, s, t) ,$$

i.e. that $f(s^t, s, t)$ can always be achieved by a linear array.

Although this conjecture may be rash, it is supported by all the available evidence. We hope that this bold conjecture will either soon be proved or else that new nonlinear codes or arrays will be found to disprove it. Such codes or arrays would be extremely interesting.

There is much more evidence for the next conjecture, which follows Chapter 11, §7 of MacWilliams and Sloane (1977), and gives the putative exact values for the function $m(t,s)$.

Research Problem 5.8. *The MDS code conjecture.* Let s be a prime power. Prove (or disprove) the following conjecture. Suppose $t \geq 2$. Then

$$m(t,s) = \max\{t, s\} + 1 ,$$

except that

$$m(3, s) = m(s-1, s) = s+2$$

if $s = 2^m$, $m \geq 2$.

If both the above conjectures are true, then we also know the exact values of the function $f(s^t, s, t)$, and it is these values that we hypothesized in Conjecture 3.11 (which is equivalent to the above conjectures).

The MDS code conjecture *is* true in a number of special cases, for example when

$$s \leq t, \qquad \text{by Corollary 2.22},$$

$$t = 2 \text{ or } 3, \qquad \text{by Corollary 3.9},$$

and

$$t = 4 \text{ and } s \text{ is 5 or a power of 2}, \qquad \text{by Corollary 3.9},$$

for in these cases we know both the exact value of $f(s^t, s, t)$ and that the linearity conjecture of Research Problem 5.5 holds. Bose (1947) had already determined $m(2, s)$ and $m(3, s)$ directly. The MDS code conjecture has also been shown to be true if

$$t = 4 \text{ or } 5, \text{ for any value of } s,$$

by Segre (1955), Casse (1969) and Gulati and Kounias (1970); or

$$\text{if } s \text{ is sufficiently large, for any } t,$$

by Thas (1968,1969); as well as in a number of other cases (see Hirschfeld, 1983).

We must emphasize that these conjectures are specifically for MDS codes and orthogonal arrays of index unity, and for s a prime power. For more general codes and orthogonal arrays over $GF(s)$ it is certainly true that nonlinear constructions can produce strictly better results than linear ones, as we shall see for example in Section 5.10.

Research Problem 5.9. Define two (possibly nonlinear) orthogonal arrays to be *formal duals* if the weight distributions of the corresponding codes satisfy 4.11. Suppose s is not a prime power. There are trivial examples of pairs of orthogonal arrays of index unity that are formal duals (see Problem 4.5 for example), but are there any nontrivial examples? Find conditions on an orthogonal array of index unity for it to have (or not to have) a formal dual.

5.7 Quadratic Residue and Golay Codes

Quadratic residue codes are another important family of cyclic codes. Several orthogonal arrays constructed from these codes are mentioned in the tables in Chapter 12. Most of these have very large numbers of runs, however, so we refer the reader to Chapter 16 of MacWilliams and Sloane (1977), for the general construction. Newhart (1988) has determined the minimal distance of many of these codes. See also Ward (1990).

The two best-known examples of quadratic residue codes are the *Golay* (1949) codes. These are $(24, 4096, 8)_2$ and $(12, 729, 6)_3$ codes, and are defined in Sections 5.11 and 5.12. They have many connections with group theory, number theory and combinatorics — see Conway and Sloane (1998).

The two Golay codes are self-dual, and in fact many extended quadratic residue codes have this property. Self-dual codes of moderate length over $GF(s)$ for $s = 2, 3, 4, 5$ and 7 have been classified — see the survey in Conway and Sloane (1998), Chapter 7, and Rains and Sloane (1998). The papers by Mallows, Pless and Sloane (1976) and MacWilliams, Odlyzko, Sloane and Ward (1978) contain a number of other self-dual codes that lead to orthogonal arrays mentioned in the tables in Chapter 12.

5.8 Reed-Muller Codes

This construction, due to Reed (1954) and Muller (1954), produces a family of binary linear codes with several useful properties:

(a) they are a generalization of Hamming codes;

(b) they are extended cyclic codes;

(c) they are easy to construct and to encode and decode;

(d) the dimension and minimal distance can be written down explicitly; and

(e) the dual codes are of the same type.

Let $x_1, \ldots, x_m$ be binary variables, that is, variables taking values in $GF(2)$. A polynomial $f(x_1, \ldots, x_m)$ can be represented by a vector $\Phi(f)$ giving its values at all 2^m possible combinations of the x_i's. For example, the function $x_1x_2x_3$ of three variables x_1, x_2, x_3 would be specified by the vector $\Phi(x_1x_2x_3)$ of length 8 shown here:

$$\begin{aligned} x_1 &= 00001111 \\ x_2 &= 00110011 \end{aligned}$$

$$\begin{aligned} x_3 &= 01010101 \\ \Phi(x_1x_2x_3) &= 00000001 \end{aligned}$$

Let $k = 2^m$, $m \geq 1$. The *r-th order Reed-Muller code* $RM(r,m)$, for $0 \leq r \leq m$, consists of all vectors $\Phi(f)$ of length 2^m, where f is any polynomial in $x_1, \ldots, x_m$ of degree at most r. Of course, since $x^2 = x$ in $GF(2)$, we can assume that the polynomials are sums of monomials of the form

$$x_1^{r_1} x_2^{r_2} \cdots x_m^{r_m}, \quad r_i \in \{0,1\} \, .$$

Reed-Muller codes are linear, since these polynomials form a vector space. As generating vectors we may take the vectors $\Phi(f)$ where f ranges over all monomials of degree at most r. There are $\binom{m}{i}$ distinct monomials of degree i, for $i = 0, \ldots, m$. Thus the rows of the generator matrix for a first-order Reed-Muller code are the vectors

$$\Phi(1), \Phi(x_1), \Phi(x_2), \ldots, \Phi(x_m) \, .$$

For a second-order Reed-Muller code we also use the vectors

$$\Phi(x_1x_2), \Phi(x_1x_3), \ldots, \Phi(x_{m-1}x_m) \, ;$$

and so on.

For example, $RM(2,3)$ has the generator matrix shown in Table 5.10.

Table 5.10. Generator matrix for second-order Reed-Muller code $RM(2,3)$.

1	1	1	1	1	1	1	1	1
x_1	0	0	0	0	1	1	1	1
x_2	0	0	1	1	0	0	1	1
x_3	0	1	0	1	0	1	0	1
x_1x_2	0	0	0	0	0	0	1	1
x_1x_3	0	0	0	0	0	1	0	1
x_2x_3	0	0	0	1	0	0	0	1

The dimension of $RM(r,m)$ is equal to the number of distinct monomials of degree at most r, which is

$$1 + \binom{m}{1} + \cdots + \binom{m}{r} \, ,$$

and the minimal distance can easily be shown by induction to be 2^{m-r} — see Problem 5.10.

Thus the parameters of the r-th order Reed-Muller code $RM(r,m)$ are

$$(2^m, 2^n, 2^{m-r})_2 , \tag{5.36}$$

where $n = 1 + \binom{m}{1} + \cdots + \binom{m}{r}$.

The dual code to $RM(r,m)$ is the code $RM(m-r-1,m)$, with parameters

$$(2^m, 2^{n'}, 2^{r+1})_2 , \tag{5.37}$$

where $n' = 1 + \binom{m}{1} + \cdots + \binom{m}{m-r-1}$.

The code $RM(m-2,m)$ is the same, apart from a rearrangement of coordinates, as an extended Hamming code, and the dual code $RM(1,m)$ is the same as the code formed by the rows of a Hadamard matrix of Sylvester type together with its negative — see Section 7.4.

In view of Theorem 4.6, the codewords of the Reed-Muller code $RM(r,m)$ form an orthogonal array with parameters

$$OA(2^n, 2^m, 2, 2^{r+1}-1) , \tag{5.38}$$

where $n = 1 + \binom{m}{1} + \cdots + \binom{m}{r}$, for any $r = 0, 1, \ldots, m$.

5.9 Codes from Finite Geometries

Reed-Muller codes can also be defined over larger fields than $GF(2)$, using as generator matrices the incidence matrices of points and flats of various dimensions in projective or affine geometries. We shall not describe the construction here, since the parameters of the resulting orthogonal arrays are either outside the range we are interested in, or else could have been obtained by other methods. There is an excellent discussion in Assmus and Key (1992a). Theorem 5.7.9 of that book summarizes the properties of this very large family of codes. Since these codes are always accompanied by a (usually tight) lower bound on the minimal distance of the dual code, that table also serves as a catalogue of orthogonal arrays.

Another important family of finite geometry codes are those based on "ovoids" in $PG(3,s)$. A set of s^2+1 points (w,x,y,z) in $PG(3,s)$ is called an *ovoid* if no three of them are collinear. For a concrete example, for any prime power s one may take the points $(w,x,y,z) \in GF(s)^4$ such that

$$xy + z^2 + azw + w^2 = 0 , \tag{5.39}$$

where $a \in GF(s)$ is such that $z^2 + az + 1 = 0$ has no root in $GF(s)$. One obtains a code with parameters $(s^2+1, s^4, s(s-1))_s$ and dual distance 4 using the vectors $(w, x, y, z)^T$ as columns of a generator matrix. From this we obtain an

$$OA(s^4, s^2 + 1, s, 3) \ . \tag{5.40}$$

For further details see Calderbank and Kantor (1986) and Dembowski (1968), page 48. A different construction for the special case $s = 2^{2e+1} > 2$, producing nonisomorphic orthogonal arrays, was found by Tits (1960/1961,1962).

The array (5.40) does not have the maximal number of factors if $s = 2$, for we already know an $OA(16, 8, 2, 3)$ (e.g. from (5.16)). For $s = 3$ we obtain an $OA(81, 10, 3, 3)$. As mentioned in Example 2.17, Seiden (1955a, 1955b) showed that 10 is indeed the maximal number of factors in this case. For larger s we can only conclude that $f(s^4, s, 3) \geq s^2 + 1$; however, it was shown by Bose (1947) for s odd and Qvist (1952) for s even that these orthogonal arrays have the maximal number of factors among *linear* arrays.

5.10 Nordstrom-Robinson and Related Codes

In this section we describe the nonlinear code and orthogonal array mentioned just below Theorem 4.15, as well as some generalizations.

The first of these codes was described by Nordstrom and Robinson (1967): it is a nonlinear $(16, 256, 6)_2$ code with many remarkable properties: for example, it has dual distance 6; furthermore any binary linear code of length 16 with minimal distance 6 can contain at most 128 codewords. By Theorem 4.9 this means that the codewords of the Nordstrom-Robinson code form an $OA(256, 16, 2, 5)$, whereas any linear array with the same number of factors and the same strength must contain at least 512 runs.

Several different infinite families of codes that generalize the Nordstrom-Robinson code were later discovered by Preparata (1968), Kerdock (1972), Goethals (1974, 1976) and Delsarte and Goethals (1975). These are also nonlinear binary codes, again usually containing more codewords than any known linear code. Their constructions were all fairly complicated, although MacWilliams and Sloane (1977), Chapter 15 gives a unified description. See also Baker, van Lint and Wilson (1983) and Hergert (1990).

It came as a considerable surprise therefore when in 1992 it was discovered that all these codes can be obtained very simply as *linear* codes over the alphabet $\mathbb{Z}_4$, the ring of integers modulo 4. We take the elements of $\mathbb{Z}_4$ to be $\{0, 1, 2, 3\}$, with arithmetic modulo 4. The $\mathbb{Z}_4$ codewords are then mapped to binary vectors by the map

$$\phi: \ 0 \mapsto 00, \quad 1 \mapsto 01, \quad 2 \mapsto 11, \quad 3 \mapsto 10 \ . \tag{5.41}$$

These discoveries are described in Forney, Sloane and Trott (1993) and Hammons, Kumar, Calderbank, Sloane and Solé (1993). In particular, the Kerdock codes now provide an easily constructable, infinite family of orthogonal arrays $OA(2^{2m}, 2^m, 2, 5)$, for all even $m \geq 4$, and the Delsarte-Goethals codes provide (for example) arrays $OA(2^{3m-1}, 2^m, 2, 7)$ for all even $m \geq 6$. These contain fewer runs than comparable orthogonal arrays obtained from the duals of BCH codes. However, if the number of factors is of the form $k = 2^m$, m odd, then the BCH arrays are still the best family known. For strength 5, for instance, duals of certain double-error-correcting BCH codes provide arrays $OA(2^{2m+1}, 2^m + 1, 2, 5)$ for both even and odd values of $m \geq 5$ — see MacWilliams and Sloane (1977), page 586.

Most of these arrays are quite large, except for the first one mentioned, that obtained from the Nordstrom-Robinson code. This now has the following extremely simple description: it is a linear code over $\mathbb{Z}_4$ of length 8, containing 4^4 codewords, with generator matrix

$$G = \begin{bmatrix} 1 & 3 & 1 & 2 & 1 & 0 & 0 & 0 \\ 1 & 0 & 3 & 1 & 2 & 1 & 0 & 0 \\ 1 & 0 & 0 & 3 & 1 & 2 & 1 & 0 \\ 1 & 0 & 0 & 0 & 3 & 1 & 2 & 1 \end{bmatrix} . \tag{5.42}$$

This code is dual to itself, so (5.42) is also a parity check matrix. The code thus consists of all vectors uG, where $u = (u_1, u_2, u_3, u_4)$ runs through $\mathbb{Z}_4^4$. By mapping these codewords to binary vectors of length 16 via (5.41), we obtain an $OA(256, 16, 2, 5)$. Note that although the code is linear over $\mathbb{Z}_4$, the final binary orthogonal array is not linear.

In fact no *linear* $OA(256, 16, 2, 5)$ can exist, for if it did then the runs of its dual would form a linear code with the same parameters as the Nordstrom-Robinson code, and this is known to be impossible.

Orthogonal arrays obtained from Preparata codes are probably less useful: they have parameters

$$OA(2^{2^m - 2m}, 2^{2m}, 2, 2^{m-1} - 2^{(m-2)/2} - 1)$$

for all even $m \geq 4$ (for $m = 4$ we get the Nordstrom-Robinson array again).

5.11 Examples of Binary Codes and Orthogonal Arrays

The next three sections give explicit constructions for most of the linear orthogonal arrays with fewer than 1000 runs that are mentioned in Tables 12.1–12.3 and 12.7, and can be obtained from codes. We also mention a few especially important arrays with larger numbers of runs.

Each entry gives the parameters of the array, generators for the array, and a brief description of either the associated code or the dual code, often using the pointed bracket notation introduced in Section 5.2. (Note that, as already mentioned in Section 5.2, the pointed bracket notation usually produces more generators than are necessary. However, the redundant generators do no harm.) In each section the arrays are arranged in order of increasing strength.

In view of Theorem 2.24, for binary orthogonal arrays we need consider only arrays of odd strength t.

$OA(2^4, 8, 2, 3)$. Extended Rao-Hamming or Hadamard array. Generators:

$$\langle 1101000 \rangle 1 \ .$$

Code is $(8, 16, 4)_2$ first-order Reed-Muller code.

$OA(2^5, 16, 2, 3)$. Extended Rao-Hamming or Hadamard array. Generators:

$$\langle 111011001010000 \rangle 1 \ .$$

Code is $(16, 32, 8)_2$ first-order Reed-Muller code.

$OA(2^6, 32, 2, 3)$. Extended Rao-Hamming or Hadamard array. Generators:

$$\langle 1110010001010111101101001100000 \rangle 1 \ .$$

Code is $(32, 64, 16)_2$ first-order Reed-Muller code.

$OA(2^7, 9, 2, 5)$. Generators

$$\langle 111000000 \rangle \ .$$

Parity checks for array:

$$\langle 110110110 \rangle \ .$$

Dual code is $(9, 4, 6)_2$ cyclic code.

$OA(2^8, 16, 2, 5)$. From Nordstrom-Robinson code: see Section 5.10.

$OA(2^{10}, 24, 2, 5)$. This is a modification of a code discovered by Wagner (1965), and has the generator matrix shown in Table 5.11.

Table 5.11. Generator matrix for $OA(1024, 24, 2, 5)$.

```
110011011001111000000000
111011110010010100000000
001101111011010010000000
101001110111110001000000
110111111110000000100000
011110110101010000010000
101111000000110000001000
101010011110000000000100
001011010100110000000010
110100011010100000000001
```

$OA(2^{11}, 32, 2, 5)$. Generators:

$$\langle 11100111000101001100100000000000 \rangle 1 \ .$$

Code is $(32, 2^{11}, 12)_2$ code, dual to $(32, 2^{21}, 6)_2$ BCH code.

$OA(2^{10}, 12, 2, 7)$. Parity checks:

$$\langle 110110110110 \rangle \ .$$

Dual code is $(12, 4, 8)_2$ cyclic code.

$OA(2^{11}, 16, 2, 7)$. Extended Rao-Hamming array: see Section 5.4.

$OA(2^{12}, 24, 2, 7)$. Generators:

$$\langle 10000000000 \rangle 0 \langle 10100011101 \rangle 1 \ , \qquad 00000000000\ 1\ 11111111111\ 0 \ .$$

Code is $(24, 2^{12}, 8)_2$ self-dual Golay code.

5.12 Examples of Ternary Codes and Orthogonal Arrays

$OA(3^2, 4, 3, 2)$. Bush array. Generators:

$$1110, \quad 0121 \ .$$

Code is $(4, 9, 3)_3$ self-dual *tetracode* — see Conway and Sloane (1998).

$OA(3^3, 13, 3, 2)$. Rao-Hamming array. Generators:

$$\langle 1011122012100 \rangle \ .$$

Code is $(13, 3^3, 9)_3$ dual Hamming code.

$OA(3^4, 40, 3, 2)$. Rao-Hamming array. Generators:

$$\langle 2121102221111010122102012001121011001000 \rangle_{-1} \ .$$

Code is $(40, 3^4, 27)_3$ dual Hamming code.

$OA(3^4, 10, 3, 3)$. From ovoid code: see Section 5.9.

$OA(3^5, 20, 3, 3)$. Generators:

$$\langle 20211021210100110000 \rangle \ .$$

Dual code is $(20, 3^{15}, 4)_3$ Kschischang-Pasupathy (1992) cyclic code.

$OA(3^6, 56, 3, 3)$. The generator matrix is shown in Table 5.12. The corresponding code was found by Hill (1973). The table shows a 6×56 matrix broken into two parts.

Table 5.12. Generator matrix for $OA(3^6, 56, 3, 3)$.

2000022	2021211	1000110	1211220
1200021	0220002	1100121	1002012
0120021	1010211	2110122	0011121
0012021	0122202	0211122	2212002
0001221	1000101	0021222	0102120
0000111	0121221	0002202	1221102

1121220	1001100	1001010	2111202
1200012	1101210	1101111	1022022
0211221	1111221	2111121	1210101
2112012	2112222	2212122	0202212
0002121	0212022	0222222	1101120
2121102	0022002	0020202	2221011

$OA(3^5, 11, 3, 4)$. Generators:

$$\langle 11011100010 \rangle \ .$$

Code is $(11, 3^5, 6)_3$ Golay code.

$OA(3^6, 14, 3, 4)$. Generators:

$$\langle 12110112100000 \rangle_{-1} \ .$$

Dual code is $(14, 3^8, 5)_3$ Kschischang-Pasupathy (1992) constacyclic code.

$OA(3^6, 12, 3, 5)$. Generators:

$$\langle 22122211121 \rangle 1 \ .$$

Code is $(12, 3^6, 6)_3$ self-dual Golay code.

5.13 Examples of Quaternary Codes and Orthogonal Arrays

As usual we denote the elements of $GF(4)$ by 0, 1, 2, 3.

$OA(4^3, 21, 4, 2)$. Rao-Hamming array. Generators:

$$\langle 131202012232221301100 \rangle_2 \ .$$

Code is $(21, 4^3, 16)_4$ dual Hamming code.

$OA(4^3, 6, 4, 3)$. Bush array. Generators:

$$100122, \quad 010212, \quad 001221 \ .$$

Code is $(6, 64, 4)_4$ self-dual hexacode — see Section 5.2 and Conway and Sloane (1998).

$OA(4^4, 17, 4, 3)$. Generators for array:

$$\langle 11201233210211000 \rangle \ .$$

Code is $(17, 4^4, 12)_4$ ovoid code.

$OA(4^5, 11, 4, 4)$. Generators for array:

$$\langle 12203310000 \rangle \ .$$

Dual code is $(11, 4^6, 5)_4$ quadratic residue code.

$OA(4^6, 21, 4, 4)$. Generators for array:

$$\langle 211212211021221000000 \rangle_2 \ .$$

Dual code is $(21, 4^{15}, 5)_4$ constacyclic code.

$OA(4^6, 12, 4, 5)$. Generators for array:

$$\langle 12203310000 \rangle 0 \ , \quad 11111111111\ 1 \ .$$

Code is $(12, 4^6, 6)_4$ quadratic residue code (compare Kschischang and Pasupathy, 1992).

The latter array is not self-dual, and in fact no linear $OA(4^6, 12, 4, 5)$ can be self-dual over $GF(4)$. However, there is a second $OA(4^6, 12, 4, 5)$, an additive (but not linear, cf. Section 4.3) array over $GF(4)$ with generators

$$\langle 310100100101 \rangle \ .$$

This is self-dual as an additive array. Code is $(12, 4^6, 6)_4$ *dodecacode*, an additive self-dual code over $GF(4)$ (Rains and Sloane, 1998).

$OA(4^{12}, 18, 4, 9)$. Generators for dual code:

$$\langle 000011 \rangle \langle 001013 \rangle \langle 011212 \rangle \ .$$

Dual code is $(18, 4^6, 10)_4$ quasicyclic code given by Gulliver and Östergård (1998).

5.14 Notes on Chapter 5

A great many other constructions for codes are known, besides those given in this chapter, for example *alternant* codes (MacWilliams and Sloane, 1977, Chapter 12, §2), *Srivastava* codes (loc. cit., §6), *two-weight codes* (Calderbank and Kantor, 1986), *algebraic geometry* codes, which are a far-reaching generalization of Bush's construction given in Theorem 3.1 (Tsfasman and Vlăduţ, 1991), to mention just a few. There is no space to discuss them here, and they are not needed to obtain the arrays mentioned in the tables in Chapter 12. Nevertheless all of them yield orthogonal arrays.

Section 5.2. While studying this section some readers may find it helpful to be told that a cyclic code is an ideal in the ring $GF(s)[X]/(X^k - 1)$, and that the generator polynomial is unique because this ring is a principal ideal domain.

Section 5.3. The earliest references for Hamming codes are Golay (1949) and Hamming (1950). See Chapter 1 of MacWilliams and Sloane (1977), for further information.

Section 5.4. One could write down formulae analogous to (5.21) and (5.24) for the parameters of e-error-correcting BCH codes, for $e = 3, 4, 5, \ldots$, and the orthogonal arrays formed by their duals. In these cases the minimal distance of the dual code is not known in general (although there are bounds on its value), so we do not have analogues of (5.22) and (5.23).

Tables of binary BCH codes can be found in Berlekamp (1968), Peterson and Weldon (1972) and MacWilliams and Sloane (1977). In Chapter 12 we will mention many other tables of codes.

Section 5.5. Reed-Solomon codes were initially described by Reed and Solomon (1960), and the extended codes were constructed by Wolf (1969), Tanaka and Nishida (1970) and Gross (1973). See also Roth (1991), Roth and Lempel (1989a, 1989b), Roth and Seroussi (1985, 1986), Seroussi and Roth (1986).

Section 5.6. As remarked in MacWilliams and Sloane (1977), the subject of MDS codes is one of the most fascinating in all of coding theory, because of their connections with other combinatorial and geometrical problems. Chapter 11 of MacWilliams and Sloane (1977) gives a comprehensive survey of what was known up to 1977, together with an extensive bibliography. We mention also the following additional references: Blokhuis, Bruen and Thas (1990), Bruen, Thas and Blokhuis (1988), Ceccherini and Tallini (1981), Gulati and Kounias (1970,

1973a), Hill (1973, 1976, 1978a,b), Hirschfeld (1979, 1983, 1985), Korchmáros (1983), Storme (1992, 1993), Storme and Szönyi (1993a, 1993b) and Storme and Thas (1991, 1992, 1993, 1994). The survey articles by Hirschfeld (1983) and Hirschfeld and Storme (1998) are especially recommended.

Sections 5.5 to 5.13. *Enumeration of orthogonal arrays.* In a few cases orthogonal arrays with specified sets of parameters have been completely enumerated. For example:

- Lam and Tonchev (1996) have classified all $OA(27, 13, 3, 2)$'s.
- Hedayat, Seiden and Stufken (1997) have classified all $OA(54, 5, 3, 3)$'s.
- It is known from coding theory that the arrays $OA(2^8, 16, 2, 5)$, $OA(2^{12}, 24, 2, 7)$, $OA(3^6, 12, 3, 5)$, etc., are unique up to isomorphism.
- As we will see in Chapter 7, orthogonal arrays based on Hadamard matrices of small orders have also been completely enumerated.
- Some further enumerations have been given by Fujii, Namikawa and Yamamoto (1989), Namikawa, Fujii and Yamamoto (1989), Yamamoto, Fujii, Hyodo and Yumiba (1992a, 1992b, 1994a, 1995) and Yamamoto, Namikawa and Fujii (1988).

It would be nice to have additional results of this type.

Research Problem 5.13. Find further classifications of orthogonal arrays whose parameter sets are not too large.

5.15 Problems

5.1. Prove Theorem 5.2.

5.2. To familiarize yourself with the pointed bracket notation introduced in Section 5.2, verify that $\langle 1110100 \rangle$ generates a binary $OA(8, 7, 2, 2)$. What can you say about the sets of vectors $\langle 1101000 \rangle$, $1\langle 1101000 \rangle$ or $\langle 110 \rangle\langle 110 \rangle\langle 110 \rangle$?

5.3. We claimed in Section 5.2 that a cyclic code can always be represented by a single generator. This problem suggests a way to prove this claim.

As in Section A.2, let $GF(s)[X]$ denote the ring of polynomials with coefficients in $GF(s)$. Let $R_k = GF(s)[X]/(X^k - 1)$ be the finite ring of residue classes of $GF(s)[X]$ modulo $X^k - 1$. Thus R_k has s^k elements, and every polynomial of degree less than or equal to $k - 1$ represents a different residue class. For ease of reference we will say that these polynomials (rather than the residue classes that they represent) are the elements of R_k.

For a $(k, s^n, d)_s$ cyclic code C, represent each codeword by a polynomial over $GF(s)$ of degree at most $k-1$ (as explained in Section 5.2). Let $\mathcal{C} \subset R_k$ consist of these s^n polynomials.

a. Among the $s^k - 1$ nonzero polynomials in $\mathcal{C}$, show that there is a unique monic polynomial of minimal degree.

b. For $f \in R_k$ and $g \in \mathcal{C}$, show that $fg \in \mathcal{C}$. (Hint: You may first want to show the validity of this statement for $f(X) = X$.)

c. Let $h \in \mathcal{C}$, $h \neq 0$, be the unique polynomial from part a. Show that for every $g \in \mathcal{C}$ there is an $f \in R_k$ such that $g = fh$. (Hint: Start by writing $g(X) = \alpha(X)h(X) + \beta(X)$ for $\alpha, \beta \in R_k$, $\deg \beta < \deg h$.)

d. Show that the codeword in C that corresponds to the polynomial h in part c is a generator for C.

5.4. Show that the dual Hamming code of length $k = (s^m - 1)/(s-1)$ and dimension m has minimal distance s^{m-1}. [Hint: see for example Assmus and Key, 1992a, p. 59.]

5.5. Show that the minimal polynomials $M^{(i)}(X)$ and $M^{(2i)}(X)$ coincide for binary codes, as stated in Section 5.4. (Hint: Use the result in Problem A.12.)

5.6. For the double-error correcting BCH code of length 31 constructed in Section 5.4, verify that if $M^{(1)}(X) = 1 + X^2 + X^5$ then $M^{(3)}(X) = 1 + X^2 + X^3 + X^4 + X^5$ as claimed.

5.7. Give generator polynomials for the following binary BCH codes:

a. Length 7, minimal distance 3.

b. Length 7, minimal distance 4.

c. Length 7, minimal distance 5.

d. What are actual minimal distances of these codes, and how many codewords do they contain?

e. Repeat for length 15 and minimal distances 1, 3, 5 and 7.

f. What are the sizes and minimal distances of the duals to the above codes?

g. What are the parameters of the orthogonal arrays formed by the codewords of each of the above codes? What are the parameters of the orthogonal arrays formed from the dual codes?

5.8. Repeat the above problem for a ternary BCH code of length 8 and minimal distance 4.

5.9. *The $(u, u+v)$ construction for codes.* Given two binary codes C_1 and C_2 of the same length, with parameters $(k, M_i, d_i)_2$, $i = 1, 2$, we may form a new code C from the set of vectors $(u, u+v)$, for $u \in C_1$, $v \in C_2$. Show that C is a $(2k, M_1M_2, d)_2$ code, where $d = \min\{2d_1, d_2\}$.

5.10. Show that the construction of the previous problem can be used to combine Reed-Muller codes $RM(r, m-1)$ and $RM(r-1, m-1)$ to produce the Reed-Muller code $RM(r, m)$. Hence show that the latter code has minimal distance 2^{m-r}.

5.11. Show that $RM(r, m)$ and $RM(m-r-1, m)$ are dual codes, for $0 \le r \le m-1$.

5.12. Give a generator matrix for the $OA(2^{10}, 12, 2, 7)$ described in Section 5.11.

5.13. Show that the $(8, 16, 4)_2$ and $(24, 2^{12}, 8)_2$ codes of lengths 8 and 24 defined in Section 5.11 are self-dual, while the other codes given there are not. What are the smallest examples of self-dual codes over the alphabets $GF(2)$, $GF(3)$, $GF(4)$ and $\mathbb{Z}_4$?

5.14. Verify the claim in Section 5.9 that (i) there are s^2+1 solutions in $PG(3, s)$ to equation (5.39), provided that a is such that $z^2 + az + 1 = 0$ has no root in $GF(s)$, and (ii) that no three of the resulting points (w, x, y, z) are collinear.

Chapter 6

Orthogonal Arrays and Difference Schemes

In this chapter we introduce the concept of a difference scheme and some of its generalizations. Difference schemes were first defined by Bose and Bush (1952), and are a simple but powerful tool for the construction of orthogonal arrays of strength two.

Difference schemes are defined in Section 6.1, and Section 6.2 shows how they are used to construct orthogonal arrays. Sections 6.3 and 6.4 present a recursive construction of Bose and Bush (1952), and constructions for difference schemes of index 2, respectively. Section 6.4 presents an alternative version of the Addelman-Kempthorne construction of Section 3.3, and extends it to cover the case when s is even. Section 6.5 contains some generalizations and variations, including the concept of a perpendicular difference array. The concluding section contains a table summarizing what is presently known about difference schemes with small numbers of levels.

6.1 Difference Schemes

Throughout this chapter, $(\mathcal{A}, +)$, or simply $\mathcal{A}$ (if the binary operation is clear from the context), will denote a finite abelian group with a binary operation $+$. The cardinality of $\mathcal{A}$ will be denoted by s, its identity element by 0, and the inverse of an element σ by $-\sigma$. In the majority of the examples the pair $(\mathcal{A}, +)$ will be taken to be the additive group associated with the Galois field $GF(s)$.

Definition 6.1. An $r \times c$ array D with entries from $\mathcal{A}$ is called a *difference scheme based on* $(\mathcal{A}, +)$ if it has the property that for all i and j with $1 \leq i, j \leq$

c, $i \neq j$, the vector difference between the ith and jth columns contains every element of $\mathcal{A}$ equally often.

Necessarily r is a multiple of s, say $r = \lambda s$, where λ is the number of times each element of $\mathcal{A}$ occurs in the difference of two columns. We will denote such an array by $D(r, c, s)$, and refer to it as a difference scheme with s levels and index λ, although of course this notation does not show the dependence on the pair $(\mathcal{A}, +)$.

Example 6.2. Any orthogonal array $OA(N, k, s, t)$ with $t \geq 2$ may be regarded as a difference scheme $D(N, k, s)$, simply by taking the levels to be integers modulo s. ∎

Example 6.3. Let $(\mathcal{A}, +)$ be the additive group associated with the field $GF(s)$, whose elements we denote by $\alpha_0, \alpha_1, \ldots, \alpha_{s-1}$. Let D be the $s \times s$ multiplication table of this field. (Thus the table contains a row and column of zeros, corresponding to multiplication by 0). Then D is a difference scheme $D(s, s, s)$. Indeed, the difference of two columns from D has the form

$$\begin{pmatrix} \beta\alpha_0 \\ \vdots \\ \beta\alpha_{s-1} \end{pmatrix} - \begin{pmatrix} \gamma\alpha_0 \\ \vdots \\ \gamma\alpha_{s-1} \end{pmatrix} = (\beta - \gamma) \begin{pmatrix} \alpha_0 \\ \vdots \\ \alpha_{s-1} \end{pmatrix},$$

where β, $\gamma \in GF(s)$, $\beta \neq \gamma$. The elements $(\beta - \gamma)\alpha_i$, $i = 0, \ldots, s-1$, include every element of $GF(s)$ exactly once. ∎

Example 6.4. By juxtaposing difference schemes $D(r_1, c, s)$ and $D(r_2, c, s)$ we obtain a difference scheme $D(r_1 + r_2, c, s)$. By taking the componentwise products of the rows of difference schemes $D(r_1, c, s_1)$ and $D(r_2, c, s_2)$ in all possible ways we obtain a difference scheme $D(r_1 r_2, c, s_1 s_2)$. ∎

Let D be a difference scheme $D(r, c, s)$, and for a fixed i_0, let C_{i_0} denote the i_0th column of D. Let D^* be the array obtained by subtracting C_{i_0} from each column of D. It is easily seen that D^* is a difference scheme with the same parameters as D, and with the additional property that one column in D^* consists entirely of the zero element of $\mathcal{A}$. Therefore, by permuting the columns if necessary, D can always be converted to a difference scheme of the form

$$\begin{bmatrix} 0_r & D^{(0)} \end{bmatrix}. \tag{6.1}$$

By considering the difference between the first column and any other column, it follows that every column of $D^{(0)}$ contains each element of $\mathcal{A}$ equally often. We say that the columns of $D^{(0)}$ are *uniform on* $\mathcal{A}$. We call D *α-resolvable* if it can be converted, by the method just described, to the form (6.1) in such a way that the rows of $D^{(0)}$ can be partitioned to form $r/(\alpha s)$ arrays, each of size $\alpha s \times c-1$ and each having its columns uniform on $\mathcal{A}$. Note that we do not require that the $\alpha s \times c-1$ subarrays should be difference schemes.

The following result is due to Jungnickel (1979). An alternative proof is given in Problem 6.4.

Theorem 6.5. *If a difference scheme $D(r,c,s)$ exists then $c \le r$.*

Proof: Let D be a difference scheme $D(r,c,s)$ based on an abelian group $\mathcal{A}$. The group $\mathcal{A}$ is a product of cyclic groups $\mathcal{B} \times \mathcal{B}' \times \mathcal{B}'' \times \cdots$, say. Let the elements of $\mathcal{B}$ be labeled $\{0,1,\ldots,b-1\}$, where $b \ge 2$, and let π denote the projection map which sends each element of $\mathcal{A}$ to the corresponding element of $\mathcal{B}$ (ignoring the other subgroups $\mathcal{B}'$, $\mathcal{B}''$, ...). Let D' be the array obtained by replacing every element of D by its image under π, so that we may regard D' as a difference scheme $D(r,c,b)$ based on the integers modulo b. We now construct an $r \times c$ complex matrix M from D' by replacing each entry $j \in \mathcal{B}$ by $e^{2\pi ij/b}$, where $i=\sqrt{-1}$. The columns of M are vectors in $\mathbb{C}^r$ (where $\mathbb{C}$ denotes the complex numbers). We define the inner product of vectors $u,v \in \mathbb{C}^r$ by $(u,v) = \sum_{j=1}^{r} u_j \bar{v}_j$, where the bar denotes complex conjugation. Since D' is a difference scheme, the inner product of any two columns of M is zero. Since $\mathbb{C}^r$ is a complex r-dimensional vector space, we must have $c \le r$. ■

The proof also shows that M satisfies $M^T\overline{M} = rI_c$, or equivalently

$$\overline{M}^T M = rI_c \; . \tag{6.2}$$

In view of (6.2), a difference scheme $D(r,c,s)$ in which $r=c$ is also called a *generalized Hadamard matrix* of order r over $(\mathcal{A},+)$ (see the definition of a Hadamard matrix in Chapter 7). In particular, any (ordinary) Hadamard matrix of order r is a difference scheme $D(r,r,2)$.

The difference schemes constructed in Example 6.3 are generalized Hadamard matrices. It also follows from the proof of Theorem 6.5 that the transpose of a generalized Hadamard matrix is again a generalized Hadamard matrix. (For if $c=r$ then M has rank r, and (6.2) implies $\overline{M}M^T = cI_c$.)

The next result, which extends the construction of Example 6.3, provides a large class of generalized Hadamard matrices.

Theorem 6.6. *A difference scheme $D(p^m,p^m,p^n)$ exists for any prime p and integers $m \ge n \ge 1$.*

Proof:
CONSTRUCTION: Let the elements of $GF(p^m)$ be represented by polynomials

$$\beta_0 + \beta_1 x + \cdots + \beta_{n-1}x^{n-1} + \cdots + \beta_{m-1}x^{m-1}, \tag{6.3}$$

where $\beta_0, \ldots, \beta_{m-1} \in GF(p)$. We may regard $GF(p^n)$ as an *additive* subgroup of $GF(p^m)$, by identifying its elements with the subset of $GF(p^m)$ consisting of elements of the form $\beta_0+\beta_1 x+\cdots+\beta_{n-1}x^{n-1}$. (Multiplication of these elements in $GF(p^n)$ is in general totally different from their multiplication in $GF(p^m)$. This does not concern us, since here we will only be using the additive structure of $GF(p^n)$.) Now let D^* be the $p^m \times p^m$ multiplication table of $GF(p^m)$. Map every entry $\beta_0+\beta_1 x+\cdots+\beta_{m-1}x^{m-1}$ in this table to $\beta_0+\beta_1 x+\cdots+\beta_{n-1}x^{n-1}$. Let D be the array obtained in this way, and view its entries as elements of $GF(p^n)$. Then D is the desired difference scheme.

VERIFICATION: Clearly D is a $p^m \times p^m$ array with entries from $GF(p^n)$. If we let ϕ denote the map from $GF(p^m)$ to $GF(p^n)$ defined in the previous paragraph, then the difference of two columns of D will have the form

$$\begin{pmatrix} \phi(\beta\alpha_0) \\ \vdots \\ \phi(\beta\alpha_{p^m-1}) \end{pmatrix} - \begin{pmatrix} \phi(\gamma\alpha_0) \\ \vdots \\ \phi(\gamma\alpha_{p^m-1}) \end{pmatrix},$$

where $\beta, \gamma \in GF(p^m)$, $\beta \neq \gamma$. From the definition of ϕ, it follows that $\phi(\beta\alpha_i) - \phi(\gamma\alpha_i) = \phi(\beta\alpha_i - \gamma\alpha_i)$, and so the above vector difference is equal to

$$\begin{pmatrix} \phi((\beta-\gamma)\alpha_0) \\ \vdots \\ \phi((\beta-\gamma)\alpha_{p^m-1}) \end{pmatrix}.$$

Since every element of $GF(p^m)$ appears once among the elements $(\beta-\gamma)\alpha_i$, $0 \le i < p^m$, every element of $GF(p^n)$ appears p^{m-n} times among the elements $\phi((\beta-\gamma)\alpha_i)$, $0 \le i < p^m$. ■

Example 6.7. We illustrate the construction for the case $p = 3$, $m = 2$, $n = 1$. This will result in a difference scheme $D(9, 9, 3)$. In this special case the field $GF(p^n)$ in the construction is actually a *subfield* of $GF(p^m)$, and multiplication of elements of $GF(p^n)$ is the same in both fields.

Table 6.8 is a multiplication table for $GF(3^2)$, based on the irreducible polynomial $f(x) = x^2+1$. We represent the nine elements of $GF(3^2)$ in a condensed notation, writing 0 as 00, 1 as 10, x as 01, $1+2x$ as 12, and so on.

Table 6.8. A multiplication table for $GF(3^2)$.

	00	10	20	01	11	21	02	12	22
00	00	00	00	00	00	00	00	00	00
10	00	10	20	01	11	21	02	12	22
20	00	20	10	02	22	12	01	21	11
01	00	01	02	20	21	22	10	11	12
11	00	11	22	21	02	10	12	20	01
21	00	21	12	22	10	01	11	02	20
02	00	02	01	10	12	11	20	22	21
12	00	12	21	11	20	02	22	01	10
22	00	22	11	12	01	20	21	10	02

Upon applying the map $\phi : \beta_0 + \beta_1 x \mapsto \beta_0$ to the entries of this table we obtain the difference scheme $D(9, 9, 3)$ exhibited in Table 6.9. ■

Table 6.9. A difference scheme $D(9, 9, 3)$ based on $(GF(3), +)$.

0	0	0	0	0	0	0	0	0
0	1	2	0	1	2	0	1	2
0	2	1	0	2	1	0	2	1
0	0	0	2	2	2	1	1	1
0	1	2	2	0	1	1	2	0
0	2	1	2	1	0	1	0	2
0	0	0	1	1	1	2	2	2
0	1	2	1	2	0	2	0	1
0	2	1	1	0	2	2	1	0

Example 6.10. To illustrate that not all difference schemes have such a simple construction as those in Theorem 6.6, Table 6.11 shows a difference scheme $D(12, 6, 12)$ found by Dulmage, Johnson and Mendelsohn (1961). This is not part of any known infinite family. The scheme is based on the abelian group $\mathbb{Z}_2 \oplus \mathbb{Z}_6$ of order 12 (where $\mathbb{Z}_m$ denotes the integers modulo m and $\oplus$ indicates a direct sum). We represent the elements of this group by pairs ab, with $0 \le a \le 1$, $0 \le b \le 5$. ■

Table 6.11. A difference scheme $D(12, 6, 12)$ based on $(\mathbb{Z}_2 \oplus \mathbb{Z}_6, +)$.

00	00	00	00	00	00
00	01	03	12	04	10
00	02	10	01	15	12
00	03	01	15	14	02
00	04	13	05	02	11
00	05	15	13	11	01
00	10	02	03	12	13
00	11	12	14	10	15
00	12	05	02	13	04
00	13	04	11	01	14
00	14	11	10	03	05
00	15	14	04	05	03

Other constructions of difference schemes will be described in later sections.

In view of Theorem 6.5, one may ask for the maximal number of columns c in a difference scheme $D(\lambda s, c, s)$ for given values of λ and s. We will return to this question in Section 6.6 (see Table 6.67).

6.2 Orthogonal Arrays Via Difference Schemes

The procedure that converts a difference scheme into an orthogonal array is very simple. If D is a difference scheme based on $(\mathcal{A}, +)$, where $\mathcal{A} = \{\sigma_0, \ldots, \sigma_{s-1}\}$, we will use D_i to denote the array obtained from D by adding σ_i to each of its entries. Obviously D_i is a difference scheme with the same parameters as D. We can juxtapose the D_i's to obtain an orthogonal array of strength two. We refer to this process as *developing* the difference scheme into an orthogonal array.

Lemma 6.12. *If D is a difference scheme $D(r, c, s)$, then*

$$A = \begin{bmatrix} D_0 \\ D_1 \\ \vdots \\ D_{s-1} \end{bmatrix}$$

is an $OA(rs, c, s, 2)$.

Proof: Select two factors from A, say F_1 and F_2, $F_1 \neq F_2$, and two elements from $\mathcal{A}$, say σ and σ', allowing the possibility that $\sigma = \sigma'$. We must show that the number of runs with factor F_1 at level σ and factor F_2 at level σ' is equal to $rs/s^2 = \lambda$.

If C_1 and C_2 denote the columns of D corresponding to the factors F_1 and F_2, respectively, then we know that λ entries in $C_1 - C_2$ are equal to $\sigma - \sigma'$. For each occurrence of $\sigma - \sigma'$ in $C_1 - C_2$ there is a unique row in a unique D_i in which F_1 is at level σ and F_2 is at level σ'. Since these are the only runs with factor F_1 at level σ and factor F_2 at level σ', we conclude that there are indeed λ such runs in A. ■

Example 6.13. The difference scheme in Table 6.11 leads to an $OA(144, 6, 12, 2)$. ■

It is possible that an orthogonal array obtained from developing a difference scheme has strength $t \geq 3$. For this to happen, however, the difference scheme D needs to possess additional properties. We will return to this in Section 6.5.

It is natural to ask if an orthogonal array of strength two obtained from developing a difference scheme can ever contain the maximal number of factors. In other words, are there cases for which a difference scheme $D(r, c, s)$ exists and $c = f(rs, s, 2)$? The answer to this question is in the negative. In order to show this, we will introduce some additional tools which will also be useful in Sections 6.3 and 6.4.

First, we introduce the concept of a-resolvability.

Definition 6.14. An orthogonal array $OA(N, k, s, 2)$ is said to be *a-resolvable* if it is statistically equivalent to the juxtaposition of N/as arrays such that each factor occurs in each of these arrays a times at each level. (Note that the subarrays are orthogonal arrays of strength 1, although not in general of strength 2.) A 1-resolvable orthogonal array is also called *completely resolvable*.

Example 6.15. The $OA(9, 3, 3, 2)$ shown in Table 6.16 is completely resolvable.

Table 6.16. A completely resolvable $OA(9, 3, 3, 2)$ obtained by the construction of Lemma 6.12.

0	0	0
0	1	2
0	2	1
1	1	1
1	2	0
1	0	2
2	2	2
2	0	1
2	1	0

The $N/as = 3$ arrays in which each factor occurs once at each level and whose juxtaposition is statistically equivalent to the $OA(9,3,3,2)$ are

$$A_1 = \begin{bmatrix} 0 & 0 & 0 \\ 1 & 1 & 1 \\ 2 & 2 & 2 \end{bmatrix}, \quad A_2 = \begin{bmatrix} 0 & 1 & 2 \\ 1 & 2 & 0 \\ 2 & 0 & 1 \end{bmatrix}, \quad A_3 = \begin{bmatrix} 0 & 2 & 1 \\ 1 & 0 & 2 \\ 2 & 1 & 0 \end{bmatrix}.$$

■

Theorem 6.17. *The existence of both an a-resolvable $OA(N,k_1,s,2)$ and an $OA(N/as,k_2,s,2)$ implies the existence of an $OA(N,k_1+k_2,s,2)$. Furthermore, if the $OA(N/as,k_2,s,2)$ is b-resolvable, there is an (abs)-resolvable $OA(N,k_1+k_2,s,2)$.*

Proof:
CONSTRUCTION: Let

$$A = [A_1^T, \ldots, A_u^T]^T$$

be the a-resolvable $OA(N,k_1,s,2)$, where $u = N/as$ and in each of $A_1, A_2, \ldots, A_u$ every factor occurs a times at each level. Let B be the $OA(N/as,k_2,s,2)$. If B is b-resolvable, we take it to be

$$B = [B_1^T, \ldots, B_v^T]^T,$$

where $v = N/(abs^2)$ and in each B_i every factor occurs b times at each level, for $i = 1,\ldots,v$. Now let C be the $N \times (k_1+k_2)$ array formed by following each run in A_i by the ith run in B, for $i = 1,\ldots,u$. We claim that C is an $OA(N,k_1+k_2,s,2)$ which is (abs)-resolvable if B is b-resolvable.

VERIFICATION: Clearly, the first k_1 factors of C form an $OA(N,k_1,s,2)$. By Property 3 of Chapter 1 the last k_2 factors of C form an $OA(N,k_2,s,2)$. Thus to verify that C is an $OA(N,k_1+k_2,s,2)$ it suffices to consider any factor F_1 out of the first k_1 factors and any factor F_2 out of the last k_2 factors and to check that every level combination occurs N/s^2 times for those two factors. For every level of F_2, there are N/as^2 runs in B for which factor F_2 is at that level. Each of these runs is appended to all the runs in some A_i, where factor F_1 occurs a times at each level. Consequently, F_1 and F_2 contain every level combination $a(N/as^2) = N/s^2$ times. Thus C is indeed an $OA(N,k_1+k_2,s,2)$.

If B is b-resolvable, the resolvability properties imply that in the first abs^2 runs of C each factor occurs abs times at each level. The same is true for every subsequent group of abs^2 runs, so C is (abs)-resolvable. ■

Provided N/as^2 is an integer, we can apply the same argument but taking B to be an $N/as \times 1$ array in which each level occurs N/as^2 times. This leads to the following corollary.

Corollary 6.18. *If a divides N/s^2, an a-resolvable $OA(N,k,s,2)$ can always be extended to an (as)-resolvable $OA(N,k+1,s,2)$. In particular, any completely resolvable $OA(N,k,s,2)$ can be extended to an s-resolvable $OA(N,k+1,s,2)$.*

We now link the notions of difference scheme and resolvability by means of the following observation.

Theorem 6.19. *An orthogonal array $OA(N,k,s,2)$ that is obtained by developing a difference scheme is completely resolvable.*

Proof: Let the orthogonal array be as in Lemma 6.12, with $N = rs$ and $k = c$. Let A_i consist of the s runs obtained by taking the ith row of each of $D_0, D_1, \ldots, D_{s-1}$, for $i = 1, \ldots, r$. Thus A_i consists of the runs formed by taking the ith row in D and adding $\sigma_0 1_c^T, \ldots, \sigma_{s-1} 1_c^T$ to it in turn. It is now clear that every factor occurs once at each level in each A_i, and that the juxtaposition of the A_i's is statistically equivalent to the $OA(rs,c,s,2)$ that we obtain by developing the difference scheme. ∎

Our initial question, whether there is a difference scheme $D(r,c,s)$ such that $c = f(rs,s,2)$, now receives a negative answer by the following corollary, which is an immediate consequence of Lemma 6.12, Theorem 6.19 and Corollary 6.18.

Corollary 6.20. *A difference scheme $D(r,c,s)$ can be used to construct an orthogonal array $OA(rs,c+1,s,2)$.*

Thus it is always true that $c+1 \le f(rs,s,2)$. Equality can indeed occur, as Example 6.21 will illustrate. However, since the $OA(rs,c+1,s,2)$ in Corollary 6.20 will still be s-resolvable, it can be anticipated from Theorem 6.17 that, depending on the value of $N = rs$, a further extension of that orthogonal array cannot be excluded, and that it is possible that $c+1 < f(rs,s,2)$. Situations of this type will be considered in Example 6.25 and in Section 6.3.

Example 6.21. Let D be a difference scheme $D(s,s,s)$ as in Example 6.3. According to Corollary 6.20, D can be used to construct an $OA(s^2,s+1,s,2)$ of index unity. We know already from Corollary 3.9(i) that $f(s^2,s,2) = s+1$ if s is a prime power. Since $r = c = s$, this is a case where $c+1 = f(rs,s,2)$.

The construction of the orthogonal arrays in this example is accomplished by developing D as in Lemma 6.12, recognizing its complete resolvability as in Theorem 6.19, and appending a symbol to each run as in Theorem 6.17. The construction is displayed in Tables 6.22 to 6.24 for the case $s = 4$. As usual we denote the elements of $GF(4)$ by $0,1,2,3$. The 4×4 arrays A_i, $i = 1, \ldots, 4$, can be formed by combining the runs that are labeled A_i in Table 6.23. The $OA(16, 5, 4, 2)$ in Table 6.24 is then obtained by appending a symbol to each run of the $OA(16, 4, 4, 2)$, as in Corollary 6.18. ∎

Table 6.22. A difference scheme $D(4, 4, 4)$.

0	0	0	0
0	1	2	3
0	2	3	1
0	3	1	2

Table 6.23. A completely resolvable $OA(16, 4, 4, 2)$ obtained by developing the difference scheme $D(4, 4, 4)$ in Table 6.22 (the array has been transposed).

0	0	0	0	1	1	1	1	2	2	2	2	3	3	3	3
0	1	2	3	1	0	3	2	2	3	0	1	3	2	1	0
0	2	3	1	1	3	2	0	2	0	1	3	3	1	0	2
0	3	1	2	1	2	0	3	2	1	3	0	3	0	2	1
A_1	A_2	A_3	A_4	A_1	A_2	A_3	A_4	A_1	A_2	A_3	A_4	A_1	A_2	A_3	A_4

Table 6.24. An $OA(16, 5, 4, 2)$ obtained by appending a symbol to each of the runs in the $OA(16, 4, 4, 2)$ of Table 6.23 (transposed).

0	0	0	0	1	1	1	1	2	2	2	2	3	3	3	3
0	1	2	3	1	0	3	2	2	3	0	1	3	2	1	0
0	2	3	1	1	3	2	0	2	0	1	3	3	1	0	2
0	3	1	2	1	2	0	3	2	1	3	0	3	0	2	1
0	1	2	3	0	1	2	3	0	1	2	3	0	1	2	3

Example 6.25. Let D be the difference scheme $D(9, 9, 3)$ exhibited in Table 6.9. By developing this we obtain a completely resolvable $OA(27, 9, 3, 2)$. From Example 6.15 we also know that a completely resolvable $OA(9, 3, 3, 2)$ exists. From Theorem 6.17 we can then deduce the existence of a 3-resolvable $OA(27, 12, 3, 2)$. By applying Corollary 6.18 we obtain an $OA(27, 13, 3, 2)$. However, from Corollary 3.21, we know already that $f(27, 3, 2) = 13$, so this array has the maximal number of factors. ∎

We shall not give details of the construction in Example 6.25, since in Section 6.3 we will describe a more general result due to Bose and Bush (1952) that includes this as a special case. However, the reader is invited to follow through the above construction and to compare the result with the arrays in Tables 3.25 and 3.26.

There is a simple generalization of Lemma 6.12. We begin with a definition.

Definition 6.26. Let $A = (a_{ij})$ and $B = (b_{ij})$ be respectively $m \times n$ and $u \times v$ matrices with entries from an abelian group $\mathcal{A}$ with binary operation $*$ (usually addition or multiplication). Their *tensor* or *Kronecker product*, denoted by $A \otimes B$, is the $mu \times nv$ matrix

$$A \otimes B = \begin{bmatrix} a_{11} * B & \cdots & a_{1n} * B \\ \vdots & & \vdots \\ a_{m1} * B & \cdots & a_{mn} * B \end{bmatrix},$$

where $a_{ij} * B$ stands for the $u \times v$ matrix with entries $a_{ij} * b_{rs}$ ($1 \le r \le u$, $1 \le s \le v$). In this chapter $*$ will always denote addition.

Using this definition we may rewrite the array A in Lemma 6.12 as

$$A = E \otimes D , \tag{6.4}$$

where $E = (\sigma_0, \sigma_1, \ldots, \sigma_{s-1})^T$. However, there are other choices for E in Equation 6.4 for which A is also an orthogonal array of strength 2.

Lemma 6.27. *If D is a difference scheme $D(r, c, s)$ and B is an $OA(N, k, s, 2)$, both based on the abelian group $\mathcal{A}$, then the array*

$$A = B \otimes D$$

is an orthogonal array $OA(Nr, kc, s, 2)$.

The proof of this result is left as an exercise for the reader (Problem 6.7). It is again always possible to add at least one factor to the array A in this lemma; we will return to this in Chapter 9 when studying methods for constructing mixed orthogonal arrays.

6.3 Bose and Bush's Recursive Construction

The construction to be discussed in this section, due to Bose and Bush (1952), is in the spirit of that in Example 6.25. It allows us to construct orthogonal arrays of strength two with a large number of factors, possibly the maximal number, provided that the number of symbols s and the index λ are powers of the same prime p. In Example 6.25 we took $s = \lambda = p = 3$ and indeed obtained an orthogonal array with the maximal number of factors.

Theorem 6.28. *Let $s = p^v$ and $\lambda = p^u$, where p is prime and u and v are integers with $u \ge 0$, $v \ge 1$. Let $d = \lfloor u/v \rfloor$. Then there exists an*

$$OA\left(\lambda s^2, \lambda \frac{s^{d+1} - 1}{s^d - s^{d-1}} + 1, s, 2\right).$$

Proof:
CONSTRUCTION: Let $D(i)$ be a difference scheme $D(\lambda s^{1-i}, \lambda s^{1-i}, s)$, for $i = 0, 1, \ldots, d$. Since s and λs^{1-i} are powers of the same prime and $\lambda s^{1-i} \geq s$, these difference schemes can be constructed as in Theorem 6.6. Develop the difference scheme $D(0)$ to obtain a completely resolvable $OA(\lambda s^2, \lambda s, s, 2)$. If $d = 0$, add one factor to this orthogonal array as in Corollary 6.18 to obtain the desired $OA(\lambda s^2, \lambda s + 1, s, 2)$. If $d \geq 1$, use $D(1)$ to construct a completely resolvable $OA(\lambda s, \lambda, s, 2)$. The completely resolvable $OA(\lambda s^2, \lambda s, s, 2)$ and the completely resolvable $OA(\lambda s, \lambda, s, 2)$ can be used as in Theorem 6.17 to obtain an s-resolvable $OA(\lambda s^2, \lambda s + \lambda, s, 2)$. If $d = 1$, we can again add one more factor as in Corollary 6.18 to obtain the desired $OA(\lambda s^2, \lambda s + \lambda + 1, s, 2)$. If $d \geq 2$, we can use the s-resolvable $OA(\lambda s^2, \lambda s + \lambda, s, 2)$ and the completely resolvable $OA(\lambda, \lambda/s, s, 2)$ which can be obtained from $D(2)$ to obtain an s^2-resolvable $OA(\lambda s^2, \lambda s + \lambda + \lambda/s, s, 2)$ by the method of Theorem 6.17.

If we continue in this way, using all the $D(i)$'s, after d applications of Theorem 6.17 and one application of Corollary 6.18 we obtain an array with $\lambda s + \lambda + \lambda/s + \cdots + \lambda/s^{d-1} + 1 = \lambda(s^{d+1} - 1)/(s^d - s^{d-1}) + 1$ factors. This is the desired orthogonal array.

VERIFICATION: It is clear that the array has the correct dimensions. Since the entire construction is based only on the results in Theorems 6.6, 6.17 and 6.19 and on Corollary 6.18, it suffices to check that these results are applicable where they are needed. This is straightforward, and will therefore be omitted. ■

As a special case of Theorem 6.28, consider the situation where λ is actually a power of s, or, equivalently, where $u \equiv 0 \pmod{v}$. Then $d = u/v$ and $\lambda = s^d$, so the number of factors in the orthogonal array in Theorem 6.28 is $(s^{d+2} - 1)/(s - 1)$. The array is thus an $OA(s^{d+2}, (s^{d+2} - 1)/(s - 1), s, 2)$, $d \geq 0$. This has the maximal number of factors, since it follows from Corollary 3.21 that $f(s^{d+2}, s, 2) = (s^{d+2} - 1)/(s - 1)$, $d \geq 0$.

Although this remark shows that the method of Theorem 6.28 is powerful, the arrays that are obtained from this theorem when λ is a power of s can also be constructed by the methods of Section 3.4. In particular, Constructions 2 and 3 of Theorem 3.20 are generally easier to implement than the method in Theorem 6.28. The reader who has worked out Example 6.25 and compared it with the construction that produced Table 3.26 will attest to this. The methods of Section 3.4, however, are applicable only when λ is a power of s. Theorem 6.28 retains its value for situations where the methods of Section 3.4 do not apply.

Whether the number of factors in the orthogonal array of Theorem 6.28 is also maximal when λ is not a power of s is unknown. There is no general method known for constructing orthogonal arrays with λs^2 runs, s levels and more factors than are given by Theorem 6.28. Also, with Theorem 2.8 in mind, it can generally be concluded that the difference between the upper bound on the maximal number of factors in an orthogonal array with λs^2 runs, s levels and strength two, and the number of factors produced by Theorem 6.28 is fairly small. If $u \not\equiv 0 \pmod{v}$, that difference is

$$\lambda(s-1)/(s^{d+1}-s^d) - 2 - \lfloor \theta \rfloor \ .$$

For example, when $s = p^2$ and $\lambda = p$, it reduces to

$$p - 2 - \lfloor (p-2)/2 \rfloor = \lfloor (p-1)/2 \rfloor \ .$$

Table 6.29. An $OA(27, 13, 3, 2)$.

0	0	0	0	0	0	0	0	0	0	0	0	0
0	1	2	0	1	2	0	1	2	0	1	2	1
0	2	1	0	2	1	0	2	1	0	2	1	2
0	0	0	2	2	2	1	1	1	1	1	1	0
0	1	2	2	0	1	1	2	0	1	2	0	1
0	2	1	2	1	0	1	0	2	1	0	2	2
0	0	0	1	1	1	2	2	2	2	2	2	0
0	1	2	1	2	0	2	0	1	2	0	1	1
0	2	1	1	0	2	2	1	0	2	1	0	2
1	1	1	1	1	1	1	1	1	0	0	0	0
1	2	0	1	2	0	1	2	0	0	1	2	1
1	0	2	1	0	2	1	0	2	0	2	1	2
1	1	1	0	0	0	2	2	2	1	1	1	0
1	2	0	0	1	2	2	0	1	1	2	0	1
1	0	2	0	2	1	2	1	0	1	0	2	2
1	1	1	2	2	2	0	0	0	2	2	2	0
1	2	0	2	0	1	0	1	2	2	0	1	1
1	0	2	2	1	0	0	2	1	2	1	0	2
2	2	2	2	2	2	2	2	2	0	0	0	0
2	0	1	2	0	1	2	0	1	0	1	2	1
2	1	0	2	1	0	2	1	0	0	2	1	2
2	2	2	1	1	1	0	0	0	1	1	1	0
2	0	1	1	2	0	0	1	2	1	2	0	1
2	1	0	1	0	2	0	2	1	1	0	2	2
2	2	2	0	0	0	1	1	1	2	2	2	0
2	0	1	0	1	2	1	2	0	2	0	1	1
2	1	0	0	2	1	1	0	2	2	1	0	2

Example 6.30. The construction of an $OA(27, 13, 3, 2)$ in Example 6.25 follows the recipe of Theorem 6.28. Since λ is a power of s (in fact $\lambda = s$), we know that this array, presented in Table 6.29, has the maximal number of factors. The levels for the first nine factors were obtained by developing $D(0)$, a

difference scheme $D(9,9,3)$. The levels for the next three factors were obtained by appending the appropriate runs of an $OA(9,3,3,2)$, which was obtained by developing $D(1)$, a difference scheme $D(3,3,3)$. The levels for the last factor were obtained as in Corollary 6.18, using the 3-resolvability of the orthogonal array formed by the first twelve factors. ■

Example 6.31. Let $p = 2$, $s = 2^v$ and $\lambda = 2s^{n-2}$. This corresponds to the parameters of an orthogonal array that was mentioned following the description of the Addelman and Kempthorne method in Theorem 3.16. When s is a power of 2, we did not give a general construction in Section 3.3 for an $OA(2s^n, 2(s^n-1)/(s-1)-1, s, 2)$, $n \geq 2$. We will now show that such arrays can be obtained from Theorem 6.28.

First, consider the case $v \geq 2$. Then $\lambda = 2s^{n-2} = 2^{(n-2)v+1}$. Hence, in notation of this section, $u = (n-2)v+1$ and $d = n-2$. Theorem 6.28 then yields an orthogonal array with parameters

$$OA\left(2s^n, 2\frac{s^n-1}{s-1} - 1, s, 2\right), \tag{6.5}$$

where $s = 2^v$, $v \geq 2$, $n \geq 2$. Theorem 6.28 provides this number of factors since $\lambda(s^{n-1}-1)/(s^{n-2}-s^{n-3})+1 = 2s(s^{n-1}-1)/(s-1)+1 = 2(s^n-1)/(s-1)-1$. This is exactly the number of factors that Addelman and Kempthorne (1961a) claimed to be possible.

If $v = 1$, and thus $s = 2$ and $u = d = n-1$, we can actually do a little better. Theorem 6.28 yields an array $OA(2^{n+1}, 2^{n+1}-1, 2, 2)$, with $n \geq 1$. This has the maximal number of factors since $u \equiv 0 \pmod{v}$. In Chapter 7 we will see that an $OA(2^{n+1}, 2^{n+1}-1, 2, 2)$ can also be obtained more easily from a Hadamard matrix of order 2^{n+1}.

An orthogonal array with parameters (6.5) in the case $s = 4$ and $n = 2$ is shown in Table 6.32. Since $d = n-2 = 0$, we need only $D(0)$, a difference scheme $D(8,8,4)$, to construct this array. An additional factor can then be added as in Corollary 6.18. Such an orthogonal array could also have been constructed by the method of Section 3.3, and those readers who have completed Problem 3.2 can compare the two methods of construction. They are likely to agree that the present method is easier to implement than that of Section 3.3.

The difference scheme $D(0)$ developed in Table 6.32 was obtained by starting with the multiplication table of $GF(2^3)$, using the irreducible polynomial $f(x) = x^3+x+1$ over $GF(2)$. The entry $\beta_0+\beta_1x+\beta_2x^2$ in this table was then mapped to $\beta_0+\beta_1x$, and these images were considered as elements of $GF(2^2)$, with x written as 2 and $x+1$ as 3. ■

Table 6.32. An $OA(32, 9, 4, 2)$ (transposed).

0	0	0	0	0	0	0	0	1	1	1	1	1	1	1	1	2	2	2	2	2	2	2	2	3	3	3	3	3	3	3	3
0	1	2	3	0	1	2	3	1	0	3	2	1	0	3	2	2	3	0	1	2	3	0	1	3	2	1	0	3	2	1	0
0	2	0	2	3	1	3	1	1	3	1	3	2	0	2	0	2	0	2	0	1	3	1	3	3	1	3	1	0	2	0	2
0	3	2	1	3	0	1	2	1	2	3	0	2	1	0	3	2	1	0	3	1	2	3	0	3	0	1	2	0	3	2	1
0	0	3	3	2	2	1	1	1	1	2	2	3	3	0	0	2	2	1	1	0	0	3	3	3	3	0	0	1	1	2	2
0	1	1	0	2	3	3	2	1	0	0	1	3	2	2	3	2	3	3	2	0	1	1	0	3	2	2	3	1	0	0	1
0	2	3	1	1	3	2	0	1	3	2	0	0	2	3	1	2	0	1	3	3	1	0	2	3	1	0	2	2	0	1	3
0	3	1	2	1	2	0	3	1	2	0	3	0	3	1	2	2	1	3	0	3	0	2	1	3	0	2	1	2	1	3	0
0	1	2	3	0	1	2	3	0	1	2	3	0	1	2	3	0	1	2	3	0	1	2	3	0	1	2	3	0	1	2	3

6.4 Difference Schemes of Index 2

In their seminal paper Bose and Bush (1952) gave an example of a difference scheme $D(6, 6, 3)$, of index 2, based on the additive group of $GF(3)$. Their scheme was apparently obtained by trial and error, rather than by any systematic construction. Clearly, using the methods of Section 6.2, this difference scheme induces an $OA(18, 7, 3, 2)$. The parameters of this orthogonal array are the special case $n = 2$, $s = 3$ of those in Section 3.3.

When $n = 2$, the Addelman and Kempthorne (1961a) construction of Section 3.3 results in an orthogonal array $OA(2s^2, 2s + 1, s, 2)$. Thus, generalizing the remark of the previous paragraph, if we can find a difference scheme $D(2s, 2s, s)$, we will have a simple construction that replaces the Addelman-Kempthorne construction when $n = 2$.

Several authors (Masuyama, 1969b; Xu, 1979; Jungnickel, 1979; Xiang, 1983) have studied the construction of index 2 difference schemes $D(2s, 2s, s)$ where s is a power of a prime. The construction given below is a modification of those of Xu (1979) and Jungnickel (1979).

If s is a power of 2, we have already seen that a difference scheme $D(2s, 2s, s)$ exists (in Theorem 6.6), and then in Example 6.31 that an $OA(2s^2, 2s+1, s, 2)$ can be obtained. We will therefore restrict our consideration to the case when s is a power of an odd prime. We write the elements of $GF(s)$ as $\alpha_0, \alpha_1, \ldots, \alpha_{s-1}$, where $\alpha_0 = 0$ and $\alpha_i = \alpha^i$, $i = 1, \ldots, s-1$, for a primitive element α. In particular, $\alpha_{s-1} = \alpha^{s-1} = 1$.

Theorem 6.33. *If s is a power of an odd prime then there exists a difference scheme $D(2s, 2s, s)$ and an orthogonal array $OA(2s^2, 2s + 1, s, 2)$.*

Proof: The method for constructing the orthogonal array from the difference scheme was described in Section 6.2. The major assertion of the theorem is the existence of the difference scheme $D(2s, 2s, s)$.

CONSTRUCTION: We construct four $s \times s$ matrices $U = (u_{ij})$, $V = (v_{ij})$, $W = (w_{ij})$, $X = (x_{ij})$, $0 \le i, j \le s-1$, whose entries are given by

$$\begin{aligned} u_{ij} &= \alpha_i \alpha_j \ , \\ v_{ij} &= \alpha_i \alpha_j + \beta \alpha_j^2 \ , \\ w_{ij} &= \alpha_i \alpha_j + \gamma \alpha_i^2 \ , \\ x_{ij} &= \nu \alpha_i \alpha_j + \delta \alpha_j^2 + \epsilon \alpha_i^2 \ , \end{aligned} \tag{6.6}$$

where β, γ, δ, ϵ, ν are any elements of $GF(s)$ that satisfy the conditions

$$\begin{aligned} &\nu \text{ is not a square in } GF(s) \ , \\ &\nu = 1 + 4\beta\epsilon = \frac{\epsilon}{\gamma} = \nu^2 - 4\delta\epsilon \ . \end{aligned} \tag{6.7}$$

In particular, we may take

$$\nu = \alpha, \quad \beta = \frac{1}{2}, \quad \gamma = \frac{\alpha - 1}{2\alpha}, \quad \delta = \frac{\alpha}{2}, \quad \epsilon = \frac{\alpha - 1}{2} \ . \tag{6.8}$$

(Many other solutions to (6.7) are possible, including more symmetrical ones — see Problem 6.18.) Then

$$D = \begin{pmatrix} U & W \\ V & X \end{pmatrix} \tag{6.9}$$

is a difference scheme $D(2s, 2s, s)$ based on the additive group of $GF(s)$.

VERIFICATION: Certainly the values specified in (6.8) satisfy (6.7), the only nontrivial assertion being that α is not a square. This holds because α is a primitive element of $GF(s)$ and s is odd.

We will now show that if conditions (6.7) hold then D is indeed a difference scheme $D(2s, 2s, s)$. Obviously D has the correct dimensions. We only have to consider the differences $C_1 - C_2$ of distinct columns C_1 and C_2 of D, and to show that every element of $GF(s)$ occurs twice in any such vector difference.

We will consider three cases. First assume that C_1 and C_2 are both among the first s columns of D. The $2s$ entries of $C_1 - C_2$ are then of the form

$$\alpha_i(\alpha_j - \alpha_{j'}) \ \text{ and } \ \alpha_i(\alpha_j - \alpha_{j'}) + \beta(\alpha_j^2 - \alpha_{j'}^2) \ , \quad 0 \le i < s \ ,$$

for some j, $j' \in \{0, 1, \ldots, s-1\}$ with $j \ne j'$. Since $\alpha_j \ne \alpha_{j'}$, it is clear that

$$\begin{aligned} \{\alpha_0, \alpha_1, \ldots, \alpha_{s-1}\} &= \{\alpha_i(\alpha_j - \alpha_{j'}) : \quad 0 \le i < s\} \\ &= \{\alpha_i(\alpha_j - \alpha_{j'}) + \beta(\alpha_j^2 - \alpha_{j'}^2) : \quad 0 \le i < s\} \ , \end{aligned}$$

independently of the value of β. Thus in this case every element of $GF(s)$ indeed occurs twice as an entry of $C_1 - C_2$.

For the second case, assume that C_1 and C_2 are both among the last s columns of D. The $2s$ entries of $C_1 - C_2$ are of the form

$$\alpha_i(\alpha_j - \alpha_{j'}) \text{ and } \nu\alpha_i(\alpha_j - \alpha_{j'}) + \delta(\alpha_j^2 - \alpha_{j'}^2), \quad 0 \le i < s,$$

with $j \neq j'$. The argument is then completed as in the previous case, using the fact that $\nu \neq 0$ (from (6.7)). The value of δ is immaterial.

For the third case, let C_1 be one of the first s columns of D and C_2 one of the last s columns. Then the $2s$ entries of $C_1 - C_2$ are of the form

$$\alpha_i(\alpha_j - \alpha_{j'}) - \gamma\alpha_i^2 \text{ and } \alpha_i(\alpha_j - \nu\alpha_{j'}) + \beta\alpha_j^2 - \delta\alpha_{j'}^2 - \epsilon\alpha_i^2, \quad 0 \le i < s,$$

for $j,\ j' \in \{0, 1, \ldots, s-1\}$. If ξ is a fixed element of $GF(s)$ then (see the Appendix) the quadratic equation in Z,

$$Z(\alpha_j - \alpha_{j'}) - \gamma Z^2 = \xi, \tag{6.10}$$

has $0, 1$ or 2 solutions in $GF(s)$ according as the discriminant

$$\Delta_1 = \alpha_j^2 - 2\alpha_j\alpha_{j'} + \alpha_{j'}^2 - 4\gamma\xi$$

is a nonresidue, zero or a quadratic residue in $GF(s)$. Similarly, the quadratic equation

$$Z(\alpha_j - \nu\alpha_{j'}) + \beta\alpha_j^2 - \delta\alpha_{j'}^2 - \epsilon Z^2 = \xi \tag{6.11}$$

has $0, 1$ or 2 solutions in $GF(s)$ according as its discriminant

$$\Delta_2 = (1 + 4\beta\epsilon)\alpha_j^2 - 2\nu\alpha_j\alpha_{j'} + (\nu^2 - 4\delta\epsilon)\alpha_{j'}^2 - 4\epsilon\xi$$

is a nonresidue, zero or a quadratic residue in $GF(s)$. From (6.7), $\Delta_2 = \nu\Delta_1$. Since ν is not a square it follows that if Δ_1 is a nonresidue then Δ_2 is a residue and so ξ is represented twice by (6.11) and not by (6.10), if Δ_1 is a residue then ξ is represented twice by (6.10) and not by (6.11), and if $\Delta_1 = \Delta_2 = 0$ then ξ is represented just once by both equations. Thus every element ξ of $GF(s)$ appears twice as an entry in $C_1 - C_2$. ■

Example 6.34. We use Theorem 6.33 to construct a difference scheme $D(10, 10, 5)$. We work modulo 5, taking $\alpha = 2$ as the primitive element of $GF(5)$, and use (6.8) to obtain the difference scheme exhibited in Table 6.35. Using the methods of Section 6.2, this can be used to construct an $OA(50, 11, 5, 2)$. ■

Table 6.35. A difference scheme $D(10, 10, 5)$.

0	0	0	0	0	0	0	0	0	0
0	4	3	1	2	1	0	4	2	3
0	3	1	2	4	4	2	0	1	3
0	1	2	4	3	1	2	3	0	4
0	2	4	3	1	4	1	3	2	0
0	2	3	2	3	0	4	1	4	1
0	1	1	3	0	2	4	4	3	2
0	0	4	4	2	3	3	1	1	2
0	3	0	1	1	2	3	2	4	4
0	4	2	0	4	3	1	2	3	1

Example 6.36. Table 6.37 shows a difference scheme $D(6, 6, 3)$ constructed in a similar way from $GF(3)$. ■

Table 6.37. A difference scheme $D(6, 6, 3)$.

0	0	0	0	0	0
0	1	2	1	2	0
0	2	1	1	0	2
0	2	2	0	1	1
0	0	1	2	2	1
0	1	0	2	1	2

Theorem 6.33 provides a simple way to construct an $OA(2s^2, 2s + 1, s, 2)$. There is an analogous construction for orthogonal arrays

$$OA(2s^n, 2(s^n - 1)/(s - 1) - 1, s, 2)$$

with $n \geq 3$ that is also simpler than that given in Section 3.3. The recursive construction that we give makes use of a series of difference schemes. This is based on the work of several authors, including Shrikhande (1964), Masuyama (1969b), Xu (1979) and Mukhopadhyay (1981).

The following lemma, due to Shrikhande (1964), is a convenient tool for recursively constructing difference schemes.

Lemma 6.38. *The tensor product of difference schemes $D(r_1, c_1, s)$ and $D(r_2, c_2, s)$ based on an abelian group $\mathcal{A}$ is a difference scheme $D(r_1r_2, c_1c_2, s)$ based on $\mathcal{A}$.*

Proof: Let $D = (d_{ij})$ be a difference scheme $D(r_1, c_1, s)$ and $D' = (d'_{lm})$ a difference scheme $D(r_2, c_2, s)$. The entries of $C_1 - C_2$, where C_1 and C_2 are distinct columns of $D \otimes D'$, are of the form

$$d_{ij} + d'_{lm} - d_{ij'} - d'_{lm'}, \quad i = 1, \ldots, r_1, \quad l = 1, \ldots, r_2 \ ,$$

for fixed $j, j' \in \{1, \ldots, c_1\}$, $m, m' \in \{1, \ldots, c_2\}$ and $(j, m) \neq (j', m')$. If $j \neq j'$ then for any fixed l we see that

$$d_{ij} + d'_{lm} - d_{ij'} - d'_{lm'}, \quad i = 1, \ldots, r_1 ,$$

contains every element of $\mathcal{A}$ equally often, because D is a difference scheme. If $j = j'$ then $m \neq m'$, and for every fixed i we see that

$$d_{ij} + d'_{lm} - d_{ij'} - d'_{lm'} = d'_{lm} - d'_{lm'}, \quad l = 1, \ldots, r_2 ,$$

contains every element of $\mathcal{A}$ equally often, because D' is a difference scheme. This shows that $D \otimes D'$ is a difference scheme. ∎

As an immediate corollary of Lemma 6.38, we obtain the following result.

Corollary 6.39. *For any $n \geq 1$ and prime power s, there exists a difference scheme $D(2s^n, 2s^n, s)$ based on the additive group associated with $GF(s)$.*

Proof: If $s = 2^v$, $v > 1$, then we know from the proof of Theorem 6.6 that there exists a difference scheme $D(2s, 2s, s)$ based on the additive group of $GF(s)$. If s is a power of an odd prime, the existence of such a difference scheme was established in the proof of Theorem 6.33. From the proof of Theorem 6.6 we also know that there exists a difference scheme $D(s, s, s)$ based on the additive group of $GF(s)$. The desired result now follows by repeatedly using these difference schemes in Lemma 6.38. ∎

This corollary enables us to establish the claim that the family of orthogonal arrays $OA(2s^n, 2(s^n - 1)/(s - 1) - 1, s, 2)$ can also be obtained via the use of difference schemes if $n \geq 3$. Of course, we knew already that this is true if s is a power of 2, from Example 6.31. But now we need not distinguish even and odd values of s.

Theorem 6.40. *If s is a power of a prime and $n \geq 2$, then an orthogonal array $OA(2s^n, 2(s^n - 1)/(s - 1) - 1, s, 2)$ can be obtained by using difference schemes.*

Proof:
CONSTRUCTION: From Corollary 6.39 and Theorem 6.19 we know how to construct a completely resolvable $OA(2s^m, 2s^{m-1}, s, 2)$, say $A(m)$, for $m \geq 2$. Applying Theorem 6.17 to $A(3)$ and $A(2)$ results in an s-resolvable $OA(2s^3, 2s^2 + 2s, s, 2)$, say $A(2, 3)$. Applying Theorem 6.17 to $A(4)$ and $A(2, 3)$ results in an s^2-resolvable $OA(2s^4, 2s^3 + 2s^2 + 2s, s, 2)$, say $A(2, 3, 4)$. Continuing in this way we eventually obtain an s^{n-2}-resolvable $OA(2s^n,\ 2s^{n-1} + 2s^{n-2} + \cdots + 2s, s, 2)$, say $A(2, 3, \ldots, n)$. As in Corollary 6.18 we can add one more factor. This gives the desired orthogonal array.

VERIFICATION: Clearly $2s^{n-1} + 2s^{n-2} + \cdots + 2s + 1 = 2(s^n - 1)/(s - 1) - 1$, so the array does have the required number of factors. The rest of the verification consists in checking that the preceding corollaries and theorems were correctly applied. The reader has undoubtedly verified this already. ∎

The construction of an $OA(54, 25, 3, 2)$, for example, is rather laborious by any method. However, the above construction is considerably easier than that of Theorem 3.3. Moreover, the arrays constructed by the current method are specified by difference schemes, which are easily constructed and determine the arrays completely.

6.5 Generalizations and Variations

Rao (1961a) introduced two variations on the concept of orthogonal arrays, which he referred to as orthogonal arrays of Types I and II. The formal definitions are as follows.

Definition 6.41. An $N \times k$ array based on s symbols, $s \geq t$, is called an *orthogonal array of Type I*, strength t and index λ if the rows of any $N \times t$ subarray consist of λ copies of all $s!/(s-t)!$ permutations of t distinct symbols.

Definition 6.42. An $N \times k$ array based on s symbols, $s \geq t$, is called an *orthogonal array of Type II*, strength t and index λ if the rows of any $N \times t$ subarray consist of λ copies of all $\binom{s}{t}$ combinations of t distinct symbols.

In both types, provided $t \geq 2$, two factors can not occur at the same level in a run. In a Type I array all t-tuples without a repeated symbol must appear equally often in any $N \times t$ subarray. In a Type II array the order in which the levels occur is irrelevant, and all subsets of t levels must appear equally often in any $N \times t$ subarray. We denote those arrays by $OA_I(N, k, s, t)$ and $OA_{II}(N, k, s, t)$ respectively. Following Rao (1973), orthogonal arrays of Type II have also been called *semibalanced* arrays.

Example 6.43. In an $OA_I(N, k, s, t)$ we have $N \equiv 0 \pmod{s!/(s-t)!}$, and in an $OA_{II}(N, k, s, t)$ we have $N \equiv 0 \left(\bmod \binom{s}{t}\right)$. Consequently, the examples in Tables 6.44 and 6.45 have the smallest possible values of N for the given s and t. They also have the largest possible values for k. ∎

Table 6.44. An $OA_I(6, 3, 3, 2)$.

0	1	2
1	2	0
2	0	1
0	2	1
1	0	2
2	1	0

Table 6.45. An $OA_{II}(3,3,3,2)$.

0	1	2
1	2	0
2	0	1

Orthogonal arrays of Types I and II have been used in the construction of various designs with applications in statistics. This includes the construction of block designs with special properties (see for example Mukhopadhyay, 1972, Morgan and Chakravarti, 1988, and Jacroux, Majumdar and Shah, 1995) and crossover designs (see for example Stufken, 1991).

However, here our interest is limited to a certain subclass of Type II orthogonal arrays, namely those of strength two and of index unity. Schellenberg, van Rees and Vanstone (1978) call these *perpendicular arrays*, a designation that we will adopt. Thus a perpendicular array is an $OA_{II}(s(s-1)/2, k, s, 2)$. A *complementable* perpendicular array is a perpendicular array for which a second perpendicular array exists with the same parameters, such that the juxtaposition of the two is an $OA_I(s(s-1), k, s, 2)$ of index unity. One array is then said to be a *complement* of the other.

As with orthogonal arrays, there are numerous ways in which one can attempt to construct a perpendicular array or a complementable perpendicular array. Our reason for considering this topic here is that one of the methods for constructing complementable perpendicular arrays is similar to the construction of orthogonal arrays from difference schemes. It requires the notion of a perpendicular difference array.

Definition 6.46. Let $(\mathcal{A},+)$ be an abelian group of odd order s. A *perpendicular difference array* D based on $(\mathcal{A},+)$ is an $(s-1)/2 \times c$ array such that for any two columns C_1 and C_2 of D every nonzero element of $\mathcal{A}$ appears once as an entry in either $C_1 - C_2$ or $C_2 - C_1$.

We will use the notation $PDA(c,s)$ to denote such a perpendicular difference array.

Example 6.47. Let s be a power of an odd prime. Let $\beta_1,\ldots,\beta_{(s-1)/2}$ be $(s-1)/2$ nonzero elements of $GF(s)$ such that if $x \in \{\beta_1,\ldots,\beta_{(s-1)/2}\}$ then $-x \notin \{\beta_1,\ldots,\beta_{(s-1)/2}\}$. Let D be the $(s-1)/2 \times s$ array obtained from the $s \times s$ multiplication table of $GF(s)$ by taking only the $(s-1)/2$ rows that correspond to multiplication by $\beta_1,\ldots,\beta_{(s-1)/2}$. The reader is invited to verify that D is a $PDA(s,s)$. Table 6.48 displays the $PDA(7,7)$ obtained in this way from $GF(7)$ with $\beta_1 = 1$, $\beta_2 = 2$, $\beta_3 = 3$. ■

Table 6.48. A perpendicular difference array $PDA(7,7)$.

0	1	2	3	4	5	6
0	2	4	6	1	3	5
0	3	6	2	5	1	4

Schellenberg, van Rees and Vanstone (1978) show that "developing" a $PDA(c, s)$, which is accomplished by a procedure analogous to that for developing a difference scheme (Section 6.2), results in a complementable perpendicular array $OA_{II}(s(s-1)/2, c, s, 2)$, say A, and that a complement for A is given by $-A$, the array obtained from A by negating each of its entries. The reader is asked to verify this in Problem 6.10.

Not only are the concepts of perpendicular difference arrays and difference schemes similar, perpendicular difference arrays can also be used to construct difference schemes.

Theorem 6.49. *A perpendicular difference array $PDA(c, s)$ can be used to construct a difference scheme $D(s, c, s)$.*

Proof:
CONSTRUCTION: If D^* is the perpendicular difference array $PDA(c, s)$, let D be the $s \times c$ array

$$\begin{pmatrix} D^* \\ -D^* \\ 0 \end{pmatrix},$$

where 0 is a row of zeros. Then D is the required difference scheme.

VERIFICATION: Take any two columns in D, and let C_1 and C_2 be the corresponding columns in D^*. The difference of the columns in D is then given by

$$\begin{pmatrix} C_1 \\ -C_1 \\ 0 \end{pmatrix} - \begin{pmatrix} C_2 \\ -C_2 \\ 0 \end{pmatrix} = \begin{pmatrix} C_1 - C_2 \\ C_2 - C_1 \\ 0 \end{pmatrix}.$$

Since D^* is a perpendicular difference array, every element of $\mathcal{A}$ occurs once as an entry in this vector difference. Thus D is a difference scheme $D(s, c, s)$. ∎

As a consequence of Theorem 6.49 and Corollary 6.20 we obtain the following result.

Corollary 6.50. *A perpendicular difference array $PDA(c, s)$ can be used to construct an orthogonal array $OA(s^2, c+1, s, 2)$.*

It is clear that any orthogonal array that can be constructed from a perpendicular difference array in this way could have been obtained from a suitable difference scheme. The advantage of using a perpendicular difference array is that it has less than half as many rows as the corresponding difference scheme.

Example 6.51. An example of an orthogonal array that had not been constructed by any other method but was obtained by use of a perpendicular difference array is an $OA(225, 6, 15, 2)$. The perpendicular difference array used to construct this orthogonal array is based on the additive group modulo 15 and was obtained by Schellenberg, van Rees and Vanstone (1978) by a computer search. It is presented in Table 6.52, while Table 6.53 presents the difference scheme obtained from this by the method used to prove Theorem 6.49. ∎

Table 6.52. A perpendicular difference array $PDA(5, 15)$ (transposed).

0	0	0	0	0	0	0
1	2	3	4	5	6	7
2	5	7	9	12	4	1
6	3	14	10	7	13	4
10	6	1	11	2	7	12

Table 6.53. A difference scheme $D(15, 5, 15)$ (transposed).

0	0	0	0	0	0	0	0	0	0	0	0	0	0	0
1	2	3	4	5	6	7	14	13	12	11	10	9	8	0
2	5	7	9	12	4	1	13	10	8	6	3	11	14	0
6	3	14	10	7	13	4	9	12	1	5	8	2	11	0
10	6	1	11	2	7	12	5	9	14	4	13	8	3	0

Example 6.54. Schellenberg, van Rees and Vanstone (1978) also constructed examples of perpendicular difference arrays $PDA(4, 33)$ and $PDA(4, 39)$, shown in Tables 6.55 and 6.56, which lead to orthogonal arrays with parameters $OA(1089, 5, 33, 2)$ and $OA(1521, 5, 39, 2)$. ∎

Table 6.55. A perpendicular difference array $PDA(4, 33)$ (transposed).

0	0	0	0	0	0	0	0	0	0	0	0	0	0	0	0
1	2	3	4	5	6	7	8	9	10	11	12	13	14	15	16
10	9	8	7	6	16	15	14	13	12	22	32	31	30	29	28
2	4	6	8	17	26	21	18	20	28	3	19	22	9	32	10

Table 6.56. A perpendicular difference array $PDA(4, 39)$.

0	1	12	3	0	6	7	18	0	11	15	32	0	16	36	22
0	2	11	1	0	7	19	26	0	12	14	27	0	17	35	30
0	3	10	6	0	8	18	24	0	13	26	5	0	18	34	11
0	4	9	8	0	9	17	20	0	14	38	23	0	19	33	2
0	5	8	10	0	10	16	35	0	15	37	25				

This chapter provides convincing evidence for the usefulness of difference schemes as a tool for constructing orthogonal arrays. However, these generally have strength two. In order to obtain orthogonal arrays of higher strength by developing a difference scheme, the scheme should possess an additional regularity property. We call such schemes "difference schemes of strength t". The case $t = 2$ will then correspond to ordinary difference schemes. These notions were first formulated by Seiden (1954).

As before, let $\mathcal{A}$ be an abelian group of order s. By $\mathcal{A}^t$, for $t \geq 1$, we will denote the abelian group of order s^t consisting of all t-tuples of elements from $\mathcal{A}$ with the usual vector addition as the binary operation. Let

$$\mathcal{A}_0^t = \{(x_1, \ldots, x_t) : x_1 = \cdots = x_t \in \mathcal{A}\} .$$

Then $\mathcal{A}_0^t$ is a subgroup of $\mathcal{A}^t$ of order s, and we will denote its cosets by $\mathcal{A}_i^t$, $i = 1, \ldots, s^{t-1} - 1$.

Definition 6.57. An $r \times c$ array D based on $\mathcal{A}$ is a *difference scheme of strength* t if for every $r \times t$ subarray each set $\mathcal{A}_i^t$, $i = 0, 1, \ldots, s^{t-1} - 1$, is represented equally often when the rows of the subarray are viewed as elements of $\mathcal{A}^t$.

It follows that r is a multiple of s^{t-1}, say $r = \lambda s^{t-1}$. We denote such an array by $D_t(r, c, s)$. For $t = 2$ this definition is equivalent to Definition 6.1.

Developing a difference scheme of strength t (as in Section 6.2) results in an orthogonal array of strength t, to which under certain conditions at least one additional factor can be added. This is formulated in the following result, the proof of which is left to the reader.

Theorem 6.58. *A difference scheme $D_t(r, c, s)$ of strength t can be used to construct an $OA(rs, c, s, t)$. If the difference scheme itself is already an orthogonal array of strength $t - 1$ or if it can be written as the juxtaposition of s difference schemes $D_{t-1}(r/s, c, s)$, then it can be used to construct an $OA(rs, c + 1, s, t)$.*

Note that Corollary 6.20 is a special case of this theorem, since any difference scheme $D_2(r, c, s)$ is the juxtaposition of s difference schemes $D_1(r/s, c, s)$.

Example 6.59. Let D be an $OA(4\lambda, k, 2, 2)$ over $(GF(2), +)$, $k \geq 3$. Then D is also a difference scheme of strength 3 (see Problem 6.14), and so we can use it to construct an $OA(8\lambda, k + 1, 2, 3)$. We will return to this in Chapter 7. For now, it suffices to give an example. Table 6.60 gives an $OA(8, 7, 2, 2)$ which is also a difference scheme $D_3(8, 7, 2)$ of strength 3 over $(GF(2), +)$. Thus among the rows of any 8×3 subarray are two occurrences of a member of $\mathcal{A}_0^3 = \{(0,0,0), (1,1,1)\}$, as well as two occurrences of a member of each of $\mathcal{A}_1^3 = \{(1,0,0), (0,1,1)\}$, $\mathcal{A}_2^3 = \{(0,1,0), (1,0,1)\}$ and $\mathcal{A}_3^3 = \{(0,0,1), (1,1,0)\}$.

Table 6.60. A difference scheme $D_3(8, 7, 2)$ of strength 3.

0	0	0	0	0	0	0
0	0	1	0	1	1	1
0	1	0	1	0	1	1
0	1	1	1	1	0	0
1	0	0	1	1	0	1
1	0	1	1	0	1	0
1	1	0	0	1	1	0
1	1	1	0	0	0	1

The resulting $OA(16, 8, 2, 3)$ is shown in Table 6.61. Note that Corollary 2.5 now tells us that $f(8, 2, 2) = 7$, while Corollary 2.6 implies $f(16, 2, 3) = 8$. ■

Table 6.61. An $OA(16, 8, 2, 3)$ (transposed).

0	0	0	0	1	1	1	1	1	1	1	1	0	0	0	0
0	0	1	1	0	0	1	1	1	1	0	0	1	1	0	0
0	1	0	1	0	1	0	1	1	0	1	0	1	0	1	0
0	0	1	1	1	1	0	0	1	1	0	0	0	0	1	1
0	1	0	1	1	0	1	0	1	0	1	0	0	1	0	1
0	1	1	0	0	1	1	0	1	0	0	1	1	0	0	1
0	1	1	0	1	0	0	1	1	0	0	1	0	1	1	0
0	0	0	0	0	0	0	0	1	1	1	1	1	1	1	1

Although difference schemes of strength t are a natural generalization of ordinary difference schemes, so far they have had limited application. The most interesting application, for schemes with $s = 2$, was given by Seiden (1954) and Seiden and Zemach (1966) (see also Problem 6.14). Other results, mostly for $t = 2$ or 3, appear in Mukhopadhyay (1981). More recently, Hedayat, Stufken and Su (1997) have studied the existence of difference schemes of strength t that can be obtained from parity check matrices.

There are several possible reasons for this neglect. It is certainly harder to construct useful schemes with large t. (Of course, since any orthogonal array of strength t is obviously a difference scheme of strength t, it is always possible to find trivial examples such as a $D_t(s^t, t+1, s)$.) Another reason is that orthogonal arrays of strength t necessarily have a large number of runs, and so are of limited value for many statistical applications. But we hope that readers will not be discouraged by this, since there are challenging unsolved mathematical problems in this area.

Research Problem 6.62. Develop new methods for the construction of difference schemes of strength t, with the ultimate goal of obtaining better orthogonal arrays of strength t.

6.6 Concluding Remarks

This chapter has shown the usefulness of difference schemes in constructing orthogonal arrays of strength two. We have shown that several families of arrays, often with the maximal number of factors, can be constructed in this way. Difference schemes are therefore an important tool to consider in cases where the maximal number of factors has not yet been determined. Another advantage is that an orthogonal array obtained in this way has a concise description (since a $D(r, c, s)$ yields an $OA(rs, c+1, s, 2)$, by Corollary 6.20, we save a factor of about s in space).

Although constructing a new difference scheme is probably easier than the direct construction of the corresponding orthogonal array, it remains a very challenging problem. No general algorithm is known. However, it seems likely that the group-theoretic approach of Kreher and Radziszowski (1986, 1987, 1990) and Kreher (1990) (see also Kreher and Stinson 1998) could be successfully applied here.

Indeed, after seeing a preliminary version of this chapter, Colbourn and Kreher (1996) used this approach to find a $D(15, 7, 3)$ and a $D(15, 7, 5)$: for the former, take 0000000 and all cyclic shifts of the vectors

$$0112122 \quad \text{and} \quad 0221211\ , \tag{6.12}$$

and for the latter, take the zero vector and all cyclic shifts of

$$0110242 \quad \text{and} \quad 0220434\ . \tag{6.13}$$

Their paper also contains many new constructions for difference schemes. Kreher's student Jun Meng later used random search to find additional examples, including a $D(15, 9, 3)$ (Meng, 1995). Unfortunately these difference schemes have no apparent structure.

Among the problems that deserve further investigation are the construction of difference schemes with more general values of s and λ. The main families of difference schemes constructed in this chapter have parameters $D(p^m, p^m, p^n)$, p prime, $m \geq n \geq 1$, with $\lambda = p^{m-n}$ (Theorem 6.6), and $D(2p^{an}, 2p^{an}, p^n)$, p prime, $n \geq 1$, $a \geq 1$, with $\lambda = 2p^{(a-1)n}$ (Theorem 6.6 if $p = 2$, Corollary 6.39 if p is odd). However, the requirement that s be a prime power is not essential for the notion of a difference scheme, and it would be nice to have more examples (like that in Example 6.51) for other values of s (such as 6 and 10).

Table 6.67 summarizes what is presently known about the existence of small difference schemes. This table is based on the results in this chapter and other sources. In particular, we make use of the following four results, which we state without proof. The first two are constructive, while the other two lead to nonexistence results. Theorem 6.63 extends Theorem 6.6 and Corollary 6.39. The sources for these results are given in the Notes at the end of the chapter.

Theorem 6.63. *A difference scheme $D(4s^m, 4s^m, s)$ exists for any prime power s and any integer $m \geq 1$.*

Using the method of Theorem 6.40, this yields an $OA(4s^{m+1}, 4(s^{m+1} - 1)/(s-1) - 3, s, 2)$ for every prime power s and integer $m \geq 1$.

Theorem 6.64. *If q is a prime power and a difference scheme $D(q+1, q+1, s)$ exists then there is a difference scheme $D(q^m(q+1), q^m(q+1), s)$ for all $m \geq 0$.*

Theorem 6.65. *Suppose c is odd and a difference scheme $D(c, c, s)$ exists. If p is an odd prime dividing s, and $b \not\equiv 0 \pmod{p}$ is an integer dividing the square-free part of c, then the multiplicative order of b modulo p is odd.*

Theorem 6.66. *If a difference scheme $D(\lambda s, 3, s)$ exists in which $s = 2u$, where u is odd, then λ is even.*

For each index $\lambda \leq 10$, Table 6.67 gives lower and upper bounds on the maximal value of c for which a difference scheme $D(\lambda s, c, s)$ exists, for $s \leq 10$ (or the exact maximal value, if that is known). The sources for some of the bounds are indicated by labels, labels on the left referring to lower bounds and labels on the right to upper bounds.

Unmarked upper bounds in the table are from Theorem 6.5. Unmarked lower bounds are from (a) the trivial observation that we always have $c \geq 2$ (take the first column to be all zeros, and the second column to contain every possible value equally often), (b) juxtaposition (Example 6.4), and (c) tensor products (Lemma 6.38). The labels on the other entries are explained below the table.

Table 6.67. Bounds on maximal c for which a difference scheme $D(\lambda s, c, s)$ exists.

$\lambda \backslash s$	2	3	4	5	6
1	2^{d}	$^{\mathrm{G}}3$	$^{\mathrm{G}}4$	$^{\mathrm{G}}5$	2^{d}
2	$^{\mathrm{H}}4$	$^{\mathrm{I2}}6$	$^{\mathrm{G}}8$	$^{\mathrm{I2}}10$	$^{\mathrm{DJ}}6-12$
3	2^{d}	$^{\mathrm{G}}9$	$^{\mathrm{b}}12$	$^{\mathrm{CK}}7-14^{\mathrm{c}}$	2^{d}
4	$^{\mathrm{H}}8$	$^{\mathrm{b}}12$	$^{\mathrm{G}}16$	$^{\mathrm{b}}20$	$6-24$
5	2^{d}	$^{\mathrm{M}}9-14^{\mathrm{c}}$	$8-20$	$^{\mathrm{G}}25$	2^{d}
6	$^{\mathrm{H}}12$	$^{\mathrm{I2}}18$	$^{\mathrm{De}}16-24$	$^{\mathrm{CK}}15-30$	$^{\mathrm{M}}7-36$
7	2^{d}	$^{\mathrm{M}}11-21$	$12-28$	$^{\mathrm{CK}}17-34^{\mathrm{c}}$	2^{d}
8	$^{\mathrm{H}}16$	$^{\mathrm{b}}24$	$^{\mathrm{G}}32$	$^{\mathrm{De}}25-40$	$^{\mathrm{Ju}}10-48$
9	2^{d}	$^{\mathrm{G}}27$	$^{\mathrm{b}}36$	$20-45$	2^{d}
10	$^{\mathrm{H}}20$	$^{\mathrm{b}}30$	$16-40$	$^{\mathrm{I2}}50$	$^{\mathrm{CK}}11-60$

$\lambda \backslash s$	7	8	9	10
1	$^{G}7$	$^{G}8$	$^{G}9$	2^{d}
2	$^{I2}14$	$^{I2}16$	$^{I2}18$	$^{M}5-20$
3	$7-20^{c}$	$8-24$	$^{G}27$	2^{d}
4	$^{a}28$	$^{G}32$	$^{a}36$	$^{AC}8-40$
5	$^{CK}11-34^{c}$	$8-40$	$18-44^{c}$	2^{d}
6	$^{im}18-42$	$16-48$	$^{a}54$	$5-60$
7	$^{G}49$	$^{b}56$	$27-63$	2^{d}
8	$28-56$	$^{G}64$	$^{b}72$	$^{AC}10-80$
9	$^{De}28-63$	$^{CK}21-72$	$^{G}81$	2^{d}
10	$18-70$	$16-80$	$36-90$	$5-100$

Key to Table 6.67

- a Theorem 6.63
- AC From difference schemes $D(40,8,40)$, $D(80,10,80)$ constructed by Abel and Cheng (1994), via Problem 6.16
- b Theorem 6.64
- c Theorem 6.65
- d Theorem 6.66
- CK Colbourn and Kreher (1996); one of their constructions is given in (6.13)
- De De Launey (1987)
- DJ Dulmage, Johnson and Mendelsohn scheme (Example 6.10), plus Problem 6.16
- G From a Galois field, see Theorem 6.6
- H From a Hadamard matrix, see Chapter 7
- im From an impulse matrix, see Colbourn and De Launey (1996)
- I2 The index 2 construction (Theorem 6.33) and its extension (Corollary 6.39)
- Ju Corollary 4.12 of Jungnickel (1979)
- M Meng (1995)

Research Problem 6.68. Remove some of the very large gaps in Table 6.67!

A much more extensive table of lower bounds is given in Colbourn and De Launey (1996).[1]

6.7 Notes on Chapter 6

Further information about difference schemes can be found in Beth, Jungnickel and Lenz (1986), Butson (1962, 1963), Dawson (1985), De Launey (1986), Drake

[1] It is claimed there that a $D(20,9,4)$ is known, but this is probably incorrect.

(1979), Dulmage, Johnson and Mendelsohn (1961), Jungnickel (1979, 1992) and Seberry (1980).

What we call a difference scheme $D(r, c, s)$ based on $\mathcal{A}$ is referred to by Beth, Jungnickel and Lenz (1986) as an $(s, c; \lambda)$ difference matrix, where $\lambda = r/s$, and by Jungnickel (1979) as an $(s, c; \lambda, \mathcal{A})$-difference matrix.

The procedure described in Section 6.2 for converting a difference scheme into an orthogonal array is an analog of the construction used to convert difference sets into balanced incomplete block designs — see Hall (1986), Chapter 11.

The sources for the theorems stated without proof in Section 6.6 are as follows: Theorem 6.63 is based on the work of Butson (1962, 1963), Dawson (1985), Drake (1979), Jungnickel (1979) and Street (1979), and is given as Theorem 2.4(i) in De Launey (1986). Theorem 6.64 is due to De Launey (1986), Theorem 2.3, generalizing an earlier result of Seberry (1980). Theorem 6.65 is due to De Launey (1986), Theorem 1.5; and Theorem 6.66 to Drake (1979) (see De Launey, 1986, Theorem 1.1, and Beth, Jungnickel and Lenz, 1986, Theorem 12.2).

De Launey (1986) gives an extensive table of generalized Hadamard matrices.

6.8 Problems

6.1. From Example 6.3 we may construct a difference scheme $D(4, 4, 4)$ based on $(GF(4), +)$. Is it possible to construct a difference scheme $D(4, 4, 4)$ based on $(\mathbb{Z}_4, +)$, where $\mathbb{Z}_4$ denotes the integers modulo 4? If so, construct it; if not, prove its nonexistence.

6.2. For each of the following schemes, verify whether it is a difference scheme based on $(GF(3), +)$. Here D_i, $i = 0, 1, 2$, denotes the 3×3 array obtained by adding i to each element of D, a difference scheme $D(3, 3, 3)$ based on $(GF(3), +)$.

(i)

$$D^* = \begin{bmatrix} D_0 \\ D_0 \end{bmatrix} .$$

(ii)

$$D^* = \begin{bmatrix} D_0 & D_0 \\ D_0 & D_1 \end{bmatrix} .$$

(iii)

$$D^* = \begin{bmatrix} D_0 & D_0 & D_0 \\ D_0 & D_1 & D_2 \\ D_0 & D_2 & D_1 \end{bmatrix} .$$

(iv)

$$D^* = \begin{bmatrix} 0 & 0 & 0 & 0 & 0 & 0 \\ 0 & 1 & 2 & 0 & 1 & 2 \\ 0 & 2 & 1 & 1 & 0 & 2 \\ 0 & 0 & 2 & 1 & 2 & 1 \\ 0 & 2 & 0 & 2 & 1 & 1 \\ 0 & 1 & 1 & 2 & 2 & 0 \end{bmatrix} .$$

(v)

$$D^* = \begin{bmatrix} 0 & 0 & 0 & 0 & 0 & 0 \\ 0 & 0 & 1 & 2 & 2 & 1 \\ 0 & 1 & 0 & 1 & 2 & 2 \\ 0 & 2 & 1 & 0 & 1 & 2 \\ 0 & 2 & 2 & 1 & 0 & 1 \\ 0 & 1 & 2 & 2 & 1 & 0 \end{bmatrix} .$$

(vi) Let $f_0(x) = x$, $f_1(x) = x + 1$, $f_2(x) = x + 2$, $f_3(x) = x^2 + 1$, $f_4(x) = x^2 + x + 2$, $f_5(x) = x^2 + 2x + 2$. Define E_1 to be the 3×3 array with (i, j)th entry equal to $f_i(j)$, for $0 \le i, j \le 2$, let E_2 have (i, j)th entry equal to $f_{3+i}(j)$, and take

$$D^* = \begin{bmatrix} E_1 & E_1 \\ E_2 & 2E_2 \end{bmatrix} .$$

6.3. For each of the following parameters, construct a difference scheme $D(r, c, s)$:

(i) $r = 8$, $c = 8$, $s = 4$;

(ii) $r = 12$, $c = 12$, $s = 4$;

(iii) $r = 12$, $c = 12$, $s = 3$;

(iv) $r = 18$, $c = 18$, $s = 3$.

6.4. a. Show that if an $OA(N, k, s, 2)$ is a-resolvable then

$$k \le N(as - 1)/(as(s - 1)) .$$

b. By combining the results of Part a and Theorem 6.19, give an alternative proof of Theorem 6.5.

6.5. a. For each of the orthogonal arrays in Tables 3.25, 3.26, 3.31 find the smallest value of a for which it is a-resolvable.

b. Delete the first factor from each of the arrays in part a and repeat the question.

6.6. a. Construct a completely resolvable $OA(16, 8, 2, 2)$.

b. Show that an $OA(12, 3, 2, 2)$ cannot be completely resolvable.

c. Construct a 2-resolvable $OA(16, 12, 2, 2)$.

d. Construct a 2-resolvable $OA(12, 4, 2, 2)$.

6.7. Prove Lemma 6.27.

6.8. Let $s = p^2$ and $\lambda = p$, where p is a prime. Below Theorem 6.28 it was stated that the number of factors in an orthogonal array obtained from that theorem is at most $\lfloor (p-1)/2 \rfloor$ less than the upper bound given by Theorem 2.8. Verify this claim.

6.9. a. Show that the existence of an $OA(s^2, k, s, 2)$ implies the existence of an $OA_I(s(s-1), k-1, s, 2)$.

b. Show that if s is even and an $OA_{II}(N, k, s, 2)$ exists with $k \geq 3$, then $N \geq s(s-1)$.

c. Show that if $s = 2^n$, $n \geq 1$, then an $OA_{II}(s(s-1), s, s, 2)$ exists.

d. Show that if $s = p^n$, where p is an odd prime and $n \geq 1$, then an $OA_{II}(s(s-1)/2, s, s, 2)$ exists.

6.10. Verify the claim in the paragraph following Example 6.47 that a perpendicular array A obtained by developing a perpendicular difference array D is a complementable perpendicular array and that $-A$ is a complement of A.

6.11. a. The definition of a perpendicular difference array requires that any nonzero element of $\mathcal{A}$ appears once in either $C_1 - C_2$ or $C_2 - C_1$ for any two columns C_1 and C_2. We could generalize this by requiring that any nonzero element of $\mathcal{A}$ appears *equally often* among the combined entries of $C_1 - C_2$ and $C_2 - C_1$. The element $0 \in \mathcal{A}$ may also appear, but at most as often as the nonzero elements, and the frequency with which it appears should not depend on C_1 and C_2. Formulate a version of Theorem 6.49 for this more general definition.

b. In this chapter we have seen examples of difference schemes with parameters $D(3, 3, 3)$, $D(9, 9, 3)$, $D(10, 10, 5)$. Could such difference schemes be obtained from the generalized perpendicular difference arrays given in part a? (Use your generalization of Theorem 6.49.)

6.12. Verify that the definitions of difference schemes of strength 2 given in Definitions 6.1 and 6.57 are equivalent.

6.13. Prove Theorem 6.58.

6.14. a. Show that any orthogonal array $OA(4\lambda, k, 2, 2)$, $k \geq 3$, is a difference scheme of strength 3.

b. As a generalization of part a, show that if t is even any orthogonal array $OA(\lambda 2^t, k, 2, t)$, $k \geq t+1$, is a difference scheme of strength $t+1$.

c. If t is even and D is an $OA(\lambda 2^t, k, 2, t)$, show that the result in part b can be used to reach the conclusion of Theorem 2.24.

6.15. Consider the orthogonal arrays in Tables 3.25, 3.26 and 6.29. Are any of them statistically equivalent? Isomorphic?

6.16. Suppose $D(r, c, s)$ is a difference scheme based on an abelian group $\mathcal{A}$ of order s. Let $\mathcal{B}$ be a subgroup of $\mathcal{A}$ of order b, $1 < b \leq s$. Show that there exists a difference scheme $D(r, c, b)$ based on $\mathcal{B}$. (This is a generalization of the argument used to prove Theorem 6.6.)

6.17. Show how to combine difference schemes $D(r_1, c, s_1)$ and $D(r_2, c, s_2)$ to produce a $D(r_1 r_2, c, s_1 s_2)$.

6.18. Other examples of $D(2s, 2s, s)$ for s an odd prime power, $s > 3$: show that it is always possible to find a solution to (6.7) in Theorem 6.33 with $\beta = \gamma$.

Chapter 7

Orthogonal Arrays and Hadamard Matrices

Hadamard matrices are square matrices of +1's and −1's whose rows are orthogonal. The study of two-level orthogonal arrays of strengths 2 and 3 is essentially equivalent to the study of these matrices. They are also the most important examples of two-level difference schemes.

In this chapter we describe the basic properties of these matrices and give a number of techniques for constructing them.

7.1 Introduction

A *Hadamard matrix of order* n is an $n \times n$ matrix H_n of +1's and −1's whose rows are orthogonal, i.e. which satisfies

$$H_n H_n^T = nI_n \ . \tag{7.1}$$

For example, here are Hadamard matrices of orders 1, 2 and 4:

$$H_1 = [1], \quad H_2 = \begin{bmatrix} 1 & 1 \\ 1 & -1 \end{bmatrix}, \quad H_4 = \begin{bmatrix} 1 & 1 & 1 & 1 \\ 1 & -1 & 1 & -1 \\ 1 & 1 & -1 & -1 \\ 1 & -1 & -1 & 1 \end{bmatrix} . \tag{7.2}$$

These matrices are named after the French mathematician Jacques Hadamard (1865–1963). In Hadamard (1893) he showed that if $A = (a_{ij})$ is any $n \times n$ matrix with $|a_{ij}| \le 1$ then

$$|\det A| \le n^{n/2} \ , \tag{7.3}$$

with equality if and only if A is what is now called a Hadamard matrix (see Problems 7.6 and 7.7).

When do Hadamard matrices exist? The answer is perhaps surprising: we don't know. It is easy to show that if an H_n exists then n is 1, 2 or a multiple of 4 (see Corollary 7.2), and everyone who has studied the problem agrees that it is almost certainly true that if n *is* a multiple of 4 then an H_n exists. However, this assertion, known as the *Hadamard conjecture*, is a basic unsolved problem in discrete mathematics.

We will discuss the connections between Hadamard matrices, orthogonal arrays and difference schemes in Section 7.3. Probably the first application of Hadamard matrices in statistics was made by Plackett and Burman (1946). These authors used certain special classes of Hadamard matrices to construct fractional factorial designs, obtaining what we now call orthogonal arrays with parameters $OA(4\lambda, 4\lambda - 1, 2, 2)$ (compare Theorem 7.5). It is not uncommon in the statistics literature to refer to all orthogonal arrays $OA(4\lambda, 4\lambda - 1, 2, 2)$ as *Plackett-Burman designs*. This is unfortunate, and incorrect, since Plackett and Burman used particular Hadamard matrices (given by the tensor product construction and the first Paley construction) to obtain their orthogonal arrays. The statistical properties of orthogonal arrays with these same parameters but obtained from other Hadamard matrices can be quite different. The term "Plackett-Burman design", as it is often used, ignores this distinction. In the mathematics literature these arrays are universally known as *Hadamard arrays* or *designs*. We recommend that the name *Hadamard array* be adopted as the standard terminology.

Hadamard matrices may be regarded as the special class of difference schemes $D(r, c, s)$ with $s = 2$, $r = c$ and index $\lambda = c/2$, achieving equality in Theorem 6.5. The study of Hadamard matrices has however received so much attention in the literature that it is easy to justify a separate chapter devoted to them. Moreover, from a practical point of view, Hadamard arrays are one of the most important classes of orthogonal arrays for statistical applications.

Section 7.2 gives the basic properties of Hadamard matrices. A very large number of techniques for constructing these matrices are known, and we present some of the most important methods in Section 7.4. Then in Section 7.5 we give constructions for Hadamard matrices of all orders up to 200 and discuss some of their properties.

7.2 Basic Properties of Hadamard Matrices

Suppose H_n is a Hadamard matrix of order n. Then $H_n^{-1} = n^{-1} H_n^T$, so

$$H_n^T H_n = nI_n \,, \tag{7.4}$$

which implies that the columns of H_n are also orthogonal.

If H_n satisfies (7.1) or (7.4), then so does any matrix obtained from H_n by permuting its rows (or columns) and negating any of its rows (or columns). All the matrices obtained in this way are said to be *isomorphic* or *equivalent* to H_n. By transformations of this kind we can always arrange that the first row and column of H_n consist entirely of +1's. Such a Hadamard matrix is said to be *normalized*. Thus any Hadamard matrix is isomorphic to a normalized one.

Lemma 7.1. *Let H_n be a normalized Hadamard matrix of order n, $n > 2$. Let $u = (u_1, \dots, u_n)$ and $v = (v_1, \dots, v_n)$ be any two distinct rows of H_n, not including the first. Then (a) there are $n/2$ coordinates i with $u_i = +1$ and $n/2$ with $u_i = -1$; (b) there are $n/4$ coordinates with $u_i = v_i = +1$, $n/4$ with $u_i = +1$, $v_i = -1$, $n/4$ with $u_i = -1$, $v_i = +1$, and $n/4$ with $u_i = v_i = -1$; (c) similar results hold for the columns of H_n.*

Proof: These are immediate consequences of the orthogonality relations (7.1) and (7.4). ■

Corollary 7.2. *If a Hadamard matrix H_n exists then n is* 1, 2 *or a multiple of* 4.

The Hadamard conjecture is that the converse to Corollary 7.2 holds. It has been said by an eminent authority that when one gives a talk it is important to repeat the central open problems in that subject. So let us state:

Research Problem 7.3. *The Hadamard conjecture.* Show that a Hadamard matrix exists if n is 1, 2 or a multiple of 4.

In other words, (by Theorem 7.5 below), show that an $OA(4\lambda, 4\lambda-1, 2, 2)$ exists for all λ.

There are a large number of techniques for constructing Hadamard matrices, and at the time this chapter was written the only orders less than 1000 for which an H_n was not known to exist were 428, 668, 716, 764 and 892. Matrices of orders 268 and 956 were found quite recently (Sawade, 1985, Djoković, 1994). There are however infinitely many orders for which Hadamard matrices are unknown.

A (possibly) easier question is:

Research Problem 7.4. Give a heuristic or probabilistic estimate for the number of distinct Hadamard matrices of order n.

We will discuss the number of nonisomorphic Hadamard matrices of small orders in Section 7.5.

7.3 The Connection Between Hadamard Matrices and Orthogonal Arrays

We saw in (2.24) and (2.25) that Rao's inequalities imply $f(4\lambda, 2, 2) \leq 4\lambda - 1$ and $f(8\lambda, 2, 3) \leq 4\lambda$. As we shall now show, equality holds in these two bounds if and only if there exists a Hadamard matrix of order 4λ. This is equivalent to the following theorem.

Theorem 7.5. *Orthogonal arrays $OA(4\lambda, 4\lambda - 1, 2, 2)$ and $OA(8\lambda, 4\lambda, 2, 3)$ exist if and only if there exists a Hadamard matrix of order 4λ.*

Proof: An $OA(4\lambda, 4\lambda - 1, 2, 2)$ exists if and only if an $OA(8\lambda, 4\lambda, 2, 3)$ exists, by Theorem 2.24. Suppose $H_{4\lambda}$ is a normalized Hadamard matrix. By Lemma 7.1, the matrix obtained by omitting the first column of $H_{4\lambda}$ is an $OA(4\lambda, 4\lambda - 1, 2, 2)$. Conversely, let A be an $OA(4\lambda, 4\lambda - 1, 2, 2)$ in which the levels are $+1$ and -1. It follows from the definition of an orthogonal array that the matrix formed by adding an initial column of $+1$'s to A satisfies (7.1). ∎

Similarly Problem 2.16 shows that $F(4\lambda - 1 - \mu, 2, 2) \geq 4\lambda$ and $F(4\lambda - \mu, 2, 3) \geq 8\lambda$, for $\lambda \geq 1$, $0 \leq \mu \leq 3$. These two inequalities become equalities if and only if a Hadamard matrix $H_{4\lambda}$ exists.

In view of Theorem 7.5, orthogonal arrays with parameters $OA(4\lambda, 4\lambda - 1, 2, 2)$ and $OA(8\lambda, 4\lambda, 2, 3)$ are called *Hadamard arrays.*

The next theorem establishes the assertion made in Section 7.1, that Hadamard matrices are a special case of the difference schemes studied in Chapter 6.

Theorem 7.6. *A Hadamard matrix H_n exists if and only if a difference scheme $D(n, n, 2)$ exists.*

Proof: By Lemma 7.1 any two distinct columns of a Hadamard matrix H_n must agree in $n/2$ places and disagree in $n/2$ places. The componentwise product of these columns therefore contains $n/2$ $+1$'s and $n/2$ -1's, and so H_n is a difference scheme $D(n, n, 2)$ based on the multiplicative group $\{+1, -1\}$. The converse assertion follows by similar arguments. ∎

7.4 Constructions for Hadamard Matrices

Several classical methods for constructing Hadamard matrices will be presented in this section. These will enable us to construct Hadamard matrices of most orders up to 200. The tensor product construction of Definition 6.26 (with multiplication as the binary operation) is very useful.

Theorem 7.7. *The tensor product $H_a \otimes H_b$ of Hadamard matrices of orders a and b is a Hadamard matrix of order ab.*

Proof: This is an immediate consequence of Lemma 6.38 and Theorem 7.6. ▪

So once we have a Hadamard matrix of order b we can immediately obtain matrices of orders $2b$, $4b$, $8b$, $16b, \ldots$ by repeatedly tensoring with the matrix $H_2 = \begin{bmatrix} 1 & 1 \\ 1 & -1 \end{bmatrix}$. The Hadamard matrices of orders 2^m obtained in this way by starting from $b = 1$ and $H_1 = [1]$ are called *Sylvester-type matrices.*

Example 7.8. Table 7.9 shows a Sylvester-type Hadamard matrix of order 8. Here and in later tables we abbreviate $+1$ by $+$ and -1 by $-$. ▪

Table 7.9. Sylvester-type Hadamard matrix of order 8.

$$\begin{matrix}
+&+&+&+&+&+&+&+\\
+&-&+&-&+&-&+&-\\
+&+&-&-&+&+&-&-\\
+&-&-&+&+&-&-&+\\
+&+&+&+&-&-&-&-\\
+&-&+&-&-&+&-&+\\
+&+&-&-&-&-&+&+\\
+&-&-&+&-&+&+&-
\end{matrix}$$

It is worth remarking that the rows of a Sylvester-type Hadamard matrix of order $n = 2^m$, together with their negatives, form the codewords of a first-order Reed-Muller code of length n (cf. Section 5.8).

The next two constructions are due to Paley (1933), and use Galois fields. Each theorem is preceded by a lemma which does the hard work.

Let s be an odd prime power, and let $\alpha_0 = 0, \alpha_1, \ldots, \alpha_{s-1}$ denote the elements of $GF(s)$. We make use of the map χ defined on $GF(s)$ by

$$\chi(\beta) = \begin{cases} 1, & \text{if } \beta \text{ is a quadratic residue in } GF(s), \\ 0, & \text{if } \beta = 0, \\ -1, & \text{otherwise.} \end{cases}$$

This map satisfies $\chi(\beta_1\beta_2) = \chi(\beta_1)\chi(\beta_2)$ for all β_1, $\beta_2 \in GF(s)$ (see Problem A.19).

Lemma 7.10. *The $s \times s$ matrix $Q = (q_{ij})$, where $q_{ij} = \chi(\alpha_i - \alpha_j)$, for $i, j = 0, 1, \ldots, s-1$, has the following properties:*

a. Q is symmetric if $s \equiv 1 \pmod 4$, skew-symmetric if $s \equiv 3 \pmod 4$,

b. $QJ_s = J_sQ = 0$.

c. $QQ^T = sI_s - J_s$,

Proof: a. $q_{ji} = \chi(\alpha_j - \alpha_i) = \chi(-1)\chi(\alpha_i - \alpha_j) = \chi(-1)q_{ij}$. Now if α is a primitive element of $GF(s)$ then $\alpha^{s-1} = 1$, $\alpha^{(s-1)/2} = -1$, and so -1 is a quadratic residue if $s \equiv 1 \pmod 4$, and a nonresidue if $s \equiv 3 \pmod 4$. In the former case Q is symmetric, in the latter case skew-symmetric.

b. This follows immediately from Theorem A.23(i).

c. Let $QQ^T = (r_{ij})$. Then

$$
\begin{aligned}
r_{ij} &= \sum_{k \neq i,j} \chi(\alpha_i - \alpha_k)\chi(\alpha_j - \alpha_k) \\
&= \sum_{k \neq i,j} \chi(\alpha_i - \alpha_k)\chi(\alpha_j - \alpha_i + \alpha_i - \alpha_k) \\
&= \sum_{k \neq i,j} \chi^2(\alpha_i - \alpha_k)\chi((\alpha_i - \alpha_k)^{-1}(\alpha_j - \alpha_i) + 1) \\
&= \sum_{k \neq i,j} \chi((\alpha_i - \alpha_k)^{-1}(\alpha_j - \alpha_i) + 1) \ . \qquad (7.5)
\end{aligned}
$$

If $i = j$, (7.5) is

$$\sum_{k \neq i} \chi(1) = s - 1 \ .$$

If $i \neq j$, we use the fact that the set

$$\{(\alpha_i - \alpha_k)^{-1}(\alpha_j - \alpha_i) : \ k \in \{0, 1, \ldots, s-1\},\ k \neq i, j\}$$

consists of all elements of $GF(s)$ except 0 and -1 to rewrite (7.5) as

$$\sum_{\substack{\beta \in GF(s) \\ \beta \neq 0,1}} \chi(\beta) = \sum_{\beta \in GF(s)} \chi(\beta) - \chi(0) - \chi(1) = 0 - 0 - 1 = -1 \ ,$$

by Theorem A.23. This proves c. ∎

Theorem 7.11. *If $n \equiv 0 \pmod 4$, $s = n - 1$ is a prime power and Q is as in Lemma 7.10 then*

$$H_n = \begin{bmatrix} 1 & -1_s^T \\ 1_s & Q + I_s \end{bmatrix}$$

is a Hadamard matrix of order n.

Proof: We use Lemma 7.10 and find that

$$
\begin{aligned}
H_n H_n^T &= \begin{bmatrix} 1 & -1_s^T \\ 1_s & Q + I_s \end{bmatrix} \begin{bmatrix} 1 & 1_s^T \\ -1_s & Q^T + I_s \end{bmatrix} \\
&= \begin{bmatrix} s+1 & 0 \\ 0 & (s+1)I_s \end{bmatrix} = (s+1)I_{s+1} \, .
\end{aligned}
$$

■

Example 7.12. Let $s = 7$ and take $\alpha_i = i$, $0 \le i \le 6$. The resulting matrix is shown in Table 7.13. It is not difficult to show (see Problem 7.2) that this matrix is equivalent to that given in Table 7.9, and in fact that all Hadamard matrices of order 8 are isomorphic. ■

Table 7.13. Hadamard matrix of order 8 obtained from first Paley construction.

$$
\begin{array}{cccccccc}
+ & - & - & - & - & - & - & - \\
+ & + & - & - & + & - & + & + \\
+ & + & + & - & - & + & - & + \\
+ & + & + & + & - & - & + & - \\
+ & - & + & + & + & - & - & + \\
+ & + & - & + & + & + & - & - \\
+ & - & + & - & + & + & + & - \\
+ & - & - & + & - & + & + & +
\end{array}
$$

Example 7.14. Let $s = 11$ and take $\alpha_i = i$, $0 \le i \le 10$. The resulting H_{12} is shown in Table 7.15. All Hadamard matrices of order 12 are isomorphic to this matrix. ■

Table 7.15. Hadamard matrix of order 12 obtained from first Paley construction.

$$
\begin{array}{cccccccccccc}
+ & - & - & - & - & - & - & - & - & - & - & - \\
+ & + & - & + & - & - & - & + & + & + & - & + \\
+ & + & + & - & + & - & - & - & + & + & + & - \\
+ & - & + & + & - & + & - & - & - & + & + & + \\
+ & + & - & + & + & - & + & - & - & - & + & + \\
+ & + & + & - & + & + & - & + & - & - & - & + \\
+ & + & + & + & - & + & + & - & + & - & - & - \\
+ & - & + & + & + & - & + & + & - & + & - & - \\
+ & - & - & + & + & + & - & + & + & - & + & - \\
+ & - & - & - & + & + & + & - & + & + & - & + \\
+ & + & - & - & - & + & + & + & - & + & + & - \\
+ & - & + & - & - & - & + & + & + & - & + & +
\end{array}
$$

Lemma 7.16. *If* $s \equiv 1 \pmod 4$ *and* Q *is as in Lemma 7.10, then the* $(s+1) \times (s+1)$ *matrix*

$$S = \begin{bmatrix} 0 & 1_s^T \\ 1_s & Q \end{bmatrix} \tag{7.6}$$

has the following properties:

$$\begin{aligned} &a. \quad S \text{ is symmetric,} \\ &b. \quad S^2 = sI_{s+1} \,. \end{aligned} \tag{7.7}$$

Proof: This follows immediately from Lemma 7.10. ∎

The matrix S defined in (7.6) is an example of what is called a *conference matrix* of order $s+1$. This is an $(s+1) \times (s+1)$ matrix with entries $-1, 0, +1$ which satisfies (7.7).

Theorem 7.17. *If* $n = 2(s+1)$ *where* $s \equiv 1 \pmod 4$ *is a prime power, then, with* S *as in Lemma 7.16,*

$$H_n = \begin{bmatrix} S + I_{s+1} & S - I_{s+1} \\ S - I_{s+1} & -S - I_{s+1} \end{bmatrix}$$

is a Hadamard matrix of order n.

Proof: This follows from Lemma 7.16 in the same way that Theorem 7.11 followed from Lemma 7.10. ∎

Theorems 7.11 and 7.17 are known as the *first* and *second Paley constructions.*

Example 7.18. Taking $n = 28$, $s = 13$ in the second Paley construction we obtain the Hadamard matrix shown in Table 7.19. ∎

Table 7.19. Hadamard matrix of order 28 obtained from second Paley construction.

\+ + + + + + + + + + + + + + − + + + + + + + + + + + + +

\+ + + − + + − − − − + + − + + − + − + + − − − − + + − +

\+ + + + − + + − − − − + + − + + − + − + + − − − − + + −

\+ − + + + − + + − − − − + + + − + − + − + + − − − − + +

\+ + − + + + − + + − − − − + + + − + − + − + + − − − − +

\+ + + − + + + − + + − − − − + + + − + − + − + + − − − −

\+ − + + − + + + − + + − − − + − + + − + − + − + + − − −

\+ − − + + − + + + − + + − − + − − + + − + − + − + + − −

\+ − − − + + − + + + − + + − + − − − + + − + − + − + + −

\+ − − − − + + − + + + − + + + − − − − + + − + − + − + +

\+ + − − − − + + − + + + − + + + − − − − + + − + − + − +

\+ + + − − − − + + − + + + − + + + − − − − + + − + − + −

\+ − + + − − − − + + − + + + + − + + − − − − + + − + − +

\+ + − + + − − − − + + − + + + + − + + − − − − + + − + −

− + + + + + + + + + + + + + − − − − − − − − − − − − − −

\+ − + − + + − − − − + + − + − − − + − − + + + + − − + −

\+ + − + − + + − − − − + + − − − − − + − − + + + + − − +

\+ − + − + − + + − − − [illegible] + + − + − − − + − − + + + + − −

\+ + − + − + − + + − − − − + − − + − − − + − − + + + + −

\+ + + − + − + − + + − − − − − − − + − − − + − − + + + +

\+ − + + − + − + − + + − − − − + − − + − − − + − − + + +

\+ − − + + − + − + − + + − − − + + − − + − − − + − − + +

\+ − − − + + − + − + − + + − − + + + − − + − − − + − − +

\+ − − − − + + − + − + − + + − + + + + − − + − − − + − −

\+ + − − − − + + − + − + − + − − + + + + − − + − − − + −

\+ + + − − − − + + − + − + − − − − + + + + − − + − − − +

\+ − + + − − − − + + − + − + − + − − + + + + − − + − − −

\+ + − + + − − − − + + − + − − − + − − + + + + − − + − −

Corollary 7.20. *If* $n = 2(s+1)$ *where* s *is an odd prime power, then there exists a Hadamard matrix of order* n.

Proof: If $s \equiv 1 \pmod 4$ this follows from Theorem 7.17, and if $s \equiv 3 \pmod 4$ from Theorems 7.7 and 7.11 and the existence of a Hadamard matrix of order 2. ∎

The next construction is essentially due to Williamson (1944). This is only the first in a series of ever more complicated constructions that replace the right-hand side of (7.10) by larger matrices with similar properties — see Seberry and Yamada (1992) for further examples.

Theorem 7.21. *If* A, B, C, D *are* $w \times w$ *matrices with entries* $+1$ *and* -1 *that satisfy*

$$XY^T = YX^T, \quad \text{for} \quad X, Y \in \{A, B, C, D\}\,, \tag{7.8}$$

and

$$AA^T + BB^T + CC^T + DD^T = 4wI_w\,, \tag{7.9}$$

then

$$H_{4w} = \begin{bmatrix} A & B & C & D \\ -B & A & -D & C \\ -C & D & A & -B \\ -D & -C & B & A \end{bmatrix} \tag{7.10}$$

is a Hadamard matrix of order $4w$.

Proof: It is straightforward to verify that the matrix H_{4w} in (7.10) satisfies (7.1). ■

A *circulant* matrix is a square matrix in which each row is obtained by cyclically shifting every entry in the previous row one place to the *right*. A *retrocirculant* matrix is a square matrix in which each row is obtained by cyclically shifting every entry in the previous row one place to the *left*. A retrocirculant matrix is therefore symmetric.

Example 7.22. Let A, B, C, D be 5×5 symmetric circulant matrices whose first rows are as follows:

$$A: \; + - - - - \,, \qquad B: \; + - - - - \,,$$
$$C: \; + + - - + \,, \qquad D: \; + - + + - \,.$$

Thus A for example is the matrix

$$\begin{bmatrix} + - - - - \\ - + - - - \\ - - + - - \\ - - - + - \\ - - - - + \end{bmatrix}.$$

The reader may check that (7.8) and (7.9) are satisfied. The resulting Hadamard matrix of order 20 is shown in Table 7.23. Many other examples of this construction are given in Tables 7.33 to 7.34. ■

The constructions given in this section provide a rich supply of Hadamard matrices, as we will now see.

Table 7.23. Hadamard matrix of order 20 from Williamson construction.

$$
\begin{array}{cccccccccccccccccccc}
+&-&-&-&-&+&-&-&-&-&+&+&-&-&+&+&-&+&+&-\\
-&+&-&-&-&-&+&-&-&-&+&+&+&-&-&-&+&-&+&+\\
-&-&+&-&-&-&-&+&-&-&-&+&+&+&-&+&-&+&-&+\\
-&-&-&+&-&-&-&-&+&-&-&-&+&+&+&+&+&-&+&-\\
-&-&-&-&+&-&-&-&-&+&+&-&-&+&+&-&+&+&-&+\\
-&+&+&+&+&+&-&-&-&-&-&+&-&-&+&+&+&-&-&+\\
+&-&+&+&+&-&+&-&-&-&+&-&+&-&-&+&+&+&-&-\\
+&+&-&+&+&-&-&+&-&-&-&+&-&+&-&-&+&+&+&-\\
+&+&+&-&+&-&-&-&+&-&-&-&+&-&+&-&-&+&+&+\\
+&+&+&+&-&-&-&-&-&+&+&-&-&+&-&+&-&-&+&+\\
-&-&+&+&-&+&-&+&+&-&+&-&-&-&-&-&+&+&+&+\\
-&-&-&+&+&-&+&-&+&+&-&+&-&-&-&+&-&+&+&+\\
+&-&-&-&+&+&-&+&-&+&-&-&+&-&-&+&+&-&+&+\\
+&+&-&-&-&+&+&-&+&-&-&-&-&+&-&+&+&+&-&+\\
-&+&+&-&-&-&+&+&-&+&-&-&-&-&+&+&+&+&+&-\\
-&+&-&-&+&-&-&+&+&-&+&-&-&-&-&+&-&-&-&-\\
+&-&+&-&-&-&-&-&+&+&-&+&-&-&-&-&+&-&-&-\\
-&+&-&+&-&+&-&-&-&+&-&-&+&-&-&-&-&+&-&-\\
-&-&+&-&+&+&+&-&-&-&-&-&-&+&-&-&-&-&+&-\\
+&-&-&+&-&-&+&+&-&-&-&-&-&-&+&-&-&-&-&+
\end{array}
$$

7.5 Hadamard Matrices of Orders up to 200

In this section we discuss Hadamard matrices of orders up to 200.

Table 7.24. Number of nonisomorphic Hadamard matrices of order $n \leq 28$.

Order n	Number	Source
1	1	Problem 7.1
2	1	Problem 7.1
4	1	Problem 7.1
8	1	Problem 7.2
12	1	Todd (1933)
16	5	Todd (1933)
20	3	Hall (1965)
24	60	Ito, Leon and Longyear (1981), Kimura (1989)
28	487	Kimura (1994a, 1994b)

The question of the number of nonisomorphic Hadamard matrices of a given order has received a great deal of attention. For small orders the matrices are unique, but as the order increases the number appears to grow very rapidly. Of course this is only an informal assertion, for if we could merely prove that the number was always greater than 1 we would have settled the Hadamard conjecture (cf. Research Problem 7.4). The numbers of nonisomorphic matrices are now known for orders up through 28, and the results are summarized in Table 7.24. The unique matrix of order 12 is shown in Table 7.15.

At order 16 there are five nonisomorphic Hadamard matrices, which can be described as follows. Let K_0 be the Sylvester-type matrix of order 8 shown in Table 7.9, and let K_1, K_2, K_3 be the three other versions of a Hadamard matrix of order 8 shown in Table 7.25. Then the five Hadamard matrices of order 16 are

$$B_0 = \begin{bmatrix} K_0 & K_0 \\ K_0 & -K_0 \end{bmatrix}, \quad B_1 = \begin{bmatrix} K_0 & K_0 \\ K_1 & -K_1 \end{bmatrix},$$

$$B_2 = \begin{bmatrix} K_0 & K_0 \\ K_2 & -K_2 \end{bmatrix}, \quad B_3 = \begin{bmatrix} K_0 & K_0 \\ K_3 & -K_3 \end{bmatrix},$$

$$B_4 = \begin{bmatrix} K_0 & K_3^T \\ K_0 & -K_3^T \end{bmatrix} = B_3^T .$$

The matrix B_0 is the Sylvester type matrix. The five matrices are shown in full (and arranged as above) in Table 7.26. The classification is due to Todd (1933), and this description follows Assmus and Key (1992b). See also Hall (1961) and Yamada (1988).

Table 7.25. Matrices used in the construction of the Hadamard matrices of order 16.

K_1	K_2	K_3
++++++++	++++++++	++++++++
+−+−+−−+	++++−−−−	+++−+−−−
++−−++−−	++−−+−+−	++−+−−−+
+−−++−+−	++−−−+−+	++−−−++−
++++−−−−	+−+−+−−+	+−++−+−−
+−+−−++−	+−+−−++−	+−+−−−++
++−−−−++	+−−+++−−	+−−++−+−
+−−+−+−+	+−−+−−++	+−−−++−+

At order 20 there are three Hadamard matrices, shown in Tables 7.27, 7.23, 7.28. The first matrix comes from the first Paley construction, the second from the Williamson construction (see Example 7.22), while the third (known as Hall's Type N) is not part of any general family. (The second Paley construction produces a matrix that is isomorphic to the Williamson matrix.) The enumeration is due to Hall (1965).

Table 7.26. The five nonisomorphic Hadamard matrices of order 16.

```
++++++++++++++++
+−+−+−+−+−+−+−+−
++−−++−−++−−++−−
+−−++−−++−−++−−+
++++−−−−++++−−−−
+−+−−+−++−+−−+−+
++−−−−++++−−−−++
+−−+−++−+−−+−++−
++++++++−−−−−−−−
+−+−+−+−−+−+−+−+
++−−++−−−−++−−++
+−−++−−+−++−−++−
++++−−−−−−−−++++
+−+−−+−+−+−++−+−
++−−−−++−−++++−−
+−−+−++−−++−+−−+
```

```
++++++++++++++++
+−+−+−+−+−+−+−+−
++−−++−−++−−++−−
+−−++−−++−−++−−+
++++−−−−++++−−−−
+−+−−+−++−+−−+−+
++−−−−++++−−−−++
+−−+−++−+−−+−++−
++++++++−−−−−−−−
+−+−+−−+−+−+−++−
++−−++−−−−++−−++
+−−++−+−−++−−+−+
++++−−−−−−−−++++
+−+−−++−−+−++−−+
++−−−−++−−++++−−
+−−+−+−+−++−+−+−
```

```
++++++++++++++++
+−+−+−+−+−+−+−+−
++−−++−−++−−++−−
+−−++−−++−−++−−+
++++−−−−++++−−−−
+−+−−+−++−+−−+−+
++−−−−++++−−−−++
+−−+−++−+−−+−++−
++++++++−−−−−−−−
++++−−−−−−−−++++
++−−+−+−−−++−+−+
++−−−+−+−−+++−+−
+−+−+−−+−+−+−++−
+−+−−++−−+−++−−+
+−−+++−−−++−−−++
+−−+−−++−++−++−−
```

```
++++++++++++++++
+−+−+−+−+−+−+−+−
++−−++−−++−−++−−
+−−++−−++−−++−−+
++++−−−−++++−−−−
+−+−−+−++−+−−+−+
++−−−−++++−−−−++
+−−+−++−+−−+−++−
++++++++−−−−−−−−
+++−+−−−−−−+−+++
++−+−−−+−−+−+++−
++−−−++−−−+++−−+
+−++−+−−−+−−+−++
+−+−−−++−+−+++−−
+−−++−+−−++−−+−+
+−−−++−+−+++−−+−
```

```
++++++++++++++++
+−+−+−+−++++−−−−
++−−++−−++−−++−−
+−−++−−++−+−+−+−
++++−−−−++−−−−++
+−+−−+−++−−++−−+
++−−−−+++−−+−++−
+−−+−++−+−+−−+−+
++++++++−−−−−−−−
+−+−+−+−−−−−++++
++−−++−−−−++−−++
+−−++−−+−+−+−+−+
++++−−−−−−++++−−
+−+−−+−+−++−−++−
++−−−−++−++−+−−+
+−−+−++−−+−++−+−
```

Table 7.27. Hadamard matrix of order 20 from first Paley construction.

```
+ − − − − − − − − − − − − − − − − − − −
+ + − + + − − − − + − + − + + + + − − +
+ + + − + + − − − − + − + − + + + + − −
+ − + + − + + − − − − + − + − + + + + −
+ − − + + − + + − − − − + − + − + + + +
+ + − − + + − + + − − − − + − + − + + +
+ + + − − + + − + + − − − − + − + − + +
+ + + + − − + + − + + − − − − + − + − +
+ + + + + − − + + − + + − − − − + − + −
+ − + + + + − − + + − + + − − − − + − +
+ + − + + + + − − + + − + + − − − − + −
+ − + − + + + + − − + + − + + − − − − +
+ + − + − + + + + − − + + − + + − − − −
+ − + − + − + + + + − − + + − + + − − −
+ − − + − + − + + + + − − + + − + + − −
+ − − − + − + − + + + + − − + + − + + −
+ − − − − + − + − + + + + − − + + − + +
+ + − − − − + − + − + + + + − − + + − +
+ + + − − − − + − + − + + + + − − + + −
+ − + + − − − − + − + − + + + + − − + +
```

Table 7.28. Hadamard matrix of order 20 of Hall's Type N.

```
+ + + + + + + + + + + + + + + + + + + +
+ + + + + + + + + + − − − − − − − − − −
+ + + + + − − − − − + + + + + − − − − −
+ − − − − − + + + + − + + + + + − − − −
+ + − − − + + + − − + + + − − − + + − −
+ − + − − + + + − − + − − + + − − − + +
+ − − + − + − − + + + + + − − − − − + +
+ − − − + + − − + + + − − + + − + + − −
− − + + − + + − + − + + − + − + + − − −
− − + + − + − + − + + − + − + + − + − −
− + − − + + + − + − + − + − + + − − + −
− + − − + + − + − + + + − + − + − − − +
− − + − + + + − − + − + + + − − − + + −
− − + − + + − + + − − + + − + − + − − +
− + − + − + − + + − − + − + + − − + + −
− + − + − + + − − + − − + + + − + − − +
− − − + + − + + + − + − + + − − − + − +
− − − + + − + + − + + + − − + − + − + −
− + + − − − + − + + + + − − + − − + − +
− + + − − − − + + + + − + + − − + − + −
```

For higher orders, the construction methods in Section 7.4 provide numerous examples. Table 7.29 shows how these methods can be used for orders up to 200. In this table S indicates a Sylvester-type matrix, $P1$ and $P2$ refer to the first and second Paley constructions, W refers to the Williamson construction

Table 7.29. Constructions for Hadamard matrices of orders up to 200.

Order	Use	Order	Use	Order	Use
4	Equation 7.2	72	$P1, T$	140	$P1$
8	Table 7.9	76	$P2, W$	144	T
12	Table 7.15	80	$P1, T$	148	$P2, W$
16	Table 7.26	84	$P1, P2, W$	152	$P1, T$
20	Tables 7.27–7.28	88	T	156	W
24	$P1, T$	92	W	160	T
28	$P1, P2$ (7.19), W	96	T	164	$P1$, $P2$
32	$S, P1$	100	$P2, W$	168	$P1, T$
36	$P2, W$	104	$P1, T$	172	W
40	T	108	$P1, P2, W$	176	T
44	$P1, W$	112	T	180	$P1, P2$
48	$P1, T$	116	W	184	T
52	$P2, W$	120	T	188	(7.11)
56	T	124	$P2, W$	192	$P1, T$
60	$P1, P2, W$	128	$S, P1$	196	$P2$
64	S	132	$P1, W$	200	$P1, T$
68	$P1, W$	136	T		

using symmetric circulant matrices A, B, C, D, as in Example 7.22, and T to the tensor product construction. The absence of a W does not mean that the Williamson construction of Theorem 7.21 does not apply, only that we do not give such a construction here. In general, of course, the matrices A, B, C, D used in the Williamson construction need not be symmetric circulant matrices. Seberry and Yamada (1992), Table A.1, give an overview of which orders can be achieved at present with the Williamson construction. As Hall (1986) remarks, this construction has succeeded in finding a Hadamard matrix of order $4m$, m odd, in every case when a complete search has been made, and it is possible that it will succeed for *every* order.

Research Problem 7.30. *Williamson construction.* Does the Williamson construction of Hadamard matrices succeed for all orders? If not, at which order does it first fail? For what fraction of orders does it succeed?

There have been extensive computer searches for symmetric circulant matrices that can be used in the Williamson construction — see for example Colbourn and Dinitz (1996a), Djoković (1992, 1994a, 1994b), Seberry and Yamada (1992).

Research Problem 7.31. Same as previous problem, but assume that A, B, C, D are symmetric circulant matrices.

The symmetric circulant matrices A, B, C, D used in the Williamson constructions mentioned in Table 7.29 are given in Tables 7.33–7.34. These tables are adapted from a much larger table (Table 9.1) in Seberry and Yamada (1992), which is itself based on discoveries made by a number of authors. The matrices A, B, C, D for order 172 are too wide to fit into Table 7.34, so we give the first rows of these circulant matrices here in smaller type (to save space, the matrices of orders 84 through 108 also appear on this page):

$n = 172$

+−−−++−−++++−+−+++−++−−++−+++−+−++++−−++−−−
++−++++++−−−−+−+−−++−++−++−−+−+−−−−++++++−+
+++−+−++−−+−+−++++−+−−−−+−++++−+−+−−++−+−++
++−−−++++−+−−+−−++−−−−−−−−++−−+−−+−++++−−−+

Table 7.32. First rows of symmetric circulant matrices A, B, C, D used in Williamson construction of Hadamard matrices in Table 7.29: (b) $n = 84$ through 108.

$n = 84$

+ + − + + + + + − + − − + − + + + + + − +
+ + − − + − + − + + − − + + − + − + − − +
+ + + + − + − − − + − − + − − − + − + + +
+ − − + + + + − − − + + − − − + + + + − −

$n = 92$

+ + + − + + + − + − − − − − − + − + + + − + +
+ + + − − − + + − + − + + − + − + + − − − + +
+ − + + − + + − − + + + + + + − − + + − + + −
+ + − − − + − − − + − + + − + − − − + − − − +

$n = 100$

+ − + + + + − − − − + + − − + + − − − − + + + + −
+ + + − + + + − − − + − + + − + − − − + + + − + +
+ − + + − + + + − + + − − − − + + − + + + − + + −
+ − − + + − + − + − − − − − − − − + − + − + + − −

$n = 108$

+ − − − + + + + + − + − + + + + − + − + + + + + − − −
+ − − − + + + + + − + − + + + + − + − + + + + + − − −
+ − − + − − + − + + + − − + + − − + + + − + − − + − −
+ + + − + + − + − − − + + − − + + − − − + − + + − + +

Table 7.33. First rows of symmetric circulant matrices A, B, C, D used in Williamson construction of Hadamard matrices H_n in Table 7.29: (a) $n = 20$ through 76.

$n = 20$

$+ - - - -$

$+ - - - -$

$+ + - - +$

$+ - + + -$

$n = 28$

$+ - - - - - -$

$+ + - - - - +$

$+ - + - - + -$

$+ - - + + - -$

$n = 36$

$+ + - - - - - - +$

$+ - + - - - - + -$

$+ - - + - - + - -$

$+ - - - + + - - -$

$n = 44$

$+ - + + - - - - + + -$

$+ + - + + - - + + - +$

$+ + - + - + + - + - +$

$+ + - - - - - - - - +$

$n = 52$

$+ - + - - + + + + - - + -$

$+ - - - + + + + + + - - -$

$+ + - + - - + + - - + - +$

$+ - - - - + - - + - - - -$

$n = 60$

$+ + - + + - + - - + - + + - +$

$+ + - + + + + - - + + + + - +$

$+ + - + + - - - - - - + + - +$

$+ - + - - - + + + + - - - + -$

$n = 68$

$+ - + + - + + + - - + + + - + + -$

$+ - - + - + + + + + + + + - + - -$

$+ + - + - - - + - - + - - - + - +$

$+ - - - + + + - - - - + + + - - -$

$n = 76$

$+ + - - + - + + + + + + + + - + - - +$

$+ - + - - - + + + + + + + + - - - + -$

$+ + - + + - - + - + + - + - - + + - +$

$+ + - - + + + - + - - + - + + + - - +$

Table 7.34. First rows of symmetric circulant matrices A, B, C, D used in Williamson construction of Hadamard matrices in Table 7.29: (c) $n = 116$ through 156.

$n = 116$

\+ + − − + − − + − + + + − + + + + − + + + − + − − + − − +
\+ + + + − + + − + − − − + + + + + + − − − + − + + − + + +
\+ − + − − − + + − − + − + + + + + + − + − − + + − − − + −
\+ + + − − − + + − − + − + − − − − + − + − − + + − − − + +

$n = 124$

\+ − − + + + − + − − + − − − − + + − − − − + − − + − + + + − −
\+ − − + + + − + − − + − − − − + + − − − − + − − + − + + + − −
\+ − − − − − + + + − + − + − − + + − − + − + − + + + − − − − −
\+ + + + + + − − − + − + − + + − − + + − + − + − − − + + + + +

$n = 132$

\+ + − − − − + − + + − − + − + − − − − + − + − − + + − + − − − − +
\+ − + − − + − − − − − + + + − − + + − − + + + − − − − − + − − + −
\+ + + + − + − + − − − + − − + − − − − + − − + − − − + − + − + + +
\+ + + − + + − − − − + + + − + + − − + + − + + + − − − − + + − + +

$n = 148$

\+ + − + − + − − − − + + − − + − − + + + + − − + − − + + − − − − + − + − +
\+ + − + − + − − − − + + − − + − − + + + + − − + − − + + − − − − + − + − +
\+ + + + − + + + + − + − − + − − − + + + + − − − + − − + − + + + + − + + +
\+ − − − + − − − − + − + + − + + + − − − − + + + − + + − + − − − − + − − −

$n = 156$

\+ + + − − + − + − − − − − + − − + + − − − − + + − − + − − − − − + − + − − + +
\+ + + + − − − + − − + + − − − − + − + − − + − + − − − − + + − − + − − − + + +
\+ + + − − + + − + − − − + − + − − + − − − − + − − + − + − − − + − + + − − + +
\+ − − − + + − + − + − − − − − + + + − + + − + + + − − − − − + − + − + + − − −

A Hadamard matrix of order 188 may be constructed as follows. (So far such a matrix has not yet been obtained by the methods of Section 7.4.) This construction is essentially due to Turyn, although the version we give is a modification of that given by Hedayat and Wallis (1978). Take the matrix

$$\begin{array}{cccc} A & B & C & D \\ -B & A & -E & F \\ -C & E & A & G \\ -D & -F & -G & A \end{array}, \tag{7.11}$$

and replace A by the circulant of order 47 whose first row is given in Table 7.35, and $B, \ldots, G$ by the retrocirculants whose first rows can also be found in that table.

Table 7.35. First rows of $A, \ldots, G$ used to construct a Hadamard matrix of order 188.

+−+−−−−+−−+−−−−++−−−++++−−−+−+−−−−++++++−−+−−−+

+−−−+−−++++++−−−−+−+−−−+−−−+++−−++++−++−++++−+−

+−−+−−+−+++−−−−−+−−−+−−−++−−+−++−+++−+−+−−+−−−+

+−−+−−+−+++−−−−−+−−−+−−−−−++−+−−+−−−+−+−++−+++−

+++−++−+−+−−−+−−+−++−−−−−+−−−+−−−−−+++−+−−+−−+−

−−−+−−+−+−+++−++−+−−++−−−+−−−+−−−−−+++−+−−+−−++

+−++++−++−++++−−+++−−−+−−−+−+−−−−++++++−−+−−−+−

7.6 Notes on Chapter 7

Section 7.1. The term “Plackett-Burman design” is used by Khuri and Cornell (1987), p. 93; Constantine (1987), p. 339; Atkinson and Donev (1992), p. 281; and others. Hughes and Piper (1985), p. 76; Beth, Jungnickel and Lenz (1986), p. 65; Seberry and Yamada (1992), p. 436, and others use “Hadamard design” or “Hadamard array”.

Hadamard matrices have also found other applications in statistics. For example, they play a prominent role in the construction of *weighing designs* and in *resampling methods.*

The use of Hadamard matrices in weighing designs began with Hotelling (1944) and Mood (1946). There is also a brief note on the subject by Rao (1945–1946). By now there is an enormous literature on weighing designs. Since we shall not discuss such designs in this book, in partial compensation we list a number of the most important references, including both classic papers and some of the most recent developments. Of particular interest are matrices that can be used as weighing designs in situations when Hadamard matrices do not exist, corresponding to cases when n is not a multiple of 4. See Banerjee (1975), Eades and Hain (1976), Ehrenfeld (1955), Farmakis (1991), Galil (1985), Galil and Kiefer (1980a, 1980b, 1980c, 1982a, 1982b), Harwit and Sloane (1979), Jacroux and Notz (1983), Kiefer (1979), Kiefer and Galil (1980), Kounias and Chadjipantelis (1983), Kounias and Farmakis (1984), Mitchell (1974a, 1974b), Moyssiadis and Kounias (1982), Raghavarao (1971), Sathe and Shenoy (1991), Shah and Sinha (1989), and Sloane and Harwit (1976).

For information about the use of Hadamard matrices in resampling methods we refer the reader to the survey by McCarthy (1969).

Another important generalization of Hadamard matrices arises from consideration of Hadamard’s inequality, (7.3). In view of this bound it is natural to ask which $n \times n$ matrix with entries $+1$ and -1 has the largest determinant, even when n is not a multiple of 4. This is often called *Hadamard’s determinant problem.* The problem is closely related to the subject of weighing designs mentioned above.

For further information about the Hadamard determinant problem, see Chadjipantelis, Kounias and Moyssiadis (1985, 1987), Ehlich (1964a, 1964b), Ehlich and Zeller (1962), Hardin and Sloane (1993) and Williamson (1946), as well as the references for weighing designs given above. In particular, Table 8 of Hardin and Sloane (1993) gives the largest determinants presently known for all $n \leq 24$.

For further information about Hadamard matrices themselves, we strongly recommend the survey by Seberry and Yamada (1992). Other surveys are Wallis, Street and Wallis (1972), Hedayat and Wallis (1978), Agaian (1985) and Hall (1986). For statistical applications, see especially Diamond (1981).

Section 7.2. It is known (J. S. Wallis, 1976) that if m is odd, then there is an integer t (in the range $2 \leq t \leq 2 \ln m$) such that a Hadamard matrix of order $2^t m$ exists. A table showing the smallest known value of t for all odd $m < 3000$ is given in Seberry and Yamada, 1996, Table A.2, together with a method for constructing a Hadamard matrix of order $2^t m$. A condensed version of this table also appears in Colbourn and Dinitz, 1995, Table 24.33.

To illustrate the use of this table, suppose we want to know if Hadamard matrices of orders 996, 1000 or 1004 are presently known. These three numbers are $2^2 \cdot 249$, $2^3 \cdot 125$ and $2^2 \cdot 251$, respectively, so we look at the entries for $m = 249$, 125 and 251, and find that the minimal values of t are 2, 2 and 3. So Hadamard matrices of orders 996 and 1000 are known, but 1004 is not.

Research Problem 7.36. A sub-Hadamard matrix of order m of a Hadamard matrix of order n is a submatrix which is a Hadamard matrix in its own right. Determine how large m can be and investigate which Hadamard matrices have large Hadamard submatrices. (It is clear that we must have $m \leq n/2$, if $m < n$. If n is a multiple of 8 this bound can be attained with the tensor product construction. So the interesting case is when $n \equiv 4 \pmod 8), n \geq 12$: how close can we get to $m = (n-4)/2$?).

Section 7.5. The complete classification of Hadamard matrices of orders 24 and 28 is due to Ito, Leon and Longyear (1981) and Kimura (1989), and Kimura (1994a, 1994b), respectively. Certain Hadamard matrices of order 28 had been classified earlier by Tonchev (1983, 1985), Kimura (1986) and Kimura and Ohmori (1986).

Most of the Hadamard matrices mentioned in Table 7.29 and the corresponding orthogonal arrays can be found in the electronic data-base with address `www.research.att.com/~njas/hadamard/`.

Yamamoto, Fujii, Hyodo and Yumiba (1992a, 1992b) investigated the question of the number of nonisomorphic $OA(4\lambda, 4\lambda - 1, 2, 2)$'s. Their results may be obtained more concisely as follows.

We must first introduce the notion of the automorphism group $Aut(H_n)$ of a Hadamard matrix H_n. Let $\mathcal{P}$ denote the set of all $n \times n$ matrices with entries from $\{-1, 0, 1\}$ containing exactly one nonzero entry in each row and column. If $P, Q \in \mathcal{P}$ then

$$P\, H_n\, Q$$

is another Hadamard matrix which by definition is isomorphic to H_n. The *automorphism group* $Aut(H_n)$ is the set of all pairs (P, Q) for which

$$P\, H_n\, Q = H_n \ .$$

Note that $(-I_n, -I_n)$ is always in $Aut(H_n)$. Leon (1979) gives an efficient algorithm for determining $Aut(H_n)$. The automorphism groups of Hadamard matrices of the first Paley type were found by Kantor (1969).

It is not difficult to show that the nonisomorphic $OA(4\lambda, 4\lambda-1, 2, 2)$'s can be obtained by taking all the nonisomorphic Hadamard matrices $H_{4\lambda}$ and taking one representative from each orbit of $Aut(H_{4\lambda})$ on the columns of $H_{4\lambda}$.

Taking the classification of Hadamard matrices of order $4\lambda \le 24$ in Table 7.24 as our starting point, it is now easy to apply Leon's algorithm to show that the automorphism group of every Hadamard matrix of order $4\lambda \le 20$ is transitive on both rows and columns, and that the columns of the 60 Hadamard matrices of order 24 fall into exactly 130 orbits.

In view of the preceding discussion, this establishes the following theorem. The cases $4\lambda = 16$ and 20 were first obtained in a different way by Yamamoto, Fujii, Hyodo and Yumiba (1992a, 1992b).

Theorem 7.37. *The number of nonisomorphic orthogonal arrays with parameters $OA(4\lambda, 4\lambda - 1, 2, 2)$ is equal to*

$$1,\ 1,\ 1,\ 5,\ 3,\ 130\ ,$$

if 4λ is equal to

$$4,\ 8,\ 12,\ 16,\ 20,\ 24\ ,$$

respectively.

Research Problem 7.38. Determine the number of nonisomorphic $OA(4\lambda, 4\lambda-1, 2, 2)$'s and $OA(8\lambda, 4\lambda, 2, 3)$'s for additional values of λ.

7.7 Problems

7.1. Show that there are unique Hadamard matrices H_1, H_2 and H_4.

7.2. (a) Show that H_8 is unique. (b) Find an explicit equivalence between the two matrices in Tables 7.9 and 7.13.

7.3. Use the second Paley construction, Theorem 7.17, to construct a Hadamard matrix of order 12. (Of course, in view of Table 7.24, this is equivalent to the matrix in Table 7.15.)

7.4. In connection with the construction of matrices satisfying (7.8), show that if A and B are arbitrary circulant matrices then $AB = BA$. [Hint: Any circulant is a polynomial function of the circulant with first row $010\ldots0$.]

7.5. Let $\begin{bmatrix} K & L \\ M & N \end{bmatrix}$ and $\begin{bmatrix} P & Q \\ R & S \end{bmatrix}$ be Hadamard matrices of orders $4a$ and $4b$, respectively, where K, L, M, N are $2a \times 2a$ matrices and P, Q, R, S are $2b \times 2b$ matrices. Show that

$$\frac{1}{2}\begin{bmatrix} (P+Q)\otimes K + (P-Q)\otimes M & (P+Q)\otimes L + (P-Q)\otimes N \\ (R+S)\otimes K + (R-S)\otimes M & (R+S)\otimes L + (R-S)\otimes N \end{bmatrix}$$

is a Hadamard matrix of order $8ab$. (This is a result of Agaian and Sarukhanyan, and Seberry and Yamada — cf. Seberry and Yamada, 1992, p. 443.)

7.6. Let $B = (b_{ij})$ be a positive definite $n \times n$ matrix. (a) Prove Hadamard's determinant inequality

$$\det\ B \le \prod_{i=1}^{n} b_{ii}\ .$$

(Hint: reduce to the case where the b_{ii}'s are all 1. Let λ_i be the eigenvalues of B. Then $\det B = \Pi\lambda_i \le \left(\frac{1}{n}\sum\lambda_i\right)^n = \left(\frac{1}{n}\ trace\ B\right)^n = 1$. Cf. Horn and Johnson, 1985, p. 477.) (b) Let $A = (a_{ij})$ satisfy $|a_{ij}| \le 1$. Take $B = AA^T$ and deduce (7.3).

7.7. Let $f(n)$, $f^*(n)$, $g(n)$, $g^*(n)$ denote the maximal determinant of any $n \times n$ matrix $A = (a_{ij})$ with entries satisfying $0 \le a_{ij} \le 1$, $a_{ij} = 0$ or 1, $-1 \le a_{ij} \le 1$, $a_{ij} = -1$ or 1, respectively. (a) Show that $f(n) = f^*(n)$ and $g(n) = g^*(n)$. (b) Show that $g(n) = 2^{n-1}f(n-1)$. (c) Show that $g(1) = 1$, $g(2) = 2$, $g(3) = 4$ and $g(4) = 16$.

Chapter 8

Orthogonal Arrays and Latin Squares

The subject of pairwise or mutually orthogonal Latin squares has fascinated researchers for many years. Although there are a number of intriguing results in this area, many open problems remain to which the answers seem as elusive as ever. The known results, however, are well documented, for example in the books by Dénes and Keedwell (1974, 1991) and Laywine and Mullen (1998), or the article by Jungnickel (1990).

Pairwise orthogonal frequency squares are a generalization of pairwise orthogonal Latin squares. Frequency squares or F-*squares* are not so well known as Latin squares, although, as reported by Dénes and Keedwell (1991) and Laywine and Mullen (1998), they have attracted considerable attention in recent years.

Some orthogonal arrays can be used to construct pairwise orthogonal Latin squares or F-squares, and conversely both pairwise orthogonal Latin squares and F-squares provide a tool for constructing orthogonal arrays. It is the latter aspect that will concern us in this chapter. We will give a number of results on the construction of orthogonal F-squares. A selection of fundamental results on pairwise orthogonal Latin squares will also be discussed, but those with a broader interest in this area should consult the more specialized references mentioned above.

8.1 Latin Squares and Orthogonal Latin Squares

A *Latin square of order* s is an $s \times s$ array with entries from a set S of cardinality s such that each element of S appears once in every row and every column. It is easily seen that a Latin square of order s exists for every positive integer s. For example, one may label the rows and columns by $0, 1, \ldots, s-1$, and take the entry in row i and column j to be $i+j$, where the addition is modulo s. The resulting Latin square for $s = 4$ is displayed in Table 8.1.

Table 8.1. A Latin square of order 4.

0	1	2	3
1	2	3	0
2	3	0	1
3	0	1	2

Two Latin squares of order s are said to be *orthogonal* to each other if when one is superimposed on the other the ordered pairs (i, j) of corresponding entries consist of all possible s^2 pairs. A collection of w Latin squares of order s, any pair of which are orthogonal, is called a set of *pairwise orthogonal Latin squares*, and denoted by $POL(s, w)$. (We allow $w = 1$: any Latin square of order s is a $POL(s, 1)$.) Such a collection is also often called a set of *mutually orthogonal Latin squares*, abbreviated $MOLS(s, w)$.

Table 8.2 exhibits three pairwise orthogonal Latin squares of order 4 forming a $POL(4, 3)$.

Table 8.2. Three pairwise orthogonal Latin squares of order 4.

0	2	3	1		0	2	3	1		0	2	3	1
3	1	0	2		1	3	2	0		2	0	1	3
1	3	2	0		2	0	1	3		3	1	0	2
2	0	1	3		3	1	0	2		1	3	2	0

Observe that none of the squares[1] in Table 8.2 is orthogonal to the square in Table 8.1. In fact, it can be shown that there exists no Latin square of order 4 that is orthogonal to the square in Table 8.1 (see Problem 8.1). We shall say that a Latin square is *orthogonally isolated* (or simply *isolated*) if there is no Latin square orthogonal to it. The square in Table 8.1 is orthogonally isolated. (In view of Theorem 8.28, we may deduce that whereas the Latin squares in Table 8.2 produce an $OA(16, 5, 4, 2)$, the square in Table 8.1 produces an $OA(16, 3, 4, 2)$—essentially that shown in Problem 1.2(d)—which cannot even be extended to an $OA(16, 4, 4, 2)$.)

[1]We often omit the adjective "Latin" when it is implied by the context.

There are certain basic operations which transform one Latin square into another. Any permutation of the rows of the array, or the columns of the array, or the elements of S gives another Latin square. One Latin square is said to be *equivalent* to another if it can be obtained from the first by a sequence of these three operations. Any two of the squares in Table 8.2 are equivalent, but none of them is equivalent to the square in Table 8.1.

In a $POL(s, w)$, a permutation of the elements of S in one or more of the Latin squares will not affect their orthogonality. A permutation of the rows or columns that is performed simultaneously on all squares of the $POL(s, w)$ also preserves the orthogonality. It is easily seen that any $POL(s, w)$ can be converted into a $POL(s, w)$ in which the entries in the first row in every square are in the natural order $0, 1, \ldots, s - 1$. Since the first entry in the second row must then be different in all the squares because of pairwise orthogonality, and cannot be 0 since all have already a 0 in the first column, it follows that $w \le s-1$ for any $POL(s, w)$. A $POL(s, s - 1)$ will therefore be referred to as a *complete set* of pairwise orthogonal Latin squares of order s.

A construction for a $POL(s, s - 1)$ is known if s is a prime power, as will be shown in the next theorem. No construction for a complete set of pairwise orthogonal Latin squares for any other value of s is known at the present time. In fact, if s is not a prime power, the largest value of w for which a $POL(s, w)$ is known to exist is usually considerably smaller than $s - 1$, and for several values of s, including those with $s \equiv 6 \pmod 8$, it is known that a $POL(s, s - 1)$ cannot exist.

Theorem 8.3. *If s is a prime power then there are $s - 1$ pairwise orthogonal Latin squares of order s.*

Proof:

CONSTRUCTION: As in previous chapters, let the elements of $GF(s)$ be denoted by $\alpha_0 = 0, \alpha_1, \ldots, \alpha_{s-1}$. For $i = 1, 2, \ldots, s - 1$ define a Latin square L_i by taking the entry in row j and column ℓ to be $\alpha_i\alpha_j + \alpha_\ell$, for $j, \ell = 0, 1, \ldots, s - 1$. This produces a set of $s - 1$ pairwise orthogonal Latin squares of order s, i.e. a $POL(s, s - 1)$.

VERIFICATION: For fixed $i \in \{1, 2, \ldots, s - 1\}$ and $j \in \{0, 1, \ldots, s - 1\}$ it is clear that the elements

$$\alpha_i\alpha_j + \alpha_\ell, \quad \ell = 0, 1, \ldots, s - 1 ,$$

comprise all of $GF(s)$. Similarly, if $i \in \{1, 2, \ldots, s - 1\}$ and $\ell \in \{0, 1, \ldots, s - 1\}$ are fixed, it is clear that

$$\alpha_i\alpha_j + \alpha_\ell, \quad j = 0, 1, \ldots, s - 1 ,$$

comprises all of $GF(s)$. This shows that each L_i is indeed a Latin square.

To verify the pairwise orthogonality, fix $i, i' \in \{1, 2, \ldots, s-1\}$, $i \neq i'$. For any $z_1, z_2 \in GF(s)$ consider the linear equations

$$\begin{aligned} \alpha_i \alpha_j + \alpha_\ell &= z_1 , \\ \alpha_{i'} \alpha_j + \alpha_\ell &= z_2 , \end{aligned}$$

in the unknowns α_j and α_ℓ. Since these are two independent equations over $GF(s)$, there is a unique solution $(\alpha_{j_0}, \alpha_{\ell_0})$. Consequently, any ordered pair (z_1, z_2) appears exactly once when L_i is superimposed on $L_{i'}$, which establishes the orthogonality of the Latin squares. ■

The $POL(4,3)$ in Table 8.2 illustrates this construction in the case $s = 4$. We define $GF(4)$ by the irreducible polynomial $f(x) = x^2 + x + 1$ over $GF(2)$, taking $\alpha_1 = x$, $\alpha_2 = x + 1$ and $\alpha_3 = 1$. Writing 2 for x and 3 for $x+1$ produces the three squares of order 4 that are exhibited in Table 8.2.

The requirements on the structure of a $POL(s, w)$ are so stringent for values of w just less than $s-1$ that such a set can often be extended to a $POL(s, s-1)$, no matter how the $POL(s, w)$ was constructed. In Problem 8.6 the reader is asked to show that any $POL(s, s-2)$, $s \geq 3$, can be extended to a $POL(s, s-1)$. In Section 8.3 we will use a relation between pairwise orthogonal Latin squares and orthogonal arrays to show that two Latin squares can be added to any $POL(s, s-3)$ to obtain a $POL(s, s-1)$, if $s \geq 5$, a result due to Shrikhande (1961).

Bruck (1963) generalized Shrikhande's result in the following way. Let us call $d = s - 1 - w$ the *deficiency* of a $POL(s, w)$. Bruck (1963) showed that a $POL(s, w)$ with deficiency d can be extended to a $POL(s, s-1)$ provided

$$s \geq (d^4 - 2d^3 + 2d^2 + d)/2 . \tag{8.1}$$

The case $d = 2$ is Shrikhande's result. Since the right-hand side of (8.1) increases rapidly with d, for large d the result provides information only for very large values of s. For deficiency $d = 2$ we know already that the bound in (8.1) cannot be improved: the $POL(4,1)$ in Table 8.1 cannot be extended to a $POL(4,3)$.

The problem of extending a $POL(s, w)$ to a $POL(s, s-1)$ is a special case of that of embedding orthogonal arrays into orthogonal arrays with a larger number of factors. For more on this problem see Shrikhande and Singhi (1979a, 1979b).

In the eighteenth century, Euler conjectured that any Latin square of order s with $s \equiv 2 \pmod 4$ is orthogonally isolated. In other words, he believed that a $POL(s, 2)$ cannot exist when $s \equiv 2 \pmod 4$. For $s = 2$ his conjecture is obviously correct. For $s = 6$, it is also correct, although a proof was not given until the beginning of the twentieth century (Tarry, 1900, 1901). It is now known that his conjecture is false for all other values of s. The work of Bose and Shrikhande (1959, 1960), Parker (1958, 1959a, 1959b) and Bose, Shrikhande and Parker (1960) established the existence of two orthogonal Latin squares of order s for any $s \equiv 2 \pmod 4$, $s \neq 2, 6$. For $s \geq 3$ and $s \not\equiv 2 \pmod 4$ the existence of a $POL(s, 2)$ was established by the following result of MacNeish (1922).

Theorem 8.4. *Let $s = p_1^{v_1} \cdots p_u^{v_u}$ be the factorization of s into primes and let $s_0 = \min\{p_i^{v_i}, i = 1, \ldots, u\}$. Then there exists a $POL(s, s_0 - 1)$.*

Proof:
CONSTRUCTION: For each $i \in \{1, \ldots, u\}$ construct a $POL(p_i^{v_i}, s_0 - 1)$, for example by the method of Theorem 8.3. Writing $s_i = p_i^{v_i}$, let

$$L_{ij} = (\alpha_{mn}^{ij})_{m=1,n=1}^{s_i \quad s_i}, \quad i = 1, \ldots, u, \quad j = 1, \ldots, s_0 - 1 ,$$

be the j-th square in the $POL(s_i, s_0 - 1)$. For each $j \in \{1, \ldots, s_0 - 1\}$ form an $s \times s$ square L_j as follows. Label its rows and columns by the u-tuples $(l_1, \ldots, l_u)$ where $l_i \in \{1, \ldots, s_i\}$, $i = 1, \ldots, u$, using the same labeling for each of the j squares. For the row and column labeled $(m_1, \ldots, m_u)$ and $(n_1, \ldots, n_u)$, respectively, define the entry of L_j to be the u-tuple

$$(\alpha_{m_1 n_1}^{1j}, \ldots, \alpha_{m_u n_u}^{uj}) .$$

The verification that the squares $L_1, \ldots, L_{s_0-1}$ form a $POL(s, s_0 - 1)$ is left to the reader (see Problem 8.7). ∎

Note that if $s \geq 3$ and $s \not\equiv 2 \pmod 4$ then $s_0 \geq 3$ and indeed a $POL(s, 2)$ exists for any such s.

The smallest integer s for which the maximal attainable value of w in a $POL(s, w)$ is unknown is $s = 10$. While Parker (1959) established the existence of a $POL(10, 2)$, showing that Euler's conjecture was false for $s = 10$, the existence of a $POL(10, 3)$ is still unknown. Franklin (1984) gives some examples of triples of "almost" $POL(10, 3)$'s, while Brouwer (1984) presents four squares that come close to a $POL(10, 4)$ except for a hole of order 2. Parker (1991) conjectures that a $POL(10, 3)$ does not exist.

In accordance with our policy of drawing attention to the most important unsolved problems in each chapter, let us highlight this:

Research Problem 8.5. Prove (or disprove) the conjecture that no three pairwise orthogonal Latin squares of order 10 exist.

For all s except $2, 3, 6, 10$ it is known that a $POL(s, 3)$ does exist (see for example Theorem 2.4.6 in Jungnickel, 1990), but our knowledge about the existence of a $POL(s, w)$ for values of s that are not prime powers is extremely limited. Abel, Brouwer, Colbourn and Dinitz (1996) give a table of lower bounds on the maximal attainable w in a $POL(s, w)$ for $s < 10000$. Part of their table is shown in Section 8.4.

An important concept for the construction of orthogonal Latin squares is that of parallel transversals. A *transversal* in a Latin square is defined to be a set of s row-column pairs such that each row and column occurs once and the entries corresponding to these row-column pairs are distinct. If we label the rows and columns of a Latin square of order s by $0, 1, \ldots, s-1$ and if we denote its entries by $\ell_{ij} \in \{0, 1, \ldots, s-1\}$, then a transversal is a set of s pairs taken from $\{0, 1, \ldots, s-1\} \times \{0, 1, \ldots, s-1\}$, say $\{(i_1, j_1), (i_2, j_2), \ldots, (i_s, j_s)\}$, such that $\{i_1, i_2, \ldots, i_s\} = \{j_1, j_2, \ldots, j_s\} = \{\ell_{i_1 j_1}, \ell_{i_2 j_2}, \ldots, \ell_{i_s j_s}\} = \{0, 1, \ldots, s-1\}$. As an example, $\{(0,0), (1,3), (2,1), (3,2)\}$ is a transversal for both the first and third squares of order 4 in Table 8.2. However, in the second square in that table these four row-column pairs correspond each time to the entry zero, so do not form a transversal for this square.

Two or more transversals of a Latin square are said to be *parallel* if the corresponding sets of row-column pairs are disjoint. Hence a Latin square of order s has at most s parallel transversals. As an example, the sets

$$\{(0,0),\quad (1,3),\quad (2,1),\quad (3,2)\}\ ,$$

$$\{(0,1),\quad (1,2),\quad (2,0),\quad (3,3)\}\ ,$$

$$\{(0,2),\quad (1,1),\quad (2,3),\quad (3,0)\}\ ,$$

$$\{(0,3),\quad (1,0),\quad (2,2),\quad (3,1)\}\ ,$$

form a complete set of parallel transversals for the first square in Table 8.2. The same set is also a complete set of parallel transversals for the third square in that table. We will say that these two Latin squares have four *common parallel transversals*. The following lemma, the proof of which is left to the reader (Problem 8.8), illustrates the importance of this concept.

Lemma 8.6. *A $POL(s, w)$ can be extended to a $POL(s, w+1)$ if and only if the w Latin squares in the $POL(s, w)$ possess s common parallel transversals.*

For example, several authors have constructed a $POL(10, 2)$. However, no $POL(10, 2)$ has yet been found in which the two Latin squares have ten common parallel transversals. The best result so far is a $POL(10, 2)$ with four common parallel transversals — see Ganter, Mathon and Rosa (1978, 1979) and Brown, Hedayat and Parker (1993). The pair from Brown, Hedayat and Parker is shown in Table 8.7. The common parallel transversals are $\{(i, j_h(i)), i = 0, 1, \ldots, 9\}$, $h = 1, 2, 3, 4$, where the $j_h(i)$ are given in Table 8.8.

Table 8.7. Two orthogonal Latin squares of order 10 with four common parallel transversals.

9	4	2	7	3	8	1	6	5	0		1	2	3	4	5	6	7	8	9	0
3	8	1	6	2	7	0	5	4	9		8	7	6	1	0	9	5	2	4	3
2	7	0	5	1	6	4	9	3	8		5	8	4	7	2	3	0	9	6	1
1	6	4	9	0	5	3	8	2	7		9	4	8	6	1	0	3	5	7	2
0	5	3	8	4	9	2	7	6	1		2	5	9	3	6	8	1	7	0	4
4	9	7	2	8	3	6	1	0	5		7	0	5	9	8	4	2	1	3	6
8	3	6	1	7	2	5	0	9	4		4	1	7	0	3	2	8	6	5	9
7	2	5	0	6	1	9	4	8	3		0	6	1	8	9	5	4	3	2	7
6	1	9	4	5	0	8	3	7	2		6	3	2	5	4	7	9	0	1	8
5	0	8	3	9	4	7	2	1	6		3	9	0	2	7	1	6	4	8	5

Table 8.8. Four common parallel transversals in the Latin squares in Table 8.7.

h	$j_h(0)$	$j_h(1)$	$j_h(2)$	$j_h(3)$	$j_h(4)$	$j_h(5)$	$j_h(6)$	$j_h(7)$	$j_h(8)$	$j_h(9)$
Transversal 1	1	5	0	4	3	9	2	6	7	8
Transversal 2	6	0	9	3	8	2	5	7	4	1
Transversal 3	8	4	5	7	9	0	1	3	2	6
Transversal 4	9	1	8	0	5	6	4	2	3	7

The concept of parallel transversals is also a useful tool when determining if a given Latin square is orthogonally isolated. The reader should keep this in mind when attempting Problem 8.1.

For further information about Latin squares see the works by Dénes and Keedwell (1974, 1991), Jungnickel (1990) and Laywine and Mullen (1998) already mentioned, and also Beth, Jungnickel and Lenz (1986), Dinitz and Stinson (1992), Raghavarao (1971), Street and Street (1987).

8.2 Frequency Squares and Orthogonal Frequency Squares

Although Latin squares have many useful properties, for some statistical applications these structures are too restrictive. The more general concepts of F-squares and orthogonal F-squares offer more flexibility. The theory pertaining to these concepts is however not nearly as well developed as that for Latin squares. In this section we will present some results on orthogonal F-squares; for further information the reader is referred to Chapter 12 of Dénes and Keedwell (1991).

An $n \times n$ array based on s symbols is called a *frequency square* or F*-square* if each symbol appears n/s times in each row and in each column. An example of an 6×6 F-square based on three symbols is given in Table 8.9. Such an array could be obtained for example by combining four, not necessarily identical, Latin squares of order 3.

Table 8.9. A 6×6 F-square based on three symbols.

0	1	2	0	1	2
1	2	0	2	0	1
2	0	1	1	2	0
0	1	2	0	1	2
1	2	0	1	2	0
2	0	1	2	0	1

Two $n \times n$ F-squares based on s symbols are said to be *orthogonal* if when one is superimposed on the other the ordered pairs of corresponding entries contain each of the s^2 possibilities n^2/s^2 times. Table 8.10 exhibits a second 6×6 F-square based on three symbols that is orthogonal to the one in Table 8.9.

Table 8.10. Another 6×6 F-square based on three symbols.

0	1	2	0	1	2
2	0	1	1	2	0
1	2	0	2	0	1
0	1	2	0	1	2
2	0	1	2	0	1
1	2	0	1	2	0

We will write $POF(n, s, w)$ for a set of w pairwise (or mutually) orthogonal $n \times n$ F-squares based on s symbols. Thus the arrays in Tables 8.9 and 8.10 form a $POF(6, 3, 2)$.

The definitions given here are in some sense unnecessarily restrictive. F-squares can be defined in a more general way by requiring that the frequency with which a symbol occurs in a row or column is independent of the selected row or column but may depend on the symbol. Thus symbol i should appear λ_i times, say, in each row and each column. Under our definition of an F-square λ_i must equal n/s for every i. Orthogonality can be defined for this wider class of squares and can even be extended to orthogonality of F-squares that are not necessarily based on the same number of symbols. However, the more restricted definition given here will suffice for our purposes.

From now on we assume that s is at least 2 and that n is a multiple of s. Since a Latin square of order s exists for any s, the construction of Table 8.9 generalizes immediately to show that an $n \times n$ F-square based on s symbols always exists.

A more interesting question is that of determining the maximal value of w for which a $POF(n, s, w)$ can be obtained. The following result provides an upper bound.

Theorem 8.11. *A necessary condition for the existence of w pairwise orthogonal $n \times n$ F-squares based on s symbols is*

$$w \leq (n-1)^2/(s-1) . \tag{8.2}$$

Proof: The following proof is essentially that of Hedayat, Raghavarao and Seiden (1975). In Chapter 9 we will see that, as pointed out by these authors, the result can also be obtained from a generalization of Corollary 2.5 to mixed orthogonal arrays.

If $F_1, \ldots, F_w$ form a $POF(n, s, w)$, let the entry $i + (j-1)n$, $i, j = 1, \ldots, n$, of the $n^2 \times 1$ vector $u_{\ell m}$, $\ell = 1, \ldots, s$, $m = 1, \ldots, w$, be defined to be 1 if the ℓ-th symbol appears in cell (i, j) of F_m and to be $-1/(s-1)$ otherwise. Let U be the vector space spanned by these sw vectors. We will show that the dimension of U is $w(s-1)$, and we will identify $2n-1$ independent vectors belonging to the orthogonal complement V of U. This will imply that

$$w(s-1) = \dim(U) = n^2 - \dim(V) \leq n^2 - 2n + 1 = (n-1)^2$$

which is the claim of Theorem 8.11.

To see that $\dim(U) = w(s-1)$, first observe that the orthogonality of the w F-squares implies that $u_{\ell m}^T u_{\ell' m'} = 0$ for any $\ell, \ell' = 1, \ldots, s$, and $m, m' = 1, \ldots, w$ with $m \neq m'$. Consequently $\dim(U) = \sum_{m=1}^{w} \dim(U_m)$, where U_m denotes the vector space spanned by $u_{1m}, \ldots, u_{sm}$. Furthermore, since F_m is an F-square, it follows that $u_{\ell m}^T u_{\ell' m} = -n^2/(s-1)^2$ and $u_{\ell m}^T u_{\ell m} = n^2/(s-1)$ for any ℓ, $\ell' = 1, \ldots, s$ with $\ell \neq \ell'$. Hence

$$[u_{1m} \cdots u_{sm}]^T [u_{1m} \cdots u_{sm}] = (n^2 s/(s-1)^2)(I_s - (1/s)J_s) .$$

Since the rank of this matrix is $s-1$, it follows that the dimension of U_m is equal to $s-1$. This shows that $\dim(U) = w(s-1)$, as claimed.

Finally, we need to identify $2n-1$ independent vectors in the orthogonal complement of U. For $h = 1, \ldots, n$, let v_h be the $n^2 \times 1$ vector with a 1 in position $h + (j-1)n$, $j = 1, \ldots, n$, and a 0 elsewhere, and let w_h be the vector with a 1 in position $i + (h-1)n$, $i = 1, \ldots, n$, and a 0 elsewhere. The $2n-1$ vectors $1_{n^2}, v_1, \ldots, v_{n-1}, w_1, \ldots, w_{n-1}$ are independent and, since $F_1, \ldots, F_w$ are F-squares, we obtain

$$1_{n^2}^T u_{\ell m} = v_h^T u_{\ell m} = w_h^T u_{\ell m} = 0$$

for any $h = 1, \ldots, n$, $\ell = 1, \ldots, s$ and $m = 1, \ldots, w$. ∎

Since a $POF(s,s,w)$ is nothing but a $POL(s,w)$, we see that Theorem 8.11 implies $w \le s-1$ for a $POL(s,w)$. We say that a $POF(n,s,w)$ with $w = (n-1)^2/(s-1)$ is *complete*.

Complete $POF(n,s,w)$'s have been constructed for various values of n, s, w, showing that in these cases the bound of Theorem 8.11 is tight. Some of these examples will be obtained as a consequence of Theorem 8.12.

The next theorem makes use of the notion of a resolvable difference scheme introduced in Section 6.1.

Theorem 8.12. *Suppose there exists a difference scheme D with parameters $D(r,c,s)$ and an orthogonal array A with parameters $OA(\lambda s^2, k, s, 2)$, such that $\lambda r s^2$ $(= n^2)$ is a perfect square, and such that one of the following conditions holds:*

(i) $r = \lambda s^2$,

(ii) $\lambda s^2 = \delta n$ *for an integer* $\delta \ge 2$ *and* A *is* (n/s)*-resolvable, or*

(iii) $r = \delta n$ *for an integer* $\delta \ge 2$ *and* D *is* (n/s)*-resolvable.*

Then there exist $k(c-1)$ pairwise orthogonal $n \times n$ F-squares based on s symbols, i.e. a $POF(n,s,k(c-1))$.

Proof:
CONSTRUCTION: Without loss of generality, the levels in A may be taken to be the elements of $\mathcal{A}$, whose binary operation will be denoted by $+$. If condition (ii) in the theorem holds, we will take A to have the form

$$A = [A_1^T \cdots A_\delta^T]^T ,$$

where each level occurs n/s times for every factor in each A_i, $i = 1, \ldots, \delta$. We suppose that D has the form (6.1). If condition (iii) of the theorem holds, we will take $D^{(0)}$ to have the form

$$D^{(0)} = [D^{(1)^T} \cdots D^{(n)^T}]^T ,$$

where each $D^{(i)}$, $i = 1, \ldots, n$, is a $\delta \times (c-1)$ array with the property that, for any $j \in \{1, \ldots, \delta\}$, combining the j-th rows in $D^{(1)}, \ldots, D^{(n)}$ results in an $n \times (c-1)$ array with columns that are uniform on $\mathcal{A}$.

Let F denote the $n^2 \times k(c-1)$ array given by

$$F = D^{(0)} \otimes A ,$$

where $\otimes$ is the tensor product operation introduced in Definition 6.26. Denoting the element in cell (ℓ, m) of F by $f(\ell, m)$, $\ell = 1, \ldots, n^2$, $m = 1, \ldots, k(c-1)$, we define the $n \times n$ arrays $F_g = (f_g(i,j))$, $g = 1, \ldots, k(c-1)$, $i, j = 1, \ldots, n$, by setting

$$f_g(i,j) = f((i-1)n + j, g) .$$

Then $F_1, \ldots, F_{k(c-1)}$ form a $POF(n, s, k(c-1))$.

VERIFICATION: To show that these squares are F-squares we must show for any $z_1 \in \mathcal{A}$ and any $g_1 \in \{1, \ldots, k(c-1)\}$ that there are exactly n/s values of j that satisfy

$$f_{g_1}(i_0, j) = z_1 , \tag{8.3}$$

for any $i_0 \in \{1, \ldots, n\}$, and exactly n/s values of i that satisfy

$$f_{g_1}(i, j_0) = z_1 , \tag{8.4}$$

for any $j_0 \in \{1, \ldots, n\}$. To show that these squares are pairwise orthogonal we must show for any z_1, $z_2 \in \mathcal{A}$ and any $g_1, g_2 \in \{1, \ldots, k(c-1)\}$, $g_1 \neq g_2$, that there are exactly n^2/s^2 pairs (i,j) that satisfy the equations

$$f_{g_1}(i,j) = z_1 , \quad f_{g_2}(i,j) = z_2 . \tag{8.5}$$

To accomplish this, we will denote the elements of A and $D^{(0)}$ by $a(\ell, m)$, $\ell = 1, \ldots, \lambda s^2$, $m = 1, \ldots, k$, and $d(\ell, m)$, $\ell = 1, \ldots, r$, $m = 1, \ldots, c-1$, respectively. We will write $g_h = k_h + (c_h - 1)k$, $k_h \in \{1, \ldots, k\}$, $c_h \in \{1, \ldots, c-1\}$, for $h \in \{1, 2\}$. We now consider the three cases of the theorem separately.

If condition (i) holds, then

$$f_{g_h}(i,j) = d(i, c_h) + a(j, k_h) ,$$

so equation (8.3) requires that there are n/s values of j that satisfy

$$a(j, k_1) = z_1 - d(i_0, c_1) .$$

This is so, since each column of the orthogonal array A contains each element of $\mathcal{A}$ exactly n/s times. For equation (8.4), we must verify that

$$d(i, c_1) = z_1 - a(j_0, k_1)$$

is satisfied by n/s values of i. This follows since the columns of $D^{(0)}$ are uniform on $\mathcal{A}$. For equations (8.5) we must verify that there are n^2/s^2 pairs (i,j) that satisfy

$$d(i, c_1) + a(j, k_1) = z_1, \quad d(i, c_2) + a(j, k_2) = z_2 . \tag{8.6}$$

If $k_1 \neq k_2$, these equations have exactly n/s^2 solutions for j for every fixed value of i, since A is an orthogonal array of strength two. Since we can choose n different values for i the result follows. If $k_1 = k_2$, and hence $c_1 \neq c_2$, there are n/s values of i that satisfy the difference of the two equations in (8.6),

$$d(i, c_1) - d(i, c_2) = z_1 - z_2 ,$$

since D is a difference scheme. For each such value of i there are n/s values of j that satisfy the first equation in (8.6), since the columns of A are uniform on $\mathcal{A}$. Hence there are also n^2/s^2 solutions (i,j) in this case.

If condition (ii) holds, then, for $i \in \{1,\ldots,n\}$, we will write $i = (i_1-1)\delta+i_2$, with $i_1 \in \{1,\ldots,r\}$, $i_2 \in \{1,\ldots,\delta\}$. Then

$$f_{g_h}(i,j) = d(i_1,c_h) + a((i_2-1)n+j,k_h) \ ,$$

so now equation (8.3) requires that there are n/s values of j that satisfy

$$a((i_{02}-1)n+j,k_1) = z_1 - d(i_{01},c_1) \ .$$

This is so since the columns of $A_{i_{02}}$ are uniform on $\mathcal{A}$. For (8.4) we must show that there are n/s pairs (i_1,i_2) such that

$$d(i_1,c_1) + a((i_2-1)n+j_0,k_1) = z_1 \ .$$

For any choice of i_2, there are r/s solutions for i_1 since the columns of $D^{(0)}$ are uniform on $\mathcal{A}$. Since there are δ possible choices for i_2, there are $r\delta/s = n/s$ solutions (i_1,i_2) to this equation. The equations in (8.5) require n^2/s^2 solutions (i_1,i_2,j) to

$$d(i_1,c_1)+a((i_2-1)n+j,k_1) = z_1 \ , \quad d(i_1,c_2)+a((i_2-1)n+j,k_2) = z_2 \, . \quad (8.7)$$

The validity of this follows as in the previous case. If $k_1 \neq k_2$, any fixed value of i_1 results in $\delta n/s^2$ solutions (i_2,j). With r possible values for i_1, we get $r\delta n/s^2 = n^2/s^2$ solutions (i_1,i_2,j). If $k_1 = k_2$, and thus $c_1 \neq c_2$, we find there are r/s values of i_1 that satisfy

$$d(i_1,c_1) - d(i_1,c_2) = z_1 - z_2 \ .$$

For each value of i_1 there are $\delta n/s$ solutions (i_2,j) to

$$a((i_2-1)n+j,k_1) = z_1 - d(i_1,c_1) \ ,$$

again resulting in n^2/s^2 solutions for (i_1,i_2,j).

Finally, if condition (iii) holds, we will write $j \in \{1,\ldots,n\}$ as $j = (j_1 - 1)(n/\delta) + j_2$, with $j_1 \in \{1,\ldots,\delta\}$, $j_2 \in \{1,\ldots,n/\delta\}$. Then

$$f_{g_h}(i,j) = d((i-1)\delta + j_1, c_h) + a(j_2,k_h) \ .$$

Equation (8.3) requires that there are n/s pairs (j_1,j_2) such that

$$d((i_0-1)\delta + j_1, c_1) + a(j_2,k_1) = z_1 \ .$$

Since every fixed value of j_1 results in $\lambda s = n/(\delta s)$ solutions for j_2, there are indeed n/s solutions (j_1,j_2). Equation (8.4) now reads

$$d((i-1)\delta + j_{01}, c_1) = z_1 - a(j_{02},k_1) \ ,$$

which indeed has n/s solutions for i since the j_{01}-th rows in $D^{(1)}, \ldots, D^{(n)}$ form an array whose columns are uniform on $\mathcal{A}$. Finally, for (8.5) we need to count the solutions (i, j_1, j_2) to the equations

$$d((i-1)\delta + j_1, c_1) + a(j_2, k_1) = z_1 \;, \quad d((i-1)\delta + j_1, c_2) + a(j_2, k_2) = z_2 \,.$$

If $k_1 \neq k_2$, then for fixed i and j_1 there are $\lambda = n/(\delta s^2)$ solutions for j_2. Since there are $n\delta$ possible choices for (i, j_1), there are n^2/s^2 solutions (i, j_1, j_2) to these equations. If $k_1 = k_2$, there are $r/s = \delta n/s$ pairs (i, j_1) such that

$$d((i-1)\delta + j_1, c_1) - d((i-1)\delta + j_1, c_2) = z_1 - z_2 \,.$$

For each of these, there are $\lambda s = n/(\delta s)$ values of j_2 such that

$$a(j_2, k_1) = z_1 - d((i-1)\delta + j_1, c_1) \;,$$

hence in this case also there are n^2/s^2 solutions (i, j_1, j_2). ■

Case (i) of Theorem 8.12 may also be found in Street (1979). The following are some consequences of this theorem.

Example 8.13. In Chapter 7 we saw that a difference scheme $D(4u, 4u, 2)$ and an orthogonal array $OA(4u, 4u-1, 2, 2)$ exist if and only if a Hadamard matrix of order $4u$ exists. For any such value of u, Theorem 8.12 allows us to obtain a $POF(4u, 2, (4u-1)^2)$ which achieves equality in (8.2). The construction is illustrated in Tables 8.14 through 8.16 for the case $u = 1$. This example was first obtained by Federer (1977). ■

Table 8.14. The ingredients: a difference scheme $D(4,4,2)$ and an orthogonal array $OA(4,3,2,2)$ based on $GF(2)$.

$$D = \begin{bmatrix} 0 & 0 & 0 & 0 \\ 0 & 0 & 1 & 1 \\ 0 & 1 & 0 & 1 \\ 0 & 1 & 1 & 0 \end{bmatrix} , \qquad A = \begin{bmatrix} 0 & 0 & 0 \\ 0 & 1 & 1 \\ 1 & 0 & 1 \\ 1 & 1 & 0 \end{bmatrix} .$$

Table 8.15. Mixing the ingredients: $F = D^{(0)} \otimes A$.

$$F = \begin{bmatrix} 0&0&0&0&0&0&0&0&0 \\ 0&1&1&0&1&1&0&1&1 \\ 1&0&1&1&0&1&1&0&1 \\ 1&1&0&1&1&0&1&1&0 \\ 0&0&0&1&1&1&1&1&1 \\ 0&1&1&1&0&0&1&0&0 \\ 1&0&1&0&1&0&0&1&0 \\ 1&1&0&0&0&1&0&0&1 \\ 1&1&1&0&0&0&1&1&1 \\ 1&0&0&0&1&1&1&0&0 \\ 0&1&0&1&0&1&0&1&0 \\ 0&0&1&1&1&0&0&0&1 \\ 1&1&1&1&1&1&0&0&0 \\ 1&0&0&1&0&0&0&1&1 \\ 0&1&0&0&1&0&1&0&1 \\ 0&0&1&0&0&1&1&1&0 \end{bmatrix}$$

Table 8.16. The final result: a $POF(4, 2, 9)$

$$F_1 = \begin{bmatrix} 0&0&1&1 \\ 0&0&1&1 \\ 1&1&0&0 \\ 1&1&0&0 \end{bmatrix} \quad F_2 = \begin{bmatrix} 0&1&0&1 \\ 0&1&0&1 \\ 1&0&1&0 \\ 1&0&1&0 \end{bmatrix} \quad F_3 = \begin{bmatrix} 0&1&1&0 \\ 0&1&1&0 \\ 1&0&0&1 \\ 1&0&0&1 \end{bmatrix}$$

$$F_4 = \begin{bmatrix} 0&0&1&1 \\ 1&1&0&0 \\ 0&0&1&1 \\ 1&1&0&0 \end{bmatrix} \quad F_5 = \begin{bmatrix} 0&1&0&1 \\ 1&0&1&0 \\ 0&1&0&1 \\ 1&0&1&0 \end{bmatrix} \quad F_6 = \begin{bmatrix} 0&1&1&0 \\ 1&0&0&1 \\ 0&1&1&0 \\ 1&0&0&1 \end{bmatrix}$$

$$F_7 = \begin{bmatrix} 0&0&1&1 \\ 1&1&0&0 \\ 1&1&0&0 \\ 0&0&1&1 \end{bmatrix} \quad F_8 = \begin{bmatrix} 0&1&0&1 \\ 1&0&1&0 \\ 1&0&1&0 \\ 0&1&0&1 \end{bmatrix} \quad F_9 = \begin{bmatrix} 0&1&1&0 \\ 1&0&0&1 \\ 1&0&0&1 \\ 0&1&1&0 \end{bmatrix}$$

Example 8.17. Let p be a prime, let $u \geq 0$, $v \geq 1$ be integers, and let $z = u + 2v$. The existence of a difference scheme $D(p^z, p^z, p^v)$ follows from Theorem 6.6; the existence of an orthogonal array $OA(p^z, k, p^v, 2)$ with $k = p^u(p^{v(d+1)} - 1)/(p^{vd} - p^{v(d-1)}) + 1$, where $d = \lfloor u/v \rfloor$, follows from Theorem 6.28. Therefore, from Theorem 8.12, there exists a $POF(p^z, p^v, k(p^z - 1))$. If $u \equiv 0 \pmod v$, so that $u = dv$ and $k = (p^z - 1)/(p^v - 1)$, this implies the existence of a $POF(p^z, p^v, (p^z - 1)^2/(p^v - 1))$. The special case $u \equiv 0 \pmod v$ thus leads to a complete set of pairwise orthogonal F-squares, the existence of which was first observed by Hedayat, Raghavarao and Seiden (1975). ∎

Example 8.18. Let p be a prime, let u and v be integers such that $v \geq 1$, $0 \leq 2u \leq v$, and let $z = -u + 2v$. If $u = 0$, we use the result of Example 8.17. For $u \geq 1$, the existence of a difference scheme $D(p^{2(v-u)}, p^{2(v-u)}, p^v)$ and a completely resolvable orthogonal array $OA(p^{2v}, p^v, p^v, 2)$ follow from Theorems 6.6 and 6.19. Condition (ii) of Theorem 8.12 is satisfied, and we obtain a $POF(p^z, p^v, (p^{2(v-u)} - 1)p^v)$. It falls short of attaining the bound in (8.2). ■

Example 8.19. The existence of a difference scheme $D(2s^u, 2s^u, s)$ and an orthogonal array $OA(2s^u, 2(s^u - 1)/(s - 1) - 1, s, 2)$, where s is a prime power and $u \geq 2$ is an integer, follows from Corollary 6.39 and Theorem 6.40. From Theorem 8.12 we obtain $(2s^u - 1)(2(s^u - 1)/(s - 1) - 1)$ pairwise orthogonal $2s^u \times 2s^u$ F-squares based on s symbols. This result was also obtained by Street (1979), and improves a result of Mandeli, Lee and Federer (1981). ■

As examples 8.17–8.19 demonstrate, Theorem 8.12 may not always produce the maximal possible value of w for a $POF(n, s, w)$. A paper of Dillon, Ferragut and Gealy (1999) describes a construction that sometimes gives better results. These authors consider the construction of so-called *linear F-squares.* This method can be applied if n and s are powers of the same prime. Suppose $n = q^z$ and $s = q^v$, where q is a prime power and z and v are relatively prime. The idea is to fill an $n \times n$ square with the s elements of $GF(q)^v$ as follows. Label the rows and columns of the square by the elements of $GF(q)^z$, say $x_1, \ldots, x_n$, and, for selected $v \times z$ matrices A, B based on $GF(q)$, set the (i, j)th entry of the square equal to

$$Ax_i + Bx_j \ .$$

It follows easily that a square constructed in this way is an F-square if and only if rank $(A) =$ rank $(B) = v$.

Two F-squares constructed in this way, using pairs of matrices (A_1, B_1) and (A_2, B_2), are orthogonal F-squares if and only if

$$\text{rank} \begin{pmatrix} A_1 & B_1 \\ A_2 & B_2 \end{pmatrix} = 2v \ , \tag{8.8}$$

or equivalently if and only if the row spaces of the matrices $(A_1 \quad B_1)$ and $(A_2 \quad B_2)$ have only the zero vector in common.

For this construction to succeed one has to find sufficiently many pairs of $v \times z$ matrices (A_i, B_i) such that rank $(A_i) =$ rank $(B_i) = v$ and Equation 8.8 is satisfied for any two of the pairs. Dillon, Ferragut and Gealy (1999) give a method for finding such pairs of matrices. Their results imply among others the existence of sets of pairwise orthogonal F-squares with parameters $POF(8, 4, 15)$, $POF(32, 4, 319)$, $POF(32, 8, 127)$, $POF(27, 9, 80)$ and $POF(64, 16, 255)$. The 15 pairs of matrices that they use for a $POF(8, 4, 15)$ are shown in Table 8.20.

Table 8.20. Fifteen pairs of 2×3 matrices over $GF(2)$ that lead to a $POF(8,4,15)$.

$$\begin{pmatrix} 110\ 101 \\ 101\ 011 \end{pmatrix}, \begin{pmatrix} 110\ 011 \\ 011\ 101 \end{pmatrix}, \begin{pmatrix} 001\ 111 \\ 111\ 110 \end{pmatrix}$$

$$\begin{pmatrix} 001\ 110 \\ 110\ 111 \end{pmatrix}, \begin{pmatrix} 010\ 101 \\ 101\ 111 \end{pmatrix}, \begin{pmatrix} 010\ 111 \\ 111\ 101 \end{pmatrix}$$

$$\begin{pmatrix} 010\ 011 \\ 011\ 001 \end{pmatrix}, \begin{pmatrix} 010\ 001 \\ 001\ 011 \end{pmatrix}, \begin{pmatrix} 100\ 101 \\ 101\ 001 \end{pmatrix}$$

$$\begin{pmatrix} 100\ 111 \\ 111\ 011 \end{pmatrix}, \begin{pmatrix} 100\ 011 \\ 011\ 111 \end{pmatrix}, \begin{pmatrix} 010\ 110 \\ 110\ 100 \end{pmatrix}$$

$$\begin{pmatrix} 100\ 001 \\ 001\ 101 \end{pmatrix}, \begin{pmatrix} 010\ 100 \\ 100\ 110 \end{pmatrix}, \begin{pmatrix} 001\ 001 \\ 010\ 010 \end{pmatrix}$$

Table 8.21 summarizes the parameters of some pairwise orthogonal F-squares that can be obtained using the methods of Examples 8.13, 8.17–8.19, and Dillon, Ferragut and Gealy (1999), the latter being denoted by "DFG" in the table. We also include an example given by Finney (1982). Entries followed by a period meet the upper bound of Theorem 8.11, and so w is as large as possible for the given values of n and s.

Table 8.21. Parameters of pairwise orthogonal F-squares $POF(n,s,w)$.

n	s	w	Method	n	s	w	Method
4	2	9.	8.13 or 8.17	27	3	338.	8.17
8	2	49.	8.13 or 8.17	8	4	15	DFG
12	2	121.	8.13	16	4	75.	8.17
16	2	225.	8.13 or 8.17	32	4	319	DFG
20	2	361.	8.13	25	5	144.	8.17
24	2	529.	8.13	50	5	539	8.19
28	2	729.	8.13	49	7	384.	8.17
32	2	961.	8.13 or 8.17	32	8	127	DFG
6	3	8	Finney	27	9	80	DFG
9	3	32.	8.17	64	16	255	DFG
18	3	119	8.19				

See Suchower (1993, 1994, 1995) for further ideas on the construction of pairwise orthogonal F-squares.

Research Problem 8.22. Find other constructions and nonexistence results for $POF(n,s,w)$'s. Determine the maximal possible value of w for the entries in Table 8.21 not followed by a period. Extend the table.

8.3 Construction of Orthogonal Arrays from Pairwise Orthogonal Latin Squares or F-Squares

There is a simple process for converting pairwise orthogonal Latin squares or F-squares into orthogonal arrays of strength 2.

Theorem 8.23. *The existence of k_1 pairwise orthogonal $n \times n$ F-squares based on s symbols, i.e. a $POF(n, s, k_1)$, implies the existence of an orthogonal array $OA(n^2, k_1 + 2, s, 2)$. If in addition an $OA(n, k_2, s, 2)$ exists then an orthogonal array $OA(n^2, k_1 + 2k_2, s, 2)$ can be constructed.*

Proof:
CONSTRUCTION: Convert each of the k_1 F-squares to an $n^2 \times 1$ array by juxtaposing the n rows of the square and transposing. Combine these arrays to form an $n^2 \times k_1$ array. Add two more columns to this array, using the symbols $\{1, \ldots, n\}$, so that the pair (i, j), $i, j \in \{1, \ldots, n\}$, is appended to that row whose entries in the first k_1 columns correspond to the symbols in cell (i, j) of the F-squares. Call the resulting $n^2 \times (k_1 + 2)$ array A^*. If an $OA(n, k_2, s, 2)$ exists, denote it by B. Otherwise, set $k_2 = 1$ and let B denote an $n \times 1$ array based on s symbols such that each symbol occurs n/s times. Now construct an $n^2 \times (k_1 + 2k_2)$ array A by starting with A^* and by replacing each symbol $i \in \{1, \ldots, n\}$ in the last two columns by the i-th row of the $n \times k_2$ array B. This gives the desired $OA(n^2, k_1 + 2k_2, s, 2)$.

VERIFICATION: Let $F_g = (f_g(i, j))$, $g = 1, \ldots, k_1$, $i, j = 1, \ldots, n$, be the F-squares in a $POF(n, s, k_1)$. With B as above, write $B = (b(u, v))$, $u = 1, \ldots, n$, $v = 1, \ldots, k_2$. Denote the entries of A by $a(\ell, m)$, $\ell = 1, \ldots, n^2$, $m = 1, \ldots, k_1 + 2k_2$. Without loss of generality, A is based on $\{0, 1, \ldots, s - 1\}$. For $\ell \in \{1, \ldots, n^2\}$ we will write $\ell = \ell_2 + n(\ell_1 - 1)$, where $\ell_1, \ell_2 \in \{1, \ldots, n\}$. The entries of A can now be expressed as

$$a(\ell, m) = \begin{cases} f_m(\ell_1, \ell_2), & \text{if } 1 \le m \le k_1 \,, \\ b(\ell_1, m - k_1), & \text{if } k_1 + 1 \le m \le k_1 + k_2 \,, \\ b(\ell_2, m - k_1 - k_2), & \text{if } k_1 + k_2 + 1 \le m \le k_1 + 2k_2 \,. \end{cases}$$

For $m < m'$, $m, m' \in \{1, \ldots, k_1 + 2k_2\}$ and $z_1, z_2 \in \{0, 1, \ldots, s - 1\}$, we must show that there are n^2/s^2 values of ℓ such that simultaneously $a(\ell, m) = z_1$ and $a(\ell, m') = z_2$. If $1 \le m < m' \le k_1$ this follows since F_m and $F_{m'}$ are orthogonal F-squares. If $1 \le m \le k_1$ and $k_1 + 1 \le m' \le k_1 + k_2$, then for every fixed value of ℓ_1 there are exactly n/s values of ℓ_2 such that $f_m(\ell_1, \ell_2) = z_1$. Since there are also n/s values of ℓ_1 such that $b(\ell_1, m' - k_1) = z_2$, the result follows. An analogous argument applies if $1 \le m \le k_1$ and $k_1 + k_2 + 1 \le m' \le k_1 + 2k_2$. If $k_1 + 1 \le m < m' \le k_1 + k_2$, a case that need be considered only if $k_2 \ge 2$, there are n/s^2 values of ℓ_1 that simultaneously satisfy $b(\ell_1, m - k_1) = z_1$ and

$b(\ell_1, m' - k_1) = z_2$. This is so because B is an orthogonal array of strength 2 and index n/s^2. Since there are no restrictions on ℓ_2, there are n^2/s^2 solutions for ℓ. The same argument, with the roles of ℓ_1 and ℓ_2 reversed, can be used if $k_1 + k_2 + 1 \leq m < m' \leq k_1 + 2k_2$. Finally, if $k_1 + 1 \leq m \leq k_1 + k_2$ and $k_1+k_2+1 \leq m' \leq k_1+2k_2$, there are n/s values of ℓ_1 such that $b(\ell_1, m-k_1) = z_1$ and n/s values of ℓ_2 such that $b(\ell_2, m' - k_1 - k_2) = z_2$, again combining to give n^2/s^2 solutions for ℓ. ∎

If $n \neq s$ the array A^* is an example of a *mixed orthogonal array*. We return to this subject in Chapter 9.

Since a $POF(s, s, k)$ is simply a $POL(s, k)$, we have:

Corollary 8.24. *If a $POL(s, k)$ exists then so does an $OA(s^2, k+2, s, 2)$.*

In particular, if s is a prime power then an $OA(s^2, s+1, s, 2)$ exists (of course we have already seen arrays with these parameters from the Rao-Hamming construction).

Example 8.25. Consider the $POF(4, 2, 9)$ in Table 8.16. The array A^* used in the proof of Theorem 8.23 is given in Table 8.26. Now take B to be the orthogonal array $OA(4, 3, 2, 2)$ in Table 2.14. Replacing the symbols in the last two columns of A^* by the corresponding rows in B results in the orthogonal array $OA(16, 15, 2, 2)$ exhibited in Table 8.27. It follows from Corollary 2.5 that this array has the maximal number of factors. In fact we saw in the previous chapter that any $OA(16, 15, 2, 2)$ can also be obtained from a Hadamard matrix of order 16, and indeed there are exactly five inequivalent arrays with these parameters (see Table 7.24). It should be observed that the first nine factors of the $OA(16, 15, 2, 2)$ in Table 8.27 are exactly the nine factors corresponding to the array in Table 8.15. The remaining six factors come from the $OA(4, 3, 2, 2)$, as described in the proof of Theorem 8.23. ∎

If $n = s$ in Theorem 8.23, and we start with a $POL(s, k_1)$, then the array A^* in the construction is itself an $OA(s^2, k_1 + 2, s, 2)$. In that case the last two rows of A^* are already based on s symbols, and use of the array B can be ignored.

Table 8.26. The array A^* induced by a $POF(4, 2, 9)$.

0	0	0	0	0	0	0	0	0	1	1
0	1	1	0	1	1	0	1	1	1	2
1	0	1	1	0	1	1	0	1	1	3
1	1	0	1	1	0	1	1	0	1	4
0	0	0	1	1	1	1	1	1	2	1
0	1	1	1	0	0	1	0	0	2	2
1	0	1	0	1	0	0	1	0	2	3
1	1	0	0	0	1	0	0	1	2	4
1	1	1	0	0	0	1	1	1	3	1
1	0	0	0	1	1	1	0	0	3	2
0	1	0	1	0	1	0	1	0	3	3
0	0	1	1	1	0	0	0	1	3	4
1	1	1	1	1	1	0	0	0	4	1
1	0	0	1	0	0	0	1	1	4	2
0	1	0	0	1	0	1	0	1	4	3
0	0	1	0	0	1	1	1	0	4	4

Table 8.27. An $OA(16, 15, 2, 2)$.

0	0	0	0	0	0	0	0	0	1	0	0	1	0	0
0	1	1	0	1	1	0	1	1	1	0	0	0	1	0
1	0	1	1	0	1	1	0	1	1	0	0	0	0	1
1	1	0	1	1	0	1	1	0	1	0	0	1	1	1
0	0	0	1	1	1	1	1	1	0	1	0	1	0	0
0	1	1	1	0	0	1	0	0	0	1	0	0	1	0
1	0	1	0	1	0	0	1	0	0	1	0	0	0	1
1	1	0	0	0	1	0	0	1	0	1	0	1	1	1
1	1	1	0	0	0	1	1	1	0	0	1	1	0	0
1	0	0	0	1	1	1	0	0	0	0	1	0	1	0
0	1	0	1	0	1	0	1	0	0	0	1	0	0	1
0	0	1	1	1	0	0	0	1	0	0	1	1	1	1
1	1	1	1	1	1	0	0	0	1	1	1	1	0	0
1	0	0	1	0	0	0	1	1	1	1	1	0	1	0
0	1	0	0	1	0	1	0	1	1	1	1	0	0	1
0	0	1	0	0	1	1	1	0	1	1	1	1	1	1

If $k_1 = (n-1)^2/(s-1)$ in Theorem 8.23, the maximal value of k_1 according to Theorem 8.11, and if $k_2 = (n-1)/(s-1)$, its maximal value according to Corollary 2.5, then the number of factors in the $OA(n^2, k_1 + 2k_2, s, 2)$ equals $(n^2-1)/(s-1)$, which is again the maximal possible value according to Corollary 2.5. Example 8.25 illustrates this situation. As a second illustration, if we start with a $POL(s, s-1)$, so that $k_1 = s-1$ attains its maximal value, then the resulting array A^* is an $OA(s^2, s+1, s, 2)$, which has the maximal number of factors according to Corollary 2.5.

Although it is not of primary interest for those who are concerned with the construction of orthogonal arrays, we should like to point out that the converse of Theorem 8.23, in other words the construction of pairwise orthogonal $n \times n$ F-squares from a given $OA(n^2, k, s, 2)$, is not so straightforward. However, a partial converse is provided by the following result.

Theorem 8.28. *An orthogonal array $OA(s^2, k+2, s, 2)$, $k \geq 2$, exists if and only if k pairwise orthogonal Latin squares of order s exist.*

Proof: That the existence of a $POL(s, k)$ is sufficient for the existence of the orthogonal array follows from Theorem 8.23. Thus it suffices to show that an $OA(s^2, k+2, s, 2)$ can be used to construct a $POL(s, k)$.

CONSTRUCTION: Let $A = (a(\ell, m))$, $\ell = 1, \ldots, s^2$, $m = 1, \ldots, k+2$ be an $OA(s^2, k+2, s, 2)$ based on $S = \{0, \ldots, s-1\}$. For $g = 1, \ldots, k$, define the squares $L_g = (L_g(i,j))$, $i, j = 0, \ldots, s-1$, by

$$L_g(i,j) = a(\ell(i,j), g) ,$$

where $\ell(i,j)$ is the unique value of ℓ that satisfies $a(\ell, k+1) = i$ and $a(\ell, k+2) = j$. Then $L_1, \ldots, L_k$ form a $POL(s, k)$.

VERIFICATION: First, the uniqueness of the solution to the preceding pair of equations follows because columns $k+1$ and $k+2$ of A form an orthogonal array of strength 2 and index unity. Thus the squares L_g are well defined. For a fixed $g_1 \in \{1, \ldots, k\}$ and a fixed $i_1 \in \{0, \ldots, s-1\}$, there is for each $z_1 \in \{0, \ldots, s-1\}$ a unique value of j such that

$$a(\ell(i_1, j), g_1) = z_1 ,$$

since columns g_1 and $k+1$ of A also form an orthogonal array of strength 2 and index unity. Similarly, by considering columns g_1 and $k+2$ of A, it follows that for every $g_1 \in \{1, \ldots, k\}$ and every $j_1 \in \{0, \ldots, s-1\}$ there is for each $z_1 \in \{0, \ldots, s-1\}$ a unique value of i such that

$$a(\ell(i, j_1), g_1) = z_1 .$$

This establishes that L_{g_1} is a Latin square of order s for any $g_1 \in \{1, \ldots, k\}$. Finally, to verify the orthogonality of any pair of squares, fix $g_1, g_2 \in \{1, \ldots, k\}$ with $g_1 \neq g_2$. For fixed $z_1, z_2 \in \{0, \ldots, s-1\}$, there is a unique pair (i, j), $i, j \in \{0, \ldots, s-1\}$, that satisfies

$$a(\ell(i,j), g_1) = z_1 \quad \text{and} \quad a(\ell(i,j), g_2) = z_2$$

since columns g_1 and g_2 of A form an orthogonal array of strength 2 and index unity. ∎

Example 8.29. In Table 6.24 we exhibited an $OA(16,5,4,2)$, which has been reproduced in Table 8.30 for ease of reference and which we will refer to as A. Using the last two rows of this $OA(16,5,4,2)$ as a frame of reference to construct L_1, L_2 and L_3 from rows 1,2 and 3, respectively, results in the $POL(4,3)$ displayed in Table 8.31.

Table 8.30. An $OA(16,5,4,2)$ (transposed).

0	0	0	0	1	1	1	1	2	2	2	2	3	3	3	3
0	1	2	3	1	0	3	2	2	3	0	1	3	2	1	0
0	2	3	1	1	3	2	0	2	0	1	3	3	1	0	2
0	3	1	2	1	2	0	3	2	1	3	0	3	0	2	1
0	1	2	3	0	1	2	3	0	1	2	3	0	1	2	3

Table 8.31. A $POL(4,3)$.

$$L_1: \begin{array}{cccc} 0 & 3 & 1 & 2 \\ 1 & 2 & 0 & 3 \\ 2 & 1 & 3 & 0 \\ 3 & 0 & 2 & 1 \end{array} \qquad L_2: \begin{array}{cccc} 0 & 2 & 3 & 1 \\ 1 & 3 & 2 & 0 \\ 2 & 0 & 1 & 3 \\ 3 & 1 & 0 & 2 \end{array} \qquad L_3: \begin{array}{cccc} 0 & 1 & 2 & 3 \\ 1 & 0 & 3 & 2 \\ 2 & 3 & 0 & 1 \\ 3 & 2 & 1 & 0 \end{array}$$

As an example of how the entries of the Latin squares are obtained, consider cell $(2,1)$. Since the pair $(2,1)$ occurs in the last two columns in row 6 of the $OA(16,5,4,2)$, we have $m(2,1) = 6$. Hence $L_1(2,1) = a(6,1) = 1$, $L_2(2,1) = a(6,2) = 0$ and $L_3(2,1) = a(6,3) = 3$. Note that any two columns of A could play the role now played by the last two columns; each of the remaining three columns of A would then be used to generate a square of a $POL(4,3)$. It should also be noted that the construction of the squares can be greatly simplified by replacing A by a statistically equivalent $OA(16,5,4,2)$ which has the pairs in its last two columns in lexicographical order $(0,0),(0,1)\ldots,(3,3)$. This array, obtained from A by permuting columns, is displayed in Table 8.32.

Table 8.32. An $OA(16,5,4,2)$ with pairs in its last two columns in lexicographical order (transposed).

0	3	1	2	1	2	0	3	2	1	3	0	3	0	2	1
0	2	3	1	1	3	2	0	2	0	1	3	3	1	0	2
0	1	2	3	1	0	3	2	2	3	0	1	3	2	1	0
0	0	0	0	1	1	1	1	2	2	2	2	3	3	3	3
0	1	2	3	0	1	2	3	0	1	2	3	0	1	2	3

The squares L_1, L_2 and L_3 are easily obtained from this new $OA(16, 5, 4, 2)$; for example, the first four elements in the g-th column, for $g = 1, 2, 3$, give the first row of L_g, the next four elements give the second row in L_g, and so on. Observe that it would have been easier to use the first and last columns of the $OA(16, 5, 4, 2)$ in Table 8.30 as the frame of reference for the construction of a $POL(4, 3)$, since these have the 16 pairs already in lexicographical order. Writing columns 2, 3 and 4 of this $OA(16, 5, 4, 2)$ in 4×4 squares, as just illustrated, gives the simplest construction for a $POL(4, 3)$. ∎

Using a given $OA(n^2, k, s, 2)$ to generate a $POF(n, s, k')$, for some k' that is preferably not too much smaller than k, is a problem that is not as well understood. Unless a certain structure can be identified in the $OA(n^2, k, s, 2)$, it is not clear how to generate the greatest number of pairwise orthogonal F-squares from a given orthogonal array. This issue will also be addressed in Problem 8.11.

Finally, now that we understand the relation between a $POL(s, k+2)$ and an $OA(s^2, k, s, 2)$, we are ready to reconsider an issue raised in Section 8.1 about embedding a $POL(s, s-3)$ into a complete $POL(s, s-1)$ by adding two Latin squares.

Theorem 8.33. *If $s \geq 5$, then any set of $s - 3$ pairwise orthogonal Latin squares of order s can be embedded into a complete set of $s-1$ pairwise orthogonal Latin squares of order s by adding two Latin squares.*

Proof:

CONSTRUCTION: Let $L_1, \ldots, L_{s-3}$ be $s \times s$ Latin squares based on $S = \{0, \ldots, s-1\}$ that form a $POL(s, s-3)$. Without loss of generality, we may assume the first row of each square is in the natural order $0, 1, \ldots, s-1$. Using the construction in the proof of Theorem 8.23 we can convert these squares to an $OA(s^2, s-1, s, 2)$, part of which is exhibited in Table 8.34. Number the runs of the array as indicated in that table.

Table 8.34. Skeleton of an $OA(s^2, s-1, s, 2)$ (transposed).

	Runs										
Factors	1	2	$\cdots$	s	$s+1$	$\cdots$	$2s$	$\cdots$	$(s-1)s+1$	$\cdots$	s^2
1	0	1	$\cdots$	$s-1$							
$\vdots$			$\vdots$								
$s-3$	0	1	$\cdots$	$s-1$							
$s-2$	0	0	$\cdots$	0	1	$1\cdots$	1	$\cdots$	$s-1$	$s-1\cdots$	$s-1$
$s-1$	0	1	$\cdots$	$s-1$	0	$1\cdots$	$s-1$	$\cdots$	0	$s-1\cdots$	$s-1$

From Theorem 8.28 we know that it suffices to show that we can embed this $OA(s^2, s-1, s, 2)$ in an $OA(s^2, s+1, s, 2)$ by adding two factors. To accomplish this, we will first fill an $s \times s$ square with the numbers $1, 2, \ldots, s^2$, in a way to be explained in the next paragraph, using each number once. With rows and columns of the square labeled $0, 1, \ldots, s-1$, if $j \in \{1, \ldots, s^2\}$ is the number that appears in cell (i_1, i_2) of the square, where $i_1, i_2 \in \{0, \ldots, s-1\}$, then we append the ordered pair (i_1, i_2) to run j of the $OA(s^2, s-1, s, 2)$. The resulting $(s+1) \times s^2$ array will be an $OA(s^2, s+1, s, 2)$.

The $s \times s$ square is to be filled with the run numbers $1, \ldots, s^2$ such that any two runs of the $OA(s^2, s-1, s, 2)$ that correspond to two run numbers assigned to cells in the same row or column of the square have no factor at a common level. To achieve this, first enter $1, 2, \ldots, s$ along the main diagonal of the square. For any pair of runs (j_1, j_2) from the first s runs, $1 \leq j_1 < j_2 \leq s$, there are exactly two runs among the last $s(s-1)$ runs of the $OA(s^2, s-1, s, 2)$ that have no factor at a common level with run j_1 nor with run j_2 (Problem 8.12). Let $m_1(j_1, j_2)$ and $m_2(j_1, j_2)$ denote these two runs. Enter $m_1(1, 2)$ in cell $(0, 1)$ of the square and $m_2(1, 2)$ in cell $(1, 0)$. If $m_1(1, 3)$ has no factor at a common level with $m_1(1, 2)$, then place $m_1(1, 3)$ in cell $(0, 2)$ and $m_2(1, 3)$ in cell $(2, 0)$; otherwise place $m_1(1, 3)$ in cell $(2, 0)$ and $m_2(1, 3)$ in cell $(0, 2)$. Continue like this, filling first the first row and column of the square and then the remaining entries. Using the square to construct an $OA(s^2, s+1, s, 2)$ as described in the previous paragraph, and using the last two columns of the $OA(s^2, s-1, s, 2)$ as a frame of reference, this $OA(s^2, s+1, s, 2)$ can now be converted to a $POL(s, s-1)$ that contains the $s-3$ squares of the original $POL(s, s-3)$.

VERIFICATION: In the $s \times s$ square we have constructed any two run numbers are in the same row or column if and only if the corresponding runs in the $OA(s^2, s-1, s, 2)$ have no factor at a common level. The validity of this statement is based on a result pertaining to L_2 association schemes and is not further pursued here. (See also Shrikhande, 1961.) We will only verify that this property of the square implies that the new $(s+1) \times s^2$ array is indeed an $OA(s^2, s+1, s, 2)$. Since it is obvious from the construction that the two new factors contain every possible level combination exactly once, it suffices to show that the same is true for one factor from the $OA(s^2, s-1, s, 2)$ and one new factor. If it is not true, then some level combination, say (i_1, i_2), must appear twice. This means that either row i_2 or column i_2 in the $s \times s$ square contains two run numbers corresponding to runs in the $OA(s^2, s-1, s, 2)$ that have at least one factor at a common level. But this contradicts the properties of the $s \times s$ square. ∎

Example 8.35. Consider the $POL(5, 2)$ shown in Table 8.36. The resulting $OA(25, 4, 5, 2)$, fleshing out the skeleton in Table 8.34, is exhibited in Table 8.37. The functions $m_1(j_1, j_2)$ and $m_2(j_1, j_2)$ defined in the proof of Theorem 8.33 are presented in Table 8.38. The resulting 5×5 square is shown in Table 8.39. This square and the $OA(25, 4, 5, 2)$ produce the $OA(25, 6, 5, 2)$ given in Table 8.40, which in turn induces the $POL(5, 4)$ in Table 8.41. ∎

Table 8.36. Two orthogonal Latin squares of order 5.

0	1	2	3	4		0	1	2	3	4
1	2	3	4	0		2	3	4	0	1
2	3	4	0	1		4	0	1	2	3
3	4	0	1	2		1	2	3	4	0
4	0	1	2	3		3	4	0	1	2

Table 8.37. An $OA(25, 4, 5, 2)$ (transposed and split into two pieces).

Runs

1	2	3	4	5	6	7	8	9	10	11	12	13
0	1	2	3	4	1	2	3	4	0	2	3	4
0	1	2	3	4	2	3	4	0	1	4	0	1
0	0	0	0	0	1	1	1	1	1	2	2	2
0	1	2	3	4	0	1	2	3	4	0	1	2

14	15	16	17	18	19	20	21	22	23	24	25
0	1	3	4	0	1	2	4	0	1	2	3
2	3	1	2	3	4	0	3	4	0	1	2
2	2	3	3	3	3	3	4	4	4	4	4
3	4	0	1	2	3	4	0	1	2	3	4

Table 8.38. Ingredients for a 5×5 square.

j_1	j_2	$m_1(j_1, j_2)$	$m_2(j_1, j_2)$
1	2	8	25
1	3	15	19
1	4	13	17
1	5	7	24
2	3	9	21
2	4	11	20
2	5	14	18
3	4	10	22
3	5	12	16
4	5	6	23

Table 8.39. The 5×5 square.

	0	1	2	3	4
0	1	8	15	17	24
1	25	2	9	11	18
2	19	21	3	10	12
3	13	20	22	4	6
4	7	14	16	23	5

Table 8.40. An $OA(25, 6, 5, 2)$ that contains the $OA(25, 4, 5, 2)$ in Table 8.37 (transposed and split into two pieces).

Runs

1	2	3	4	5	6	7	8	9	10	11	12	13
0	1	2	3	4	1	2	3	4	0	2	3	4
0	1	2	3	4	2	3	4	0	1	4	0	1
0	0	0	0	0	1	1	1	1	1	2	2	2
0	1	2	3	4	0	1	2	3	4	0	1	2
0	1	2	3	4	3	4	0	1	2	1	2	3
0	1	2	3	4	4	0	1	2	3	3	4	0

14	15	16	17	18	19	20	21	22	23	24	25
0	1	3	4	0	1	2	4	0	1	2	3
2	3	1	2	3	4	0	3	4	0	1	2
2	2	3	3	3	3	3	4	4	4	4	4
3	4	0	1	2	3	4	0	1	2	3	4
4	0	4	0	1	2	3	2	3	4	0	1
1	2	2	3	4	0	1	1	2	3	4	0

Table 8.41. Four pairwise orthogonal Latin squares of order 5 that include the squares in Table 8.36.

0	1	2	3	4		0	1	2	3	4		0	1	2	3	4		0	1	2	3	4
1	2	3	4	0		2	3	4	0	1		3	4	0	1	2		4	0	1	2	3
2	3	4	0	1		4	0	1	2	3		1	2	3	4	0		3	4	0	1	2
3	4	0	1	2		1	2	3	4	0		4	0	1	2	3		2	3	4	0	1
4	0	1	2	3		3	4	0	1	2		2	3	4	0	1		1	2	3	4	0

8.4 Concluding Remarks

F-squares and pairwise orthogonal F-squares, with Latin squares and pairwise orthogonal Latin squares as special cases, present an interesting and challenging area of research. However, although the connections between orthogonal arrays and pairwise orthogonal F-squares are fascinating, it is debatable how important these connections are for the construction of new orthogonal arrays. In previous chapters where various types of combinatorial structures were used to construct orthogonal arrays, these structures were typically smaller than the arrays and their construction was perceived as a more tractable problem than the direct construction of the orthogonal arrays. This reduction in complexity is no longer apparent for the construction discussed in this chapter. After all, each of the F-squares presents the levels for a factor in an orthogonal array, merely displaying them in a square instead of a vector.

We have already mentioned (in Research Problem 8.5) the problem of deciding if a $POL(10,3)$ exists. At the other extreme, Lam, Thiel and Swiercz (1989) have given a computer-assisted proof of the nonexistence of a $POL(10,9)$. (An earlier attack on the nonexistence of a $POL(10,9)$ was made by MacWilliams, Sloane and Thompson, 1973, using error-correcting codes.) In view of Theorem 8.33, this implies that $w \leq 6$ in a $POL(10,w)$.

Research Problem 8.42. Find a proof that no set of nine pairwise orthogonal Latin squares of order 10 exists that does not depend on a computer search.

The papers of MacWilliams, Sloane and Thompson and Lam, Thiel and Swiercz actually consider a geometrical formulation of the same problem. A *projective plane of order* n is a set of n^2+n+1 "points" and a set of n^2+n+1 "lines" with an "incidence" relation between them with the properties that (a) every point is incident with $n+1$ lines and (b) every line is incident with $n+1$ points (see Hughes and Piper, 1973). The connection with pairwise orthogonal Latin squares arises because of the following result.

Theorem 8.43. *If any one of the following exists then they all do:*

(a) a projective plane of order n,

(b) a $POL(n, n-1)$,

(c) an $OA(n^2, n+1, n, 2)$.

The equivalence of (a) and (b) is due to Bose (1938). For a proof see for example Ryser (1963), page 92. Certainly this theorem shows that determining the existence of orthogonal arrays is a nontrivial problem. The Lam, Thiel and Swiercz (1989) result is:

Theorem 8.44. *There is no projective plane of order 10.*

At the present time it is known that projective planes of order n exist whenever n is a prime power, and no example is known in which n is not a prime power. Whether such planes ever exist is another basic unsolved question of discrete mathematics, which we emphasize by stating:

Research Problem 8.45. Construct a projective plane of order 12 or 18, or prove that no such planes exist.

Research Problem 8.46. *Finite projective planes.* Determine for which orders n there exist a projective plane of order n. On the basis of our limited present knowledge, some people guess that the answer is that n must be a prime power. This problem has been proposed as a candidate for the next "Fermat's Last Theorem" (Mullen, 1995).

For enumerating the Latin squares of a given order, the concept of a reduced Latin square plays an important role. A Latin square of order s is said to be *reduced* (or *normalized*) if its first row and column are both in the natural order $0, 1, \ldots, s-1$. It is not hard to show (see Problem 8.13) that the number of Latin squares of order s is equal to the number of reduced Latin squares of order s times $s!(s-1)!$. Table 8.47 gives the numbers of reduced Latin squares of order s, $R(s)$, and the number of inequivalent squares, $I(s)$, for the first few values of s. The value of $I(8)$ was found by Kolesova, Lam and Thiel (1990), and $R(10)$ by McKay and Rogoyski (1995). (Incorrect versions of this table have been published.) For the most up-to-date information about these two sequences see Sloane (1999).

Table 8.47. The numbers of reduced and inequivalent Latin squares of order s.

s	Reduced squares $R(s)$	Inequivalent squares $I(s)$
2	1	1
3	1	1
4	4	2
5	56	2
6	9,408	22
7	16,942,080	564
8	535,281,401,856	1,676,257
9	377,597,570,964,258,816	??
10	7,580,721,483,160,132,811,489,280	??

As can be seen, these numbers grow very rapidly. There does not seem to be a good asymptotic estimate known for either sequence, but a crude estimate is given by

$$\frac{1}{s^2} \log R(s) \sim \log s \ .$$

Research Problem 8.48. Find asymptotic estimates for $R(s)$ and $I(s)$. (Compare Godsil and McKay, 1990.)

The enumeration of all Latin squares of a given order — provided the order is not too large — is of interest for statistical applications; it enables the random selection of a square from all possible squares of the specified size. For details see Hinkelmann and Kempthorne (1994). For larger orders there are too many squares for a complete enumeration to be useful. As a substitute we are satisfied with a process which generates Latin squares at random. An algorithm which generates Latin squares approximately at random has been described by Jacobson and Mattheus (1996).

Research Problem 8.49. Find a polynomial-time algorithm that generates Latin squares of order s at random.

For $s < 10,000$ Abel, Brouwer, Colbourn and Dinitz (1996) present a table with the largest known value for w in a $POL(s, w)$. Table 8.50 gives these numbers for the values of $s \leq 100$ that are not prime powers.

Table 8.50. Largest known value for w in a $POL(s, w)$.

s	w	s	w	s	w	s	w
6	1	36	5	58	5	82	8
10	2	38	4	60	4	84	6
12	5	39	4	62	4	85	6
14	3	40	7	63	6	86	6
15	4	42	5	65	7	87	6
18	3	44	5	66	5	88	7
20	4	45	6	68	5	90	6
21	5	46	4	69	6	91	7
22	3	48	5	70	6	92	6
24	5	50	6	72	7	93	6
26	4	51	5	74	5	94	6
28	5	52	5	75	5	95	6
30	4	54	4	76	6	96	7
33	5	55	5	77	6	98	6
34	4	56	7	78	6	99	8
35	5	57	7	80	9	100	8

The results for $s = 6$ and 10 in Table 8.50 have already been discussed in this chapter. The existence of an orthogonal array $OA(144, 7, 12, 2)$, and thus a $POL(12, 5)$, follows from the difference scheme $D(12, 6, 12)$ in Table 6.11. Example 6.51 implies the existence of a $POL(15, 4)$. Some of the other results in Table 8.50, namely those for $s = 60$, 63, 66, 72, 77, 88 and 99, follow either directly from Theorem 8.4 or, in the case of $s = 60$, from an obvious extension of the theorem using the existence of a $POL(12, 5)$ and a $POL(5, 4)$.

Research Problem 8.51. For the entries in Table 8.50, find better lower bounds, or prove that the entries shown cannot be improved. At present only the first entry in the table is known to be the best possible value.

A *Latin square of order s with a subsquare of order* s_0 is an $s \times s$ Latin square that contains an $s_0 \times s_0$ Latin square, based on a subset of cardinality

s_0 of the symbols, as a subsquare. The rows and columns of the subsquare need not be contiguous in the original square. Table 8.52 shows a square of order 6 with a subsquare of order 2.

Table 8.52. A Latin square of order 6 with a subsquare of order 2.

5	6	3	4	1	2
2	1	6	5	3	4
6	5	1	2	4	3
4	3	5	6	2	1
1	4	2	3	5	6
3	2	4	1	6	5

The subsquare identified in Table 8.52 is only one of several subsquares of order 2 in this Latin square.

An *incomplete Latin square of order s with a hole of order* s_0 is an (incomplete) square obtained from a Latin square of order s with a subsquare of order s_0 by deleting the subsquare. If $S_0 \subset S = \{0, 1, \ldots, s-1\}$ denotes the set of symbols for the deleted subsquare, we will say that a set of w incomplete Latin squares of order s with holes of order s_0 (appearing at the same place in each of the squares) are pairwise orthogonal if, for any two of the incomplete squares, when one is superimposed on the other the ordered pairs of corresponding entries consist of all $s^2 - s_0^2$ pairs from $(S \times S) \setminus (S_0 \times S_0)$. We denote such a set by $IPOL(s, s_0, w)$. An $IPOL(6, 2, 2)$ based on the square in Table 8.52 is shown in Table 8.53. The first such pair appeared in Euler (1782).

Table 8.53. An $IPOL(6, 2, 2)$.

5	6	3	4	1	2		1	2	5	6	3	4
2	1	6	5	3	4		6	5	1	2	4	3
6	5	1	2	4	3		4	3	6	5	1	2
4	3	5	6	2	1		5	6	4	3	2	1
1	4	2	3				2	4	3	1		
3	2	4	1				3	1	2	4		

The "almost" $POL(10, 4)$ found by Brouwer (1984), alluded to in Section 8.1, is actually an $IPOL(10, 4, 2)$. A necessary condition for the existence of an $IPOL(s, s_0, w)$ is $s \geq (w+1)s_0$ (see Problem 8.14). Abel, Colbourn and Dinitz (1996) give a table for $1 \leq s \leq 1000$, $1 \leq s_0 \leq 50$ with the largest known value of w in an $IPOL(s, s_0, w)$.

Maurin (1985, 1996) and Hedayat and Stufken (1986, 1992) considered the closely related concept of an incomplete orthogonal array. An *incomplete orthogonal array* $IOA(N, k, (s, s_0), t)$ is an $N \times k$ array in which, using the above notation, in any $N \times t$ subarray no row has all its entries from S_0 while all other ordered t-tuples appear equally often as a row, namely $N/(s^t - s_0^t)$ times. It can be shown (Problem 8.15) that such an array must satisfy $s \geq (k-t+1)s_0$; for $t = 2$ this gives the preceding necessary condition for the existence of an $IPOL(s, s_0, w)$ via the obvious relation between an $IPOL$ and an IOA.

8.5 Problems

8.1. Show that the Latin square in Table 8.1 is orthogonally isolated.

8.2. To generalize the result in Problem 8.1, define $L = (\ell_{ij})$, where $\ell_{ij} = i+j$, $i, j \in \{0, 1, \ldots, s-1\}$, with addition modulo s.

a. Show that L is a Latin square of order s.

b. Show that L is not isolated if s is odd.

c. Show that L does not possess any transversals if s is even. (As a consequence, L is isolated if s is even.)

8.3. Let L be an $s \times s$ Latin square based on the symbols $\{0, 1, \ldots, s-1\}$. We say that L has a Latin subsquare of order u if there are u rows and u columns in L such that the $u \times u$ subsquare induced by these rows and columns is a Latin square based on a subset of size u of $\{0, 1, \ldots, s-1\}$.

a. Verify that the 4×4 Latin square in Table 8.1 has a subsquare of order 2.

b. Provide an example of a 5×5 Latin square with a subsquare of order 2.

c. Show that if a Latin square of order s has a subsquare of order u, then $u \leq \lfloor s/2 \rfloor$.

d. Show that if a Latin square of order $s = 4m + 2$ has a subsquare of order $u = 2m + 1$, it has not even one transversal and is therefore isolated.

8.4. a. Prove the following statement or provide a counterexample: If L_1 and L_2 form a $POL(s, 2)$, then L_1 and L_2 must be inequivalent Latin squares.

b. Show that if L_1 and L_2 are two equivalent Latin squares then L_1 has a set of n parallel transversals if and only if L_2 has a set of n parallel transversals.

8.5. Show that there are four reduced and two inequivalent Latin squares of order 4, as claimed in Table 8.47.

8.6. Give a simple construction to show how a $POL(s, s-2)$ can be extended to a $POL(s, s-1)$.

8.7. Show that the squares $L_1, \ldots, L_{s_0-1}$ in the proof of Theorem 8.4 form a $POL(s, s_0 - 1)$, and use this result to construct a $POL(12, 2)$. (As seen in Section 8.4, it is actually possible to obtain a $POL(12, 5)$.)

8.8. Prove Lemma 8.6.

8.9. Show that the following Latin square is isolated:

$$\begin{array}{ccccc} 3 & 4 & 2 & 0 & 1 \\ 1 & 3 & 4 & 2 & 0 \\ 0 & 2 & 1 & 3 & 4 \\ 4 & 0 & 3 & 1 & 2 \\ 2 & 1 & 0 & 4 & 3 \end{array}$$

8.10. a. Some reflection reveals that the square in Problem 8.9 contains a subsquare of order 2. Prove that any 5×5 Latin square that contains a subsquare of order 2 is isolated.

b. As a generalization of part a, show that if $s \equiv 1 \pmod 4$ any $s \times s$ Latin square that contains a subsquare of order $(s-1)/2$ is isolated.

8.11. Let A be an $OA(n^2, k, s, 2)$ with runs labeled $1, 2, \ldots, n^2$. Let $\cup_{i=1}^n A_i$ and $\cup_{i=1}^n B_i$ be two partitions of $\{1, 2, \ldots, n^2\}$ such that:

(i) $|A_i| = |B_i| = n$ for all $i = 1, \ldots, n$;
(ii) $|A_i \cap B_{i'}| = 1$ for all $i, i' = 1, \ldots, n$;
(iii) all $2n$ subarrays of A obtained by taking the runs corresponding to one of the A_i's or B_i's are orthogonal arrays $OA(n, k, s, 1)$.

Show that A can be used to construct a $POF(n, s, k)$. (Comment: For an arbitrary $OA(n^2, k, s, 2)$ there is no guarantee that the partitions needed for this problem exist. If they do not exist, one should try to find an $n^2 \times k'$ subarray of A, with k' as large as possible, for which such partitions do exist.)

8.12. Show that for any two runs in an $OA(s^2, s-1, s, 2)$, say u_1 and u_2, that have one factor at the same level, there are precisely two runs in the array that have no factor at a common level with either u_1 or u_2. (This result is used in the construction part of Theorem 8.33.)

8.13. If $R(s)$ denotes the number of reduced Latin squares of order s, show that the number of Latin squares of order s is equal to $s!(s-1)!R(s)$. (Hint: For each of the reduced squares, consider the $s!(s-1)!$ squares obtained by using one of the $s!$ possible column permutations and one of the $(s-1)!$ possible permutations of the last $s-1$ rows.)

8.14. Show that $s \geq (w+1)s_0$ for an $IPOL(s, s_0, w)$.

8.15. Show that $s \geq (k-t+1)s_0$ for an $IOA(N, k, (s, s_0), t)$.

Chapter 9

Mixed Orthogonal Arrays

In this chapter we investigate orthogonal arrays in which the various factors may have different numbers of levels — these are called *mixed* or *asymmetrical* orthogonal arrays.

9.1 Introduction

In earlier chapters we assumed that all the factors had the same number of levels. For statistical applications (discussed in Chapter 11) things are not always so simple and it may be desirable to use arrays in which some factors have three levels (for example) while others have only two levels. Such "mixed" orthogonal arrays were already studied by Addelman and Kempthorne (1961b) and others, and were called "asymmetrical" orthogonal arrays by Rao (1973). Rao also used the terminology "orthogonal arrays with variable numbers of symbols". Neither of these terms is completely satisfactory, and we will use a third term that has appeared in the literature, namely "mixed" orthogonal arrays.

In this chapter we give the appropriate version of the Rao inequalities for mixed orthogonal arrays, Theorem 9.4 (there is also a linear programming bound), and then describe some constructions. Several other constructions are known, as can be seen from Table 12.7, which provides a list of the best (mixed- or fixed-level) arrays of strength 2 known to us with up to 100 runs.

Before giving the definition of a mixed-level array, we will slightly modify our notation for *fixed-level* orthogonal arrays (i.e. those with the same number of levels for all factors). Whereas in previous chapters we used the symbol

$$OA(N, k, s, t) \ ,$$

we will now write this as

$$OA(N, s^k, t) ,$$

where "s^k" indicates that there are k factors each at s levels.

The following definition generalizes this to allow the factors to have different levels.

Definition 9.1. A *mixed orthogonal array* $OA(N, s_1^{k_1} s_2^{k_2} \cdots s_v^{k_v}, t)$ is an array of size $N \times k$, where $k = k_1 + k_2 + \cdots + k_v$ is the total number of factors, in which the first k_1 columns have symbols from $\{0, 1, \ldots, s_1 - 1\}$, the next k_2 columns have symbols from $\{0, 1, \ldots, s_2 - 1\}$, and so on, with the property that in any $N \times t$ subarray every possible t-tuple occurs an equal number of times as a row. The symbol $L_N(s_1^{k_1} s_2^{k_2} \cdots s_v^{k_v})$ has also been used in the literature.

Note that it is not required that s_1, $s_2, \ldots$ be distinct. This is convenient for the general constructions in Sections 9.3 and 9.4, when we may not know if s_1 and s_2 (for example) are equal. Normally we combine terms with equal s_l's, so that $2^2 2^3 3^2$ would be replaced by $2^5 3^2$. Of course if all s_l's are equal we do not usually refer to the array as mixed.

Example 9.2. The array in Table 9.3 is an $OA(12, 2^4 3^1, 2)$. The first four factors are at two levels, the fifth at three levels. ■

Table 9.3. A mixed orthogonal array $OA(12, 2^4 3^1, 2)$ (transposed).

0	0	1	1	0	0	1	1	0	0	1	1
0	1	0	1	0	1	0	1	0	1	0	1
0	0	1	1	1	1	0	0	1	0	0	1
0	1	0	1	1	0	0	1	1	0	1	0
0	0	0	0	1	1	1	1	2	2	2	2

The number of possible t-tuples that occur in the $N \times t$ subarray now depends on which t columns are chosen. For the array in Table 9.3, in which $t = 2$, the possible pairs in the last two factors are

$$00, 01, 02, 10, 11, 12,$$

and each occurs twice, whereas the possible pairs in the first two factors are

$$00, 01, 10, 11,$$

and each occurs three times. Thus the number of runs must be divisible by both 6 and by 4. In general, N must be a multiple of every number

$$s_1^{i_1} s_2^{i_2} \cdots s_v^{i_v} , \tag{9.1}$$

where $0 \le i_1 \le k_1, \ldots, 0 \le i_v \le k_v$ and $i_1 + i_2 + \cdots + i_v \le t$. If N_0 denotes the least common multiple of all these numbers, it follows that N must be divisible by N_0. In the above example, $N_0 = \text{l.c.m.}\{6, 4\} = 12$.

In the special case where all s_i's are equal, Definition 9.1 reduces to the one given in Chapter 1.

9.2 The Rao Inequalities for Mixed Orthogonal Arrays

The Rao bounds on the number of runs in fixed orthogonal arrays (see Theorem 2.1) can easily be extended to mixed orthogonal arrays.

To state this bound we define the set $I_m(v)$, where $m \ge 0$ and $v \ge 1$ are integers, by

$$I_m(v) = \{(i_1, i_2, \ldots, i_v) : i_1 \ge 0, \ldots, i_v \ge 0, \sum_{l=1}^{v} i_l = m\} .$$

The notation $\sum_{I_m(v)}$ will denote a sum over all v-tuples in $I_m(v)$.

Theorem 9.4. *Consider an* $OA(N, s_1^{k_1} s_2^{k_2} \cdots s_v^{k_v}, t)$ *where, without loss of generality, we assume* $s_1 \le s_2 \le \cdots \le s_v$. *The parameters of the array satisfy:*

$$N \ge \sum_{m=0}^{u} \sum_{I_m(v)} \binom{k_1}{i_1}\binom{k_2}{i_2} \cdots \binom{k_v}{i_v} (s_1-1)^{i_1}(s_2-1)^{i_2}\cdots(s_v-1)^{i_v}, \quad \text{if} \quad t = 2u\ , \tag{9.2}$$

$$\begin{aligned} N \ge & \sum_{m=0}^{u} \sum_{I_m(v)} \binom{k_1}{i_1}\binom{k_2}{i_2} \cdots \binom{k_v}{i_v} (s_1-1)^{i_1}(s_2-1)^{i_2}\cdots(s_v-1)^{i_v} \\ & + \sum_{I_u(v)} \binom{k_1}{i_1} \cdots \binom{k_{v-1}}{i_{v-1}}\binom{k_v - 1}{i_v} (s_1-1)^{i_1}\cdots(s_{v-1}-1)^{i_{v-1}}(s_v-1)^{i_v+1}, \\ & \text{if} \quad t = 2u+1\ , \end{aligned} \tag{9.3}$$

for $u \ge 0$.

(For $v = 1$ the second summation in (9.3) reduces to $\binom{k_1-1}{u}(s_1-1)^{u+1}$.) This result can be established in the same way as Theorem 2.1, except that we need to use a different $(s_l - 1) \times s_l$ matrix B_l for each $l \in \{1, 2, \ldots, v\}$ instead of a single matrix B. Details of the proof are omitted.

The reader can verify that the Theorem 9.4 reduces to Theorem 2.1 if $v = 1$ or $s_1 = s_2 = \cdots = s_v$.

An inequality similar to (9.3) holds if we let one of $s_1, \ldots, s_{v-1}$ play the role of s_v. Thus, for example,

$$\begin{aligned} N \quad \geq \quad & \sum_{m=0}^{u} \sum_{I_m(v)} \binom{k_1}{i_1}\binom{k_2}{i_2}\cdots\binom{k_v}{i_v}(s_1-1)^{i_1}(s_2-1)^{i_2}\cdots(s_v-1)^{i_v} \\ & + \sum_{I_u(v)} \binom{k_1-1}{i_1}\binom{k_2}{i_2}\cdots\binom{k_v}{i_v}(s_1-1)^{i_1+1}(s_2-1)^{i_2}\cdots(s_v-1)^{i_v}, \end{aligned}$$

if $t = 2u+1$. But if $s_1 \leq s_2 \leq \cdots \leq s_v$, (9.3) gives the best inequality among these different choices (Problem 9.1).

Example 9.5. The Rao bound states that

$$N \geq 1 + k_1(s_1-1) + k_2(s_2-1) \tag{9.4}$$

for an $OA(N, s_1^{k_1} s_2^{k_2}, 2)$,

$$N \geq 1+k_1(s_1-1)+k_2(s_2-1)+\binom{k_1}{2}(s_1-1)^2+k_1k_2(s_1-1)(s_2-1)+\binom{k_2}{2}(s_2-1)^2$$

for an $OA(N, s_1^{k_1} s_2^{k_2}, 4)$, and, assuming without loss of generality that $s_1 \leq s_2$,

$$N \geq 1 + k_1(s_1-1) + k_2(s_2-1) + k_1(s_1-1)(s_2-1) + (k_2-1)(s_2-1)^2$$

for an $OA(N, s_1^{k_1} s_2^{k_2}, 3)$. ∎

Example 9.6. A $POF(n, s, w)$ is equivalent to an $OA(n^2, s^w n^2, 2)$ (see A^* in the proof of Theorem 8.23). Rao's inequality states that $n^2 \geq 1 + w(s-1) + 2(n-1)$, or $w \leq (n-1)^2/(s-1)$. This is precisely the result of Theorem 8.11. The parameters in Table 8.21 give various possibilities for n, s and w, although n is fairly large in some cases. ∎

We will see that the Rao bound for mixed orthogonal arrays is sometimes sharp. Example 9.19, for instance, contains an $OA(36, 2^{11}3^{12}, 2)$, which satisfies (9.4) with equality. On the other hand there are again numerous parameters for which the Rao bound can be improved. Mukerjee and Wu (1995) showed that for certain sets of parameters satisfying both the Rao inequalities and the divisibility conditions in (9.1), no mixed orthogonal arrays exist. Sloane and Stufken (1996) showed how Delsarte's theory (see Chapter 4) also applies to mixed orthogonal arrays, resulting in a **linear programming bound** for the number of runs in these arrays. Further results were obtained in Sloane and Young (1999). With N^*_{Rao} and N^*_{LP} denoting the values obtained by rounding the Rao bound N_{Rao} and the linear programming bound N_{LP} up to the nearest integer that satisfies the divisibility conditions of (9.1), Table 9.7 presents sets of parameters for an $OA(N, 2^{k_1}3^{k_2}, t)$ for which $N^*_{LP} > N^*_{Rao}$. Not all of these arrays $OA(N^*_{LP}, 2^{k_1}3^{k_2}, t)$ are known to exist though, so that additional improvements will certainly be possible for some values of k_1, k_2 and t.

Table 9.7. Improvements to the Rao bound for an $OA(N, 2^{k_1}3^{k_2}, t)$.

t	k_1	k_2	N^*_{LP}	Applies for
2	1	$9m-1$	$18(m+1)$	All $m \geq 1$
2	$12m-3$	1	$12(m+1)$	All $m \geq 1$
2	$2m+1$	$17-m$	72	$m = 1..15, m \neq 5$
2	$2m+1$	$35-m$	108	$m = 1..29, m \neq$ 5,11,12,14,17,18,20,23
2	30	2	72	
3	1	$9m$	$54(m+1)$	$m = 1..4$
3	$24m-3$	2	$72(m+1)$	$m = 1, 2$
3	$2m+1$	$36-m$	432	$m = 1..29, m \neq 12, 18$
4	m	1	$48\lfloor m/3 \rfloor$	$m = 6, 7, 9, 10, 11, 14, 16, 17, 19$
4	20	1	336	

Research Problem 9.8. Of the putative mixed arrays $OA(N^*_{LP}, 2^{k_1}3^{k_2}, t)$ mentioned in Table 9.7, which actually exist? Many further open questions of this type can be found in Sloane and Young (1999).

9.3 Constructing Mixed Orthogonal Arrays

Most of the methods presently known for constructing mixed orthogonal arrays apply only to arrays of strength 2. Some basic constructions will be discussed in this section, with additional methods being covered in Section 9.4.

The first method is based on replacing one or more factors in a given orthogonal array, fixed or mixed, by one or more factors with different numbers of levels. To make this statement more precise we will distinguish between two types of replacements.

Let A be an orthogonal array of strength 2, fixed or mixed, in which factor 1, say, has s_1 levels. Let B also be an orthogonal array of strength 2, either fixed or mixed, with s_1 runs. After making a one-to-one correspondence between the levels of factor 1 in A and the runs of B, if we replace each level of factor 1 in A by the corresponding run from B, the resulting array is again an orthogonal array of strength 2 and has at least as many factors as A. We will refer to this method of construction as the *expansive replacement* method.

Example 9.9. Starting with an $OA(16, 4^5, 2)$ and an $OA(4, 2^3, 2)$, the expansive replacement method, applied to m factors of the $OA(16, 4^5, 2)$, for $0 \leq m \leq 5$, yields an $OA(16, 2^{3m}4^{5-m}, 2)$. ∎

The expansive replacement method allows us to take the symbol for an orthogonal array, and to replace any 4^1 by 2^3, any 8^1 by 2^7 or $2^4 4^1$ (using the $OA(8, 2^4 4^1, 2)$ constructed in the next example), any 12^1 by 2^{11}, $2^4 3^1$ (using the $OA(12, 2^4 3^1, 2)$ in Table 9.3), $2^2 6^1$ or $3^1 4^1$, etc. Table 12.7 provides a much more extensive list of such substitutes.

The reader is invited to investigate whether the expansive replacement method can be extended to arrays of strength $t \geq 3$ (see Problem 9.2).

Table 9.10. (a) An $OA(8, 2^7, 2)$, and (b) the mixed $OA(8, 2^4 4^1, 2)$ obtained by contracting the first three factors.

0	0	0	0	0	0	0
0	0	0	1	1	1	1
0	1	1	0	0	1	1
0	1	1	1	1	0	0
1	0	1	0	1	0	1
1	0	1	1	0	1	0
1	1	0	0	1	1	0
1	1	0	1	0	0	1

(a)

0	0	0	0	0
1	1	1	1	0
0	0	1	1	1
1	1	0	0	1
0	1	0	1	2
1	0	1	0	2
0	1	1	0	3
1	0	0	1	3

(b)

Conversely, it may sometimes be possible for a judiciously selected set of factors in an orthogonal array of strength 2 to be replaced by a single factor with a larger number of levels, in such a way that the resulting array is still an orthogonal array of strength 2. In particular, the following method, which we will call the *contractive replacement* method, can be used if the original array possesses the requisite structure. Let A be an orthogonal array $OA(N, s_1 s_2 \cdots s_k, 2)$, where the s_i's are not necessarily all distinct, such that for a subset of u factors, say the first u, the runs of the array obtained from A by removing the other $k-u$ factors consist of N/N_1 copies of each of the runs of an $OA(N_1, s_1 s_2 \cdots s_u, 2)$, say B, that is tight. (Extending the definition of tight arrays given in Chapter 2, an orthogonal array of strength 2 is said to be *tight* if $N_1 = 1 + \sum_{i=1}^{u}(s_i - 1)$.) After labeling the runs of B by $0, 1, \ldots, N_1 - 1$, replacing each level combination of the first u factors in A by the corresponding label in B results in an $OA(N, N_1 s_{u+1} \cdots s_k, 2)$. (The reader is asked to verify this assertion in Problem 9.3.)

We will describe this by saying that the first u factors have been *contracted* to form an N_1-level factor. Application of the contractive replacement method is obviously somewhat limited by the requirement that the initial array A possesses a subarray that can be contracted.

Example 9.11. Consider the $OA(8, 2^7, 2)$ in Table 9.10 (a). When restricted to the first three factors, the runs consist of two copies of each of 000, 011, 101 and 110, which form a tight $OA(4, 2^3, 2)$. Contracting these three factors results in the $OA(8, 2^4 4^1, 2)$ shown in Table 9.10 (b). ■

Example 9.12. Consider the $OA(16, 2^{15}, 2)$ in Table 9.13, which was obtained from the first Hadamard matrix of order 16 in Table 7.26.

When restricted to the first seven factors, the array in Table 9.13 consists of two copies of each of the runs of a tight $OA(8, 2^7, 2)$. Therefore replacing the first seven factors by the column $(0, 1, \ldots, 7, 0, 1, \ldots, 7)^T$ produces an $OA(16, 2^8 8^1, 2)$. Alternatively, consider the partition $\{1, 2, 3\}$, $\{4, 8, 12\}$, $\{5, 11, 14\}$, $\{6, 9, 15\}$ and $\{7, 10, 13\}$ of the factors of the array in Table 9.13. When restricted to any one of these triples, the array consists of four copies of a tight $OA(4, 2^3, 2)$. Hence, for $0 \le m \le 5$, the contractive replacement method can transform this $OA(16, 2^{15}, 2)$ into an $OA(16, 2^{15-3m} 4^m, 2)$, the same parameters that were obtained in Example 9.9.

Table 9.13. An $OA(16, 2^{15}, 2)$.

1	2	3	4	5	6	7	8	9	10	11	12	13	14	15
1	1	1	1	1	1	1	1	1	1	1	1	1	1	1
0	1	0	1	0	1	0	1	0	1	0	1	0	1	0
1	0	0	1	1	0	0	1	1	0	0	1	1	0	0
0	0	1	1	0	0	1	1	0	0	1	1	0	0	1
1	1	1	0	0	0	0	1	1	1	1	0	0	0	0
0	1	0	0	1	0	1	1	0	1	0	0	1	0	1
1	0	0	0	0	1	1	1	1	0	0	0	0	1	1
0	0	1	0	1	1	0	1	0	0	1	0	1	1	0
1	1	1	1	1	1	1	0	0	0	0	0	0	0	0
0	1	0	1	0	1	0	0	1	0	1	0	1	0	1
1	0	0	1	1	0	0	0	0	1	1	0	0	1	1
0	0	1	1	0	0	1	0	1	1	0	0	1	1	0
1	1	1	0	0	0	0	0	0	0	0	1	1	1	1
0	1	0	0	1	0	1	0	1	0	1	1	0	1	0
1	0	0	0	0	1	1	0	0	1	1	1	1	0	0
0	0	1	0	1	1	0	0	1	1	0	1	0	0	1

Table 9.14. Another $OA(16, 2^{15}, 2)$.

1	2	3	4	5	6	7	8	9	10	11	12	13	14	15
1	1	1	1	1	1	1	1	1	1	1	1	1	1	1
0	1	0	1	0	1	0	1	0	1	0	1	0	1	0
1	0	0	1	1	0	0	1	1	0	0	1	1	0	0
0	0	1	1	0	0	1	1	0	0	1	1	0	0	1
1	1	1	0	0	0	0	1	1	1	1	0	0	0	0
0	1	0	0	1	0	1	1	0	1	0	0	1	0	1
1	0	0	0	0	1	1	1	1	0	0	0	0	1	1
0	0	1	0	1	1	0	1	0	0	1	0	1	1	0
1	1	1	1	1	1	1	0	0	0	0	0	0	0	0
1	1	0	1	0	0	0	0	0	0	1	0	1	1	1
1	0	1	0	0	0	1	0	0	1	0	1	1	1	0
1	0	0	0	1	1	0	0	0	1	1	1	0	0	1
0	1	1	0	1	0	0	0	1	0	0	1	0	1	1
0	1	0	0	0	1	1	0	1	0	1	1	1	0	0
0	0	1	1	0	1	0	0	1	1	0	0	1	0	1
0	0	0	1	1	0	1	0	1	1	1	0	0	1	0

In contrast, to illustrate the difficulty with the contractive replacement method, consider the $OA(16, 2^{15}, 2)$ in Table 9.14, which was obtained from the second Hadamard matrix of order 16 in the second row of Table 7.26. It can be shown that there is now no pair of disjoint sets of 3 factors which can both be contracted to a 4-level factor (Problem 9.4). ∎

The expansive replacement method has been used extensively in the literature, for example in Addelman and Kempthorne (1961b), Wang and Wu (1991) and Hedayat, Pu and Stufken (1992); see also Dey (1985). The contractive replacement method has also been used successfully in some situations, especially by Addelman (1962a), Wu (1989) and Wu, Zhang and Wang (1992).

Difference schemes are another powerful tool for constructing mixed orthogonal arrays of strength 2. They are heavily used by Wang and Wu (1991), Hedayat, Pu and Stufken (1992), Mandeli (1995), among others. The principal idea is essentially that used in Chapter 6 for constructing fixed orthogonal arrays, with Lemma 6.27 playing an important role. The fixed orthogonal arrays constructed in that lemma have certain resolvability properties (see Chapter 6 for the definition), and allow the addition of more factors, possibly making them into mixed arrays. This is precisely the key tool in Wang and Wu (1991). The following is a slightly more general formulation due to Dey and Midha (1996).

Theorem 9.15. *Let B be an orthogonal array $OA(N, s_1^{k_1} \cdots s_v^{k_v}, 2)$. Suppose we can write B as*

$$B = \begin{bmatrix} B_{11} & \cdots & B_{1v} \\ \vdots & & \vdots \\ B_{u1} & \cdots & B_{uv} \end{bmatrix}, \tag{9.5}$$

where each B_{hj} is an $(N/u) \times k_j$ orthogonal array of strength (at least) 1 for k_j factors with s_j levels. If, for some M, there are difference schemes $D(M, c_j, s_j)$ (of strength 2), for $1 \le j \le v$, then there exists an $OA(MN,(Mu)^1 s_1^{k_1 c_1} \cdots s_v^{k_v c_v}, 2)$.

Proof:
CONSTRUCTION: For $j = 1, \ldots, v$, let $D(j)$ denote a difference scheme $D(M, c_j, s_j)$ based on an abelian group $\mathcal{A}_j$. Using the elements of $\mathcal{A}_j$ also as entries in $B_{1j}, \ldots, B_{uj}$, define

$$A = \begin{bmatrix} B_{11} \otimes D(1) & \cdots & B_{1v} \otimes D(v) & \\ \vdots & & \vdots & A_0 \\ B_{u1} \otimes D(1) & \cdots & B_{uv} \otimes D(v) & \end{bmatrix},$$

where A_0, which represents an (Mu)-level factor, is equal to

$$A_0 = (A_{01}^T, \ldots, A_{01}^T, \ldots, A_{0u}^T, \ldots, A_{0u}^T)^T ,$$

with each

$$A_{0h} = ((h-1)M, \ldots, hM - 1)^T$$

appearing N/u times in A_0. Then A is the desired orthogonal array.

VERIFICATION: It follows from Lemma 6.27 that

$$\begin{bmatrix} B_{1j} \\ \vdots \\ B_{uj} \end{bmatrix} \otimes D(j) = \begin{bmatrix} B_{1j} \otimes D(j) \\ \vdots \\ B_{uj} \otimes D(j) \end{bmatrix}$$

is an orthogonal array of strength 2.

Further, for $h \in \{1, \ldots, u\}$ and $l \in \{1, \ldots, M\}$, when the entry in A_0 is $(h-1)M + l - 1$, the entries for the s_j-level factors in A are given by

$$B_{hj} \otimes d_l(j)^T , \tag{9.6}$$

where

$$D(j) = \begin{bmatrix} d_1(j)^T \\ \vdots \\ d_M(j)^T \end{bmatrix} .$$

Since each B_{hj} is an orthogonal array of strength 1, it follows that for each s_j-level factor all of its levels appear equally often in (9.6).

Finally, consider the factors at s_{j_1} and s_{j_2} levels, $j_1 \neq j_2$, and the $MN \times (k_{j_1}c_{j_1}+k_{j_2}c_{j_2})$ subarray of A generated by these factors. For each $l \in \{1,\ldots,M\}$, N rows of this subarray are given by

$$\left[\begin{bmatrix} B_{1j_1} \\ \vdots \\ B_{uj_1} \end{bmatrix} \otimes d_l(j_1)^T \quad \begin{bmatrix} B_{1j_2} \\ \vdots \\ B_{uj_2} \end{bmatrix} \otimes d_l(j_2)^T\right], \tag{9.7}$$

where $d_l(j_1)^T$ and $d_l(j_2)^T$ again represent rows of $D(j_1)$ and $D(j_2)$. Each of the first $k_{j_1}c_{j_1}$ columns in (9.7) can be obtained by a permutation of levels from a column in $[B_{1j_1}^T \cdots B_{uj_1}^T]^T$; similarly, each of the last $k_{j_2}c_{j_2}$ columns can be obtained in this way from $[B_{1j_2}^T \cdots B_{uj_2}^T]^T$. That every level combination for one factor from the first $k_{j_1}c_{j_1}$ and one from the last $k_{j_2}c_{j_2}$ appears equally often now follows from the fact that B is an orthogonal array of strength 2 and the invariance property 5 of Chapter 1, which continues to hold for mixed orthogonal arrays. ∎

A few remarks concerning Theorem 9.15 are in order. First, if we take $v = 1$ and $k_1 = 1$, the orthogonal array B has only one factor. The tensor product $B \otimes D(1)$ then takes the form of the array in Lemma 6.12, possibly with the D_l repeated. Second, we can always take $u = 1$ and then do not need orthogonal arrays with special properties to apply Theorem 9.15. But if the parameters of the orthogonal array B allow a value of $u \geq 2$, the conclusion will of course be stronger. Third, the number of levels Mu for the last factor in A can be quite large. In such cases the expansive replacement method can be used to replace this factor by one or more factors with a smaller number of levels.

The following are some examples of the application of Theorem 9.15. For further examples the reader may consult Wang and Wu (1991), Dey and Midha (1996) and Table 12.7. It should however be noted that not all of the examples in these papers lead to arrays with the largest possible numbers of factors, a comment that applies especially to Dey and Midha (1996).

Example 9.16. Let $B = \left[\begin{smallmatrix} 0 \\ 1 \end{smallmatrix}\right]$, so that $v = k_1 = u = 1$, $s_1 = 2$. Using a Hadamard difference scheme $D(4\lambda, 4\lambda, 2)$ (with levels 0,1) in Theorem 9.15, we obtain an $OA(8\lambda, 2^{4\lambda}(4\lambda)^1, 2)$. The 4λ-level factor can, if desired, be replaced by a 4λ-run orthogonal array of strength 2 via the expansive replacement method. In this way we can obtain arrays $OA(8, 2^4 4^1, 2)$, $OA(16, 2^8 8^1, 2)$, $OA(16, 2^{12} 4^1, 2)$, $OA(24, 2^{12} 12^1, 2)$, $OA(24, 2^{14} 6^1, 2)$, $OA(24, 2^{16} 3^1, 2)$, etc. An $OA(16, 2^{12} 4^1, 2)$ can of course also be obtained as in Examples 9.9 and 9.12. ∎

Example 9.17. Again we use a Hadamard difference scheme $D(4\lambda, 4\lambda, 2)$, but now for B we take an orthogonal array $OA(4\mu, 2^{4\mu-1}, 2)$. This is possible provided Hadamard matrices of orders 4λ and 4μ exist (see Section 7.3); the choices $\lambda = 1/2$ and $\mu = 1/2$ are also allowed. Theorem 9.15 then provides an orthogonal array $OA(16\lambda\mu, 2^{4\lambda(4\mu-1)}(4\lambda)^1, 2)$, which can be converted to an $OA(16\lambda\mu, 2^{16\lambda\mu-1}, 2)$ by the expansive replacement method. The existence of this last array is of course not breathtaking; it can also be obtained directly from a Hadamard matrix of order $16\lambda\mu$, which exists under the assumption that Hadamard matrices of order 4λ and 4μ exist. The above construction of an $OA(16\lambda\mu, 2^{4\lambda(4\mu-1)}(4\lambda)^1, 2)$ can however be deployed to obtain various other arrays (Wang and Wu, 1991). To illustrate this, without loss of generality we can write

$$D(1) = [0_{4\lambda}\ d_1 \cdots d_f\ \bar{D}(1)]$$

and

$$B = [b_1 \cdots b_f\ \bar{B}]\ ,$$

where $0 \leq f \leq min(4\lambda - 1, 4\mu - 1)$. The $OA(16\lambda\mu, 2^{4\lambda(4\mu-1)}(4\lambda)^1, 2)$ obtained from Theorem 9.15 is now of the form

$$[B \otimes D(1) \quad 0_{4\mu} \otimes R_{4\lambda}]\ ,$$

where $R_{4\lambda} = (0, 1, \ldots, 4\lambda - 1)^T$. Using the expansive replacement method, the column $0_{4\mu} \otimes R_{4\lambda}$ can be replaced by the orthogonal array

$$0_{4\mu} \otimes [d_1 \cdots d_f\ \bar{D}(1)]\ ,$$

an $OA(16\lambda\mu, 2^{4\lambda-1}, 2)$. The resulting $OA(16\lambda\mu, 2^{16\lambda\mu-1}, 2)$ contains the columns

$$b_i \otimes 0_{4\lambda}, \quad b_i \otimes d_i, \quad 0_{4\mu} \otimes d_i, \quad i = 1, \ldots, f\ .$$

Since $b_i \otimes 0_{4\lambda} + b_i \otimes d_i \equiv 0_{4\mu} \otimes d_i \pmod 2$ for each i, the contractive replacement method can be used to form f 4-level factors to obtain an array with parameters $OA(16\lambda\mu, 2^{16\lambda\mu-3f-1}4^f, 2)$. An $OA(16, 2^6 4^3, 2)$ obtained by this method, using $\lambda = \mu = 1$ and $f = 3$, is shown in Table 9.18. ∎

Table 9.18. An $OA(16, 2^6 4^3, 2)$.

0	0	0	0	0	0	0	0	0
1	1	0	1	0	1	0	2	2
0	1	1	1	1	0	2	0	2
1	0	1	0	1	1	2	2	0
0	0	1	1	1	1	0	3	3
1	1	1	0	1	0	0	1	1
0	1	0	0	0	1	2	3	1
1	0	0	1	0	0	2	1	3
1	1	0	0	1	1	3	0	3
0	0	0	1	1	0	3	2	1
1	0	1	1	0	1	1	0	1
0	1	1	0	0	0	1	2	3
1	1	1	1	0	0	3	3	0
0	0	1	0	0	1	3	1	2
1	0	0	0	1	0	1	3	2
0	1	0	1	1	1	1	1	0

Example 9.19. With $u = v = k_1 = 1$ and $s_1 = 3$, let $B = (0, 1, 2)^T$. If $D(1)$ is a difference scheme $D(M, c_1, 3)$, Theorem 9.15 yields an orthogonal array $OA(3M, M^1 3^{c_1}, 2)$. From the results in Table 6.67, we thus obtain arrays with parameters $OA(9, 3^4, 2)$, $OA(18, 3^6 6^1, 2)$ (shown in Table 9.20), $OA(27, 3^9 9^1, 2)$, $OA(36, 3^{12} 12^1, 2)$, $OA(45, 3^9 15^1, 2)$, $OA(54, 3^{18} 18^1, 2)$, etc. Using the expansive replacement method we can then also obtain arrays $OA(18, 2^1 3^7, 2)$, $OA(27, 3^{13}, 2)$, $OA(36, 2^4 3^{13}, 2)$, $OA(36, 2^{11} 3^{12}, 2)$, $OA(54, 3^{24} 6^1, 2)$ and $OA(54, 2^1 3^{25}, 2)$, among others. Using the same ideas, starting with a difference scheme $D(36, 36, 3)$, we obtain arrays $OA(108, 3^{36} 36^1, 2)$, $OA(108, 2^{35} 3^{36}, 2)$, $OA(108, 2^{11} 3^{48}, 2)$ and $OA(108, 2^4 3^{49}, 2)$. More generally, if a difference scheme $D(r, c, s)$ exists then so does an $OA(rs, s^c r^1, 2)$. ∎

Table 9.20. An $OA(18, 3^6 6^1, 2)$ (transposed); replacing the 6^1 by $2^1 3^1$ produces an $OA(18, 2^1 3^7, 2)$.

0	0	0	0	0	0	1	1	1	1	1	1	2	2	2	2	2	2
0	1	2	1	2	0	1	2	0	2	0	1	2	0	1	0	1	2
0	2	1	1	0	2	1	0	2	2	1	0	2	1	0	0	2	1
0	2	2	0	1	1	1	0	0	1	2	2	2	1	1	2	0	0
0	0	1	2	2	1	1	1	2	0	0	2	2	2	0	1	1	0
0	1	0	2	1	2	1	2	1	0	2	0	2	0	2	1	0	1
0	1	2	3	4	5	0	1	2	3	4	5	0	1	2	3	4	5

9.4 Further Constructions

In this section we describe two additional methods for constructing mixed orthogonal arrays. Both methods are of interest but are fairly difficult to apply at this time.

The first method is due to Wang (1996a). The key result for his method is presented in Theorem 9.21. If $\mathcal{A} = \mathcal{A}_1 \times \mathcal{A}_2$ is the direct product of two abelian groups $\mathcal{A}_1$ and $\mathcal{A}_2$, we define the group homomorphisms $\phi_i : \mathcal{A} \to \mathcal{A}_i$, $i = 1, 2$, by

$$\phi_1((\alpha_1, \alpha_2)) = \alpha_1, \quad \phi_2((\alpha_1, \alpha_2)) = \alpha_2 .$$

If A is an array with entries from $\mathcal{A}$, and ϕ is a map defined on $\mathcal{A}$, we write $\phi(A)$ for the array obtained by applying ϕ to every element of A.

Theorem 9.21. *Let $\mathcal{A}$ be the direct product of two abelian groups $\mathcal{A}_1$ and $\mathcal{A}_2$ of orders s_1 and s_2, respectively. Let D be a difference scheme $D(r, c, s_1 s_2)$ based on $\mathcal{A}$, and let $D(i)$ be a difference scheme $D(r, c_i, s_i)$ based on $\mathcal{A}_i$, $i = 1, 2$. If the arrays*

$$[\phi_1(D)\ D(1)] \quad \text{and} \quad [\phi_2(D)\ D(2)]$$

are difference schemes based on $\mathcal{A}_1$ and $\mathcal{A}_2$, respectively, then an orthogonal array $OA(rs_1s_2, r^1(s_1s_2)^c s_1^{c_1} s_2^{c_2}, 2)$ can be constructed.

Proof:
CONSTRUCTION: Let $E_{s_1s_2}$ be an $s_1s_2 \times 1$ vector whose entries are precisely the elements of $\mathcal{A}$, and let $R_r = (0, 1, \ldots, r-1)^T$. With $0_{s_1s_2}$ as the $s_1s_2 \times 1$ vector of zeros, define

$$A = [E_{s_1s_2} \otimes D \quad \phi_1(E_{s_1s_2}) \otimes D(1) \quad \phi_2(E_{s_1s_2}) \otimes D(2) \quad 0_{s_1s_2} \otimes R_r] \ . \tag{9.8}$$

Then A is the desired orthogonal array.

VERIFICATION: That each of the arrays $E_{s_1s_2} \otimes D$, $\phi_1(E_{s_1s_2}) \otimes D(1)$, and $\phi_2(E_{s_1s_2}) \otimes D(2)$ is an orthogonal array of strength 2 follows from Lemma 6.12. That the column $0_{s_1s_2} \otimes R_r$ can be added to each of these arrays to obtain a larger orthogonal array of strength 2 follows as in Corollary 6.20. Thus it suffices to show that all possible level combinations for a pair of columns appear equally often if (i) one column is taken from $E_{s_1s_2} \otimes D$ and the other from $\phi_1(E_{s_1s_2}) \otimes D(1)$ or $\phi_2(E_{s_1s_2}) \otimes D(2)$; and (ii) one column is taken from $\phi_1(E_{s_1s_2}) \otimes D(1)$ and the other from $\phi_2(E_{s_1s_2}) \otimes D(2)$. Part (ii) is by far the easiest to establish. If $d_i = (d_{i1}, \ldots, d_{ir})^T$ is a column in $D(i)$, $i = 1, 2$, then the rows of

$$[\phi_1(E_{s_1s_2}) \otimes d_1 \quad \phi_2(E_{s_1s_2}) \otimes d_2]$$

are, except for their order, the same as those in

$$\begin{bmatrix} d_{11} + \phi_1(E_{s_1 s_2}) & d_{21} + \phi_2(E_{s_1 s_2}) \\ \vdots & \vdots \\ d_{1r} + \phi_1(E_{s_1 s_2}) & d_{2r} + \phi_2(E_{s_1 s_2}) \end{bmatrix},$$

where $+$ is used generically to denote addition in both $\mathcal{A}_1$ and $\mathcal{A}_2$. That these two columns contain all possible level combinations equally often follows immediately by observing that

$$[\phi_1(E_{s_1 s_2}) \quad \phi_2(E_{s_1 s_2})]$$

contains each of the possible level combinations precisely once as a row.

For part (i), without loss of generality we may assume that the second column, say $\phi_1(E_{s_1 s_2}) \otimes d_1$ where $d_1 = (d_{11}, \ldots, d_{1r})^T$, is taken from $\phi_1(E_{s_1 s_2}) \otimes D(1)$. Write the column from $E_{s_1 s_2} \otimes D$ as $E_{s_1 s_2} \otimes d_0$, where $d_0 = (d_{01}, \ldots, d_{0r})^T$. For fixed h_0, $i_0 \in \mathcal{A}_1$, and $j_0 \in \mathcal{A}_2$ we need to show that the level combination $((i_0, j_0), h_0)$ appears r/s_1 times as a row in the $rs_1 s_2 \times 2$ array

$$[E_{s_1 s_2} \otimes d_0 \quad \phi_1(E_{s_1 s_2}) \otimes d_1] .$$

Hence we need to show that there are r/s_1 solutions $(i, j, l) \in \mathcal{A}_1 \times \mathcal{A}_2 \times \{1, \ldots, r\}$ to the equations

$$\begin{aligned} i &+ \phi_1(d_{0l}) = i_0 , \\ j &+ \phi_2(d_{0l}) = j_0 , \\ i &+ d_{1l} = h_0 . \end{aligned} \tag{9.9}$$

The first and last of these equations imply that $\phi_1(d_{0l}) - d_{1l} = i_0 - h_0$. Since $[\phi_1(D) \ D(1)]$ is a difference scheme $D(r, c + c_1, s_1)$, there are precisely r/s_1 values of $l \in \{1, \ldots, r\}$ that satisfy this. For each such l the equations in (9.9) determine unique values for $i \in \mathcal{A}_1$, and $j \in \mathcal{A}_2$ that make (i, j, l) a solution. Hence there are precisely r/s_1 solutions to (9.9) for any possible choice of i_0, j_0 and h_0. ∎

Observe that either (or both) of $D(1)$ and $D(2)$ can be taken to be the empty array. If $D(1)$ is the empty array, then so is $\phi_1(E_{s_1 s_2}) \otimes D(1)$ in Equation (9.8). If both $D(1)$ and $D(2)$ are taken to be the empty array, the result in Theorem 9.21 is a special case of Theorem 9.15.

The question of finding difference schemes D, $D(1)$ and $D(2)$ that satisfy the hypotheses of Theorem 9.21 requires further attention. Wang (1996a) demonstrates the potential of the method by various examples, including the following.

Example 9.22. We will use Theorem 9.21 to construct an $OA(36, 3^4 6^3, 2)$. Our starting point is the difference scheme $D(6, 6, 3)$ in Table 6.37, which is for ease of reference repeated in Table 9.23. We take the difference scheme D needed for Theorem 9.21 to be the scheme $D(6,2,6)$ shown in Table 9.24. Note that $\phi_1(D)$ consists precisely of the first two columns of the scheme in Table 9.23. For $D(1)$ we take the scheme $D(6,4,3)$ consisting of the last four columns of the scheme in Table 9.23, while for $D(2)$ we take the empty array.

Table 9.23. A difference scheme $D(6,6,3)$.

0	0	0	0	0	0
0	1	2	1	2	0
0	2	1	1	0	2
0	2	2	0	1	1
0	0	1	2	2	1
0	1	0	2	1	2

Table 9.24. A difference scheme $D(6,2,6)$ based on $\mathbb{Z}_3 \times \mathbb{Z}_2$

(0,0)	(0,0)
(0,0)	(1,0)
(0,0)	(2,0)
(0,0)	(2,1)
(0,0)	(0,1)
(0,0)	(1,1)

Since

$$[\phi_1(D)\ D(1)]$$

is a difference scheme $D(6,6,3)$, it follows from Theorem 9.21 that

$$A = [E_6 \otimes D \quad \phi_1(E_6) \otimes D(1) \quad 0_6 \otimes R_6]$$

is an $OA(36, 3^4 6^3, 2)$. This array, with the levels of the 6-level factors written as $i + 3j$ instead of (i, j), is shown in Table 9.25. (Note that Finney, 1982, has found an $OA(36, 3^7 6^3, 2)$—see Table 12.7.) ■

Table 9.25. An $OA(36, 3^4 6^3, 2)$ (transposed).

```
021210102021210102021210102021210102
011022122100200211011022122100200211
020121101202212010020121101202212010
002112110220221001002112110220221001
000000111111222222333333444444555555
012534120345201453345201453012534120
012345012345012345012345012345012345
```

The second construction to be described in this section is due to Zhang, Lu and Pang (1999). Their method is based on decomposing an orthogonal projection matrix into the sum of lower rank orthogonal projection matrices, and facilitates (in the presence of some creativity) the construction of larger orthogonal arrays from smaller arrays.

In order to describe this construction we will need to introduce some new concepts and notation.

We will continue to use $A \otimes B$ to denote the tensor product of the matrices A and B (see Definition 6.26), but with multiplication of real numbers (rather than in a finite field) as the underlying binary operation. Thus if $A = (a_{ij})$ is an $m \times n$ matrix, then

$$A \otimes B = \begin{bmatrix} a_{11}B & \cdots & a_{1n}B \\ \vdots & & \vdots \\ a_{m1}B & \cdots & a_{mn}B \end{bmatrix} .$$

Some elementary properties of the tensor product operator are:

$$\begin{aligned} (A \otimes B) \otimes C &= A \otimes (B \otimes C) , \\ (A \otimes B)^T &= A^T \otimes B^T , \\ (A \otimes B)(C \otimes D) &= (AC) \otimes (BD) , \end{aligned}$$

where the latter requires that the various matrix products are well defined.

Orthogonal projection matrices also play an important role in what follows. A matrix A is said to be an *orthogonal projection matrix* if it is idempotent ($A^2 = A$) and symmetric ($A^T = A$). For an arbitrary matrix X, the matrix

$$P_X = X(X^TX)^-X^T$$

is idempotent and symmetric: this is the orthogonal projection matrix onto the column space of X. If A is an orthogonal projection matrix, then it follows immediately that $P_A = A$, so any orthogonal projection matrix is of the form P_X for some matrix X.

We will make use of the following well known result concerning orthogonal projection matrices (see for example Searle, 1971, p. 62).

Theorem 9.26. *Let $X_1, \ldots, X_n$ be $m \times m$ symmetric matrices with the property that $X = X_1 + \cdots + X_n$ is idempotent and* $\text{rank}(X) = \sum_{i=1}^n \text{rank}(X_i)$. *Then each X_i is idempotent and $X_iX_j = 0$ for $i \neq j$.*

Let x be an $N \times 1$ vector with entries from $\{0, 1, \ldots, s-1\}$, and let X be the $N \times s$ incidence matrix of x. In other words, $X = (x_{ij})$, $i = 1, \ldots, N$,

$j = 0, \ldots, s-1$, where

$$x_{ij} = \begin{cases} 1 & \text{if the } i\text{-th entry of } x \text{ is } j, \\ 0 & \text{otherwise.} \end{cases}$$

Our interest will be in the orthogonal projection matrix $P_X - P_{1_N}$, which has rank $s-1$ provided that each element from $\{0, 1, \ldots, s-1\}$ appears at least once in x.

If we start with an $N \times k$ array, say

$$A = [x_1 \cdots x_k] \,, \tag{9.10}$$

where the entries in x_l belong to $\{0, 1, \ldots, s_l - 1\}$, there is an incidence matrix corresponding to each column, and we obtain k orthogonal projection matrices $P_{X_l} - P_{1_N}$, $l = 1, \ldots, k$. It is relatively easy to characterize an orthogonal array of strength 2 in terms of these matrices.

Theorem 9.27. *The array A in (9.10) is an orthogonal array of strength 2 if and only if (i) it is an orthogonal array of strength 1, and (ii) $\sum_{l=1}^{k}(P_{X_l} - P_{1_N})$ is an idempotent matrix of rank $\sum_{l=1}^{k}(s_l - 1)$.*

The proof of this result is left as an exercise for the reader (Problem 9.8). It follows from Theorem 9.26 that condition (ii) in Theorem 9.27 implies $(P_{X_l} - P_{1_N})(P_{X_{l'}} - P_{1_N}) = 0$ for all $l \neq l'$. (The latter condition can in fact be used instead of (ii) in Theorem 9.27.)

It now follows that if A in (9.10) is an orthogonal array of strength 2, then

$$(I_N - P_{1_N}) - \sum_{l=1}^{k}(P_{X_l} - P_{1_N})$$

is an orthogonal projection matrix of rank $N - 1 - \sum_{l=1}^{k}(s_l - 1)$. (Since this is a nonnegative integer, we may deduce Rao's inequality for mixed orthogonal arrays of strength 2.) Consequently, any orthogonal array of strength 2 induces a decomposition of I_N into a sum of orthogonal projection matrices that project into orthogonal spaces.

Example 9.28. Consider the orthogonal array $OA(4, 3, 2, 2)$ in Table 9.29. With the notation just introduced, the matrices P_{X_1}, P_{X_2}, P_{X_3} are respectively

$$\frac{1}{2}\begin{bmatrix} 1 & 1 & 0 & 0 \\ 1 & 1 & 0 & 0 \\ 0 & 0 & 1 & 1 \\ 0 & 0 & 1 & 1 \end{bmatrix}, \quad \frac{1}{2}\begin{bmatrix} 1 & 0 & 1 & 0 \\ 0 & 1 & 0 & 1 \\ 1 & 0 & 1 & 0 \\ 0 & 1 & 0 & 1 \end{bmatrix}, \quad \frac{1}{2}\begin{bmatrix} 1 & 0 & 0 & 1 \\ 0 & 1 & 1 & 0 \\ 0 & 1 & 1 & 0 \\ 1 & 0 & 0 & 0 \end{bmatrix}.$$

Hence $\sum_{l=1}^{3}(P_{X_l} - P_{1_4}) = I_4 - P_{1_4}$. The reason for equality is that an $OA(4, 3, 2, 2)$ is a tight orthogonal array. ■

Table 9.29. An $OA(4,3,2,2)$.

0	0	0
0	1	1
1	0	1
1	1	0

For the remainder of this section we will write Q_N to denote the matrix $I_N - P_{1_N}$. Further, for an orthogonal array A as in (9.10) we will write $[A]$ to denote $\sum_{l=1}^{k}(P_{X_l} - P_{1_N})$. If C is an $N \times N$ permutation matrix, then CA is again an orthogonal array, and $[CA] = C[A]C^T$. Also, $1_m \otimes A$ is an orthogonal array for every positive integer m, and $[1_m \otimes A] = P_{1_m} \otimes [A]$.

While we now know how to decompose Q_N when we start with a given N-run orthogonal array, it is of much greater interest to reverse this procedure. How should we proceed to find a suitable decomposition of Q_N that can be used for the construction of an orthogonal array? There is unfortunately no simple general answer to this question. There are infinitely many ways to write Q_N as the sum of orthogonal projection matrices whose ranks add to $N-1$, but if these orthogonal projection matrices are not of the form $P_{X_l} - P_{1_N}$, or a sum of such matrices, the decomposition is of no help for the construction of orthogonal arrays. The strategy deployed by Zhang, Lu and Pang (1999) is to decompose Q_N into a sum of orthogonal projection matrices that can be easily related to known orthogonal arrays. While they do not present any general method for doing this, the basic ideas are demonstrated in the following example.

Example 9.30. We will construct an $OA(36, 2^{27}3^1, 2)$ and an $OA(36, 2^{20}3^2, 2)$ from a judiciously selected decomposition of Q_{36} and by using various smaller orthogonal arrays. Let A_1 be the $OA(4,3,2,2)$ from Table 9.29. The first column of this array, say x_1, is equal to $(0,1)^T \otimes (1,1)^T$, so that $P_{X_1} - P_{1_4} = Q_2 \otimes P_{1_2}$. The second and third columns of the array are of the form $C_l x_1$ for 4×4 permutation matrices C_l, $l = 2, 3$. Hence

$$P_{X_l} - P_{1_4} = C_l(Q_2 \otimes P_{1_2})C_l^T, \quad l = 2,3,$$

and

$$Q_4 = [A_1] = \sum_{l=1}^{3} C_l(Q_2 \otimes P_{1_2})C_l^T\ ,$$

where C_1 can be taken to be I_4.

Similarly, using an orthogonal array $OA(9,4,3,2)$, say A_2, we can write

$$Q_9 = [A_2] = \sum_{l=1}^{4} D_l(P_{1_3} \otimes Q_3)D_l^T$$

for 9×9 permutation matrices D_l, $l = 1, 2, 3, 4$.

Furthermore, let A_3 denote any 12-run orthogonal array of strength 2 obtained by deleting the columns $1_6 \otimes (0,1)^T$ and $1_3 \otimes (0,1)^T \otimes (0,1)^T$ from a 12-run orthogonal array of strength 2 that contains these columns. We will use an $OA(12, 9, 2, 2)$ and an $OA(12, 2^2 3^1, 2)$ for A_3. Using the notation $A \leq B$ to indicate that the difference $B - A$ is a nonnegative definite matrix, we have

$$[A_3] \leq Q_{12} - P_{1_6} \otimes Q_2 - P_{1_3} \otimes Q_2 \otimes Q_2 = Q_3 \otimes I_4 + P_{1_3} \otimes Q_2 \otimes P_{1_2} \,. \quad (9.11)$$

(Problem 9.10 asks the reader to prove the final equality in (9.11).)

The next step consists in finding an appropriate decomposition of Q_{36}. With C_l, $l = 1, 2, 3$, and D_l, $l = 1, 2, 3, 4$, as the permutation matrices in the expressions for $[A_1]$ and $[A_2]$, respectively, we have

$$\begin{aligned}
Q_{36} &= I_9 \otimes I_4 - P_{1_9} \otimes P_{1_4} = Q_9 \otimes I_4 + P_{1_9} \otimes Q_4 \\
&= \left(\sum_{l=1}^{4} D_l(P_{1_3} \otimes Q_3) D_l^T\right) \otimes I_4 + P_{1_9} \otimes \left(\sum_{l=1}^{3} C_l(Q_2 \otimes P_{1_2}) C_l^T\right) \\
&= \sum_{l=1}^{3} (D_l(P_{1_3} \otimes Q_3) D_l^T) \otimes (C_l I_4 C_l^T) \\
&\quad + \sum_{l=1}^{3} (D_l P_{1_9} D_l^T) \otimes (C_l(Q_2 \otimes P_{1_2}) C_l^T) \\
&\quad + D_4(P_{1_3} \otimes Q_3) D_4^T) \otimes I_4 \\
&= \sum_{l=1}^{3} (D_l \otimes C_l)(P_{1_3} \otimes Q_3 \otimes I_4 + P_{1_9} \otimes Q_2 \otimes P_{1_2})(D_l^T \otimes C_l^T) \\
&\quad + (D_4 \otimes I_4)(P_{1_3} \otimes Q_3 \otimes P_{1_4})(D_4^T \otimes I_4) \\
&\quad + (D_4 \otimes I_4)(P_{1_3} \otimes Q_3 \otimes Q_4)(D_4^T \otimes I_4) \\
&= \sum_{l=1}^{3} (D_l \otimes C_l)(P_{1_3} \otimes \{Q_3 \otimes I_4 + P_{1_3} \otimes Q_2 \otimes P_{1_2}\})(D_l^T \otimes C_l^T) \\
&\quad + (D_4 \otimes I_4)(P_{1_3} \otimes Q_3 \otimes P_{1_4})(D_4^T \otimes I_4) \\
&\quad + (D_4 \otimes I_4)(P_{1_3} \otimes Q_3 \otimes Q_4)(D_4^T \otimes I_4) \,.
\end{aligned} \quad (9.12)$$

We recognize several terms in this decomposition. Let A_l^*, $l = 1, 2, 3$, denote the 36-run orthogonal array of strength 2 defined by

$$A_l^* = (D_l \otimes C_l)(1_3 \otimes A_3) \,.$$

Then, using (9.11) and Problem 9.9,

$$\begin{aligned}
[A_l^*] &= (D_l \otimes C_l)(P_{1_3} \otimes [A_3])(D_l^T \otimes C_l^T) \\
&\leq (D_l \otimes C_l)(P_{1_3} \otimes \{Q_3 \otimes I_4 + P_{1_3} \otimes Q_2 \otimes P_{1_2}\})(D_l^T \otimes C_l^T) \,.
\end{aligned}$$

Also, with

$$A_4^* = (D_4 \otimes I_4)(1_3 \otimes (0,1,2)^T \otimes 1_4) \ ,$$

a vector of length 36 with entries 0, 1 and 2,

$$[A_4^*] = (D_4 \otimes I_4)(P_{1_3} \otimes Q_3 \otimes P_{1_4})(D_4^T \otimes I_4) \ .$$

Since the ranks of the five matrices in the decomposition of Q_{36} in (9.12) are 9, 9, 9, 2 and 8, respectively, which sum to the rank of Q_{36}, it follows from Theorems 9.26 and 9.27 that the array

$$A = [A_1^* \ \ A_2^* \ \ A_3^* \ \ A_4^*]$$

is an orthogonal array of strength 2. If we take A_3 to be an $OA(12,9,2,2)$ for each A_i^*, then A is an $OA(36, 2^{27}3^1, 2)$. If we make this choice for A_3 only for two of the A_i^*'s, while taking the third A_3 to be an $OA(12, 2^2 3^1, 2)$, A is an $OA(36, 2^{20}3^2, 2)$.

Table 9.32 shows an $OA(36, 2^{27}3^1, 2)$ constructed by this method. For A_1 we used the $OA(4,3,2,2)$ in Table 9.29; for A_2 the $OA(9,4,3,2)$ in Table 2.3; and for A_3 the last 9 columns of the $OA(12,11,2,2)$ shown in Table 9.31.

In the same way, taking A_3 to be an $OA(12,1,6,2)$ for one or two of the A_i^*'s, we can construct 36-run arrays with factors $2^9 3^1 6^2$, $2^{11}3^2 6^1$, $2^{18}3^1 6^1$, $2^4 3^3 6^1$ and $2^2 3^2 6^2$. ∎

Table 9.31. An $OA(12,11,2,2)$.

0	0	0	0	0	0	0	0	0	0	0
1	0	1	0	0	0	1	1	1	0	1
0	1	1	0	1	0	0	0	1	1	1
1	1	0	1	0	0	0	1	1	1	0
0	0	1	1	1	0	1	1	0	1	0
1	0	1	1	0	1	0	0	0	1	1
0	1	1	1	0	1	1	0	1	0	0
1	1	0	1	1	0	1	0	0	0	1
0	0	0	1	1	1	0	1	1	0	1
1	0	0	0	1	1	1	0	1	1	0
0	1	0	0	0	1	1	1	0	1	1
1	1	1	0	1	1	0	1	0	0	0

Table 9.32. An $OA(36, 2^{27}3^1, 2)$.

```
0000000000000000000000000000
1000111010100011101010001110
1010001111000111010100011100
0100011101010001111000111010
0000000001110110101110110101
1000111010110100011101101001
1010001111101000110110100011
0100011101101101001101000111
0000000000111011010111011012
1000111011011010000001110112
1010001110011101101011010002
0100011100000111011001110102
1110110100000000000111011011
1101000110100011100001110111
1101101001000011101101101000 1
0110100011010001110011101101
1110110101110110100000000002
1101000110110100011010001112
1101101001101000110100011102
0110100011101101001000111012
1110110100111011011110110100
1101000111011010001101101000
1101101000011101100110100010
0110100010001110111101000110
0111011010000000000111011010 2
0011101100100011101101101002
0001110111000111010110100012
1011010001010001111101000112
0111011011110110100111011010
0011101100110100010001110110
0001110111101000111011010000
1011010001101101000011101100
0111011010111011010000000001
0011101101011010001010001111
0001110110011101100100011101
1011010000000111011100011101 1
```

9.5 Notes on Chapter 9

Mixed orthogonal arrays have not nearly received the same attention in the literature as fixed orthogonal arrays. In statistical applications however the use of mixed arrays is often quite natural: when one goal of an experiment is to compare three catalysts, why should all other factors of possible interest (such as temperature and pressure) also be used with precisely three levels?

A considerable portion of the work that has been done on mixed orthogonal arrays applies specifically to arrays of strength 2. But even in that case, our knowledge about the existence and nonexistence of arrays is rather limited. And while $t = 2$ is arguably the most important case for statistical applications, there is an urgent need for better methods for $t \geq 3$.

Research Problem 9.33. Develop better methods and tools for the construction of mixed orthogonal arrays and for establishing their nonexistence.

Although *codes* over mixed alphabets have been investigated in several papers—see in particular Brouwer, Hämäläinen, Östergård and Sloäne (1998)—usually they do not form orthogonal arrays of strength 2, and do not even have the appropriate number of vectors.

Thus up to now we have been deprived of one of the most powerful construction methods that we had for fixed-level arrays.

Research Problem 9.34. Find a good way to use error-correcting codes to construct mixed orthogonal arrays. In particular, redo the investigation of mixed binary/ternary codes carried out in the above reference, concentrating on dual distance rather than distance, and on codes in which the number of vectors would be appropriate for an $OA(N, 2^{k_1}3^{k_2}, 2)$.

Finally, mixed orthogonal arrays will reappear in both Chapters 10 and 11.

9.6 Problems

9.1. Consider an $OA(N, s_1^{k_1} s_2^{k_2} \cdots s_v^{k_v}, 2u+1)$, $u \geq 0$. For $l \in \{1, 2, \ldots, v\}$ define

$$b_l = \sum_{I_u(v)} \left\{ \frac{k_l - i_l}{k_l} (s_l - 1) \prod_{j=1}^{v} \left(\binom{k_j}{i_j} (s_j - 1)^{i_j} \right) \right\} .$$

If $s_1 \leq s_2 \leq \cdots \leq s_v$, show that $b_1 \leq b_2 \leq \cdots \leq b_v$. (Note that b_v is precisely the last term in (9.3).)

9.2. In the expansive replacement method (Section 9.3) we constructed an orthogonal array of strength 2 from two other arrays of strength 2, called A and B. If A and B are both orthogonal arrays of strength t, $t \geq 3$, can we conclude that the resulting array is also an orthogonal array of strength 3? Explain your answer.

9.3. (i). Show that the $N \times (k-u+1)$ array obtained from the contractive replacement method (see Section 9.3) is indeed an $OA(N, N_1 s_{u+1} \cdots s_k, 2)$.

(ii). If the arrays A and B in the description of the contractive replacement method are both of strength t, and if B is tight (i.e., it gives equality in (9.2) or (9.3), depending on whether t is even or odd), does the contractive replacement method result in an orthogonal array of strength t? Explain your answer.

9.4. For the $OA(16, 2^{15}, 2)$ in Table 9.14 it was asserted that there are no two 16×3 subarrays, based on two disjoint sets of three columns, such that the runs for each subarray consist of four copies of the runs of an $OA(4, 2^3, 2)$. Verify that this claim is correct.

9.5. Table 9.25 shows an $OA(36, 3^4 6^3, 2)$, whereas Table 12.7 asserts that an $OA(36, 3^7 6^3, 2)$ exists. Investigate whether an $OA(36, 3^k 6^3, 2)$ with $k > 4$ can be obtained from the array in Table 9.25 by adjoining one or more 3-level factors.

9.6. Use the method of Theorem 9.21 to construct an $OA(72, 2^{17} 3^3 6^6, 2)$. (Hint: Use $r = 12$ for the difference schemes D, $D(1)$ and $D(2)$.)

9.7. Show that the column-spaces of two matrices X and Y (of sizes $m \times n_1$ and $m \times n_2$ respectively) coincide if and only if the orthogonal projection matrices P_X and P_Y are identical.

9.8. Prove Theorem 9.27.

9.9. Establish the following properties of an N-run orthogonal array A that were stated in Section 9.4. Here m is any positive integer.

(i) If C is a $N \times N$ permutation matrix, then $[CA] = C[A]C^T$.

(ii) $[1_m \otimes A] = P_{1_m} \otimes [A]$.

9.10. Verify that

$$Q_{12} - P_{1_6} \otimes Q_2 - P_{1_3} \otimes Q_2 \otimes Q_2 = Q_3 \otimes I_4 + P_{1_3} \otimes Q_2 \otimes P_{1_2} \,.$$

(This identity was used in (9.11). One way to establish it is of course to simply compute both sides of the equation; there are however more elegant ways that make use of the properties of tensor products.)

9.11. By examining Table 9.3, show that if a factor at s_1 levels and a factor at s_2 levels are combined in an orthogonal array of strength 2 to form one factor at $s_1 s_2$ levels, then the resulting array may no longer be an orthogonal array of strength 2. Do this both for the case that $s_1 = s_2$ and $s_1 \neq s_2$. (The contractive replacement method of Section 9.3 gives conditions under which such combinations are justified.)

9.12. (In this problem s_1, s_2, s_3 need not be distinct.)

(i) Show that an $OA(N, s_1^1 s_2^1, 2)$ exists if and only if N is divisible by $s_1 s_2$.

(ii) Show that an $OA(N, s_1^1 s_2^1 s_3^1, 2)$ exists if and only if N is divisible by l.c.m. $\{s_1 s_2, s_1 s_3, s_2 s_3\}$.

Research Problem 9.35. Using the results of Hanani (see Hanani, 1975), Colbourn (1996a) and others mentioned in Section 10.9, extend the results in Problem 9.12 to classify mixed orthogonal arrays with 4, 5, 6, ... factors.

Chapter 10

Further Constructions and Related Structures

This chapter discusses a number of different topics that do not quite fit into any of the earlier chapters. These are

- Constructions for orthogonal arrays inspired by coding theory
- Bounds on the size of orthogonal arrays with many factors
- Compound orthogonal arrays
- Orthogonal multi-arrays
- Transversal designs, resilient functions and nets
- Orthogonal arrays and association schemes

10.1 Constructions Inspired by Coding Theory

We begin with some general remarks about analogies between constructions for codes and for orthogonal arrays. There are many constructions that take one or more codes and perform some operation to produce a new code (cf. Chapter 17 of MacWilliams and Sloane, 1977). We believe that corresponding to every one of these constructions there should be an analogous construction for orthogonal arrays! That is probably too strong a statement, but it serves as a useful guide.

If the construction applies specifically to linear codes, it is straightforward to find an analogue for orthogonal arrays (see Fig. 10.1). One starts with one or more linear orthogonal arrays, takes their duals, regards them as codes, applies the construction, and again takes the dual to produce the new orthogonal array. The juxtaposition construction described in Section 10.2 is of this type; so is the "residual code" construction mentioned in Chapter 12.

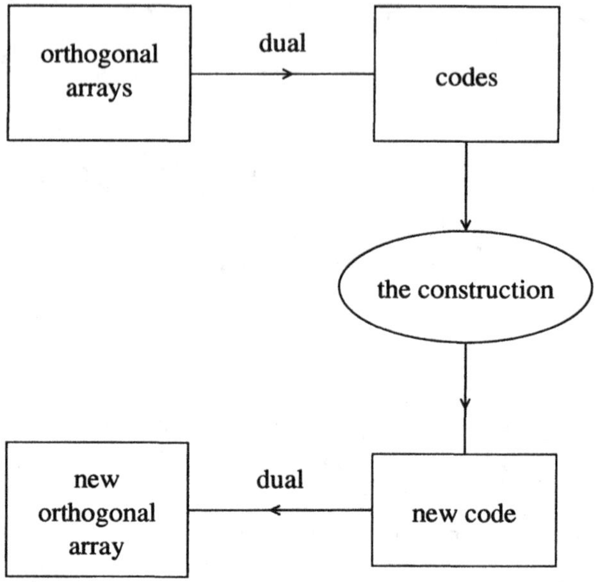

Figure 10.1: Converting a construction for linear codes into a construction for linear orthogonal arrays.

If the construction involves nonlinear codes, however, we cannot follow this simple procedure. In this situation we should look for an analogous construction that applies directly to nonlinear orthogonal arrays. To find this construction, it is sometimes enough to pretend temporarily that the codes are linear, and to examine how the orthogonal arrays formed from the dual codes get transformed in the construction. Two such constructions are described in Sections 10.3 and 10.4.

10.2 The Juxtaposition Construction

The following "juxtaposition" construction is exceedingly simple when considered in coding terms, less so in terms of orthogonal arrays. It combines two linear orthogonal arrays to produce a third. The reason for the name will become clear from the proof. The analogous construction for codes is described in MacWilliams and Sloane (1977), page 49, Figure 2.6. (Note that this construction is different from the "juxtaposition construction" described in Raktoe,

Hedayat and Federer, 1981, §13.4.)

Theorem 10.1. *Let A_i be a linear orthogonal array $OA(N_i, k_i, s, t_i)$, $i = 1, 2$, where $N_1 = s^{k_1-n_1}$, $N_2 = s^{k_2-n_1}$. Then there exists an $OA(s^{k_1+k_2-n_1}, k_1 + k_2, s, t_1 + t_2 + 1)$.*

The crucial condition is that $k_1 - \log_s N_1$ and $k_2 - \log_s N_2$ must be equal.

Proof: Let H_1 and H_2 be parity check matrices for the two arrays. The hypotheses imply that H_1 and H_2 have the same number of rows. By juxtaposing H_1 and H_2 we obtain a matrix $[H_1\ H_2]$ which we take as a parity check matrix for a new array, A_3 say. To determine the strength of A_3, we note that the code with generator matrix H_i has minimal distance at least $t_i + 1$ ($i = 1, 2$) so $[H_1\ H_2]$ is a generator matrix for a code with minimal distance at least $t_1 + t_2 + 2$, and so A_3 has strength at least $t_1 + t_2 + 1$. ∎

An explicit generator matrix for A_3 is given in Problem 10.1.

Example 10.2. By combining an $OA(2^4, 7, 2, 3)$ with an $OA(2^8, 11, 2, 5)$ we obtain an $OA(2^{15}, 18, 2, 9)$. ∎

10.3 The $(u, u+v)$ Construction

The $(u, u+v)$ construction is a simple yet powerful method for combining two possibly nonlinear codes to produce a third code (Sloane and Whitehead, 1970; MacWilliams and Sloane, 1977, Chapter 2). There is an analogous construction for OA's. We begin by describing the coding version. Suppose C_i is a $(k_i, N_i, d_i)_s$ code over the field $S = GF(s)$, for $i = 1, 2$. Then the set of vectors

$$C_3 = \{(u, u+v) :\ u \in C_1,\ v \in C_2\}$$

is a code of length $k_3 = \max\{2k_1, k_1 + k_2\}$, size $N_1 N_2$ and minimal distance $d = \min\{2d_1, d_2\}$.

When forming $u + v$, if the length of u is different from the length of v, we simply append zeros to the shorter vector. The special case $k_1 = k_2$ of this construction has already been seen in Problem 5.9.

For example, by taking C_1 to be the $(6, 8, 3)_2$ code with generator matrix

$$\begin{bmatrix} 100011 \\ 010101 \\ 001110 \end{bmatrix}$$

and C_2 to be the $(6,2,6)_2$ trivial code we find that C_3 is a $(12,16,6)_2$ code. Taking C_1 to be the latter code and C_2 to be the $(12,2,12)_2$ trivial code we obtain a $(24,32,12)_2$ code, and hence an $OA(2^{19},24,2,11)$ and an $OA(2^{18},23,2,10)$. The latter appears in Table 12.1.

There is an analogous construction for orthogonal arrays We state it here for the case $k_1 = k_2$ and postpone the case of unequal numbers of factors to Problem 10.2. In this theorem and the next, s is arbitrary.

Theorem 10.3. *Let A_i be an $OA(N_i, k, s, t_i)$, for $i = 1, 2$. Then the set of all vectors $(u, u+v)$, where u is a row of A_2 and v is a row of A_1, forms an $OA(N_1N_2, 2k, s, t_3)$ with $t_3 = \min\{2t_1+1, t_2\}$.*

Proof: We apply Theorem 3.30, and will show that, for all $x, y \in S^k$ with $0 < w(x) + w(y) \le t_3$, the sum

$$\sum_{u\in A_2, v\in A_1} \zeta^{(u,u+v)(x,y)^T} = \sum_{v\in A_1} \zeta^{vy^T} \sum_{u\in A_2} \zeta^{u(x+y)^T} \tag{10.1}$$

is equal to zero. From the hypotheses and Theorem 3.30 we know that

$$\sum_{v\in A_1} \zeta^{vp^T} = 0\,, \quad \text{for } 0 < w(p) \le t_1\,, \tag{10.2}$$

$$\sum_{u\in A_2} \zeta^{uq^T} = 0\,, \quad \text{for } 0 < w(q) \le t_2\,. \tag{10.3}$$

If $x + y \ne 0$ then $w(x+y) \le w(x) + w(y) \le t_3 \le t_2$, and (10.1) vanishes by (10.3). On the other hand if $x + y = 0$ then $w(x) = w(y)$, $2w(y) \le t_3 \le 2t_1+1$, $w(y) \le t_1$, and (10.1) vanishes by (10.2). ∎

Example 10.4. If A_1 is an $OA(2,4,2,1)$ and A_2 is an $OA(8,4,2,3)$ then the construction produces an $OA(16,8,2,3)$; if A_1 is an $OA(16,8,2,3)$ and A_2 is an $OA(128,8,2,7)$ the result is an $OA(2^{11},16,2,7)$, and so on. ∎

10.4 Construction X4

Construction X4 combines four orthogonal arrays to produce a fifth. This is an adaption for orthogonal arrays of a construction for codes given by Sloane, Reddy and Chen (1972). The ingredients are four orthogonal arrays,

$$A_1 : OA(N_1, k_1, s, t_1)\,, \quad A_3 : OA(N_3, k_3, s, t_3)\,,$$

$$A_2 : OA(N_2, k_1, s, t_2)\,, \quad A_4 : OA(N_4, k_3, s, t_4)\,,$$

with the property that

$$A_2 \text{ is a union of } m \text{ disjoint translates of } A_1 ,$$

$$A_4 \text{ is a union of } m \text{ disjoint translates of } A_3 ,$$

and $t_4 \geq \min\{t_2, k_3\}$, $t_1 + t_3 \geq t_2 - 1$. Suppose in fact that

$$\begin{aligned} A_2 &= \bigcup_{i=1}^{m} (u_i + A_1) , \\ A_4 &= \bigcup_{i=1}^{m} (v_i + A_3) , \end{aligned}$$

for suitable vectors $u_1, \ldots, u_m, v_1, \ldots, v_m$.

Theorem 10.5. *Suppose $A_1, \ldots, A_4$ are orthogonal arrays as defined above. Then the set of runs*

$$A_5 = \bigcup_{i=1}^{m} \bigcup_{a \in A_1} \bigcup_{b \in A_3} (u_i + a,\ v_i + b)$$

forms an $OA(mN_1N_2, k_1 + k_3, s, t_2)$.

Proof: Again we apply Theorem 3.30, and will show that, for all $x \in S^{k_1}$, $y \in S^{k_3}$ with $1 \leq w(x) + w(y) \leq t_2$, the sum

$$\sum_{(p,q) \in A_5} \zeta^{(p,q)(x,y)^T} = \sum_{i=1}^{m} \zeta^{u_i x^T + v_i y^T} \sum_{a \in A_1} \zeta^{a x^T} \sum_{b \in A_3} \zeta^{b y^T}$$

is equal to zero. There are four cases to consider:

$$w(x) = 0, \quad 1 \leq w(y) \leq t_2 ,$$

$$1 \leq w(x) \leq t_2, \quad w(y) = 0 ,$$

$$1 \leq w(x) \leq t_1, \quad 1 \leq w(y) \leq t_2 - w(x) ,$$

$$t_1 + 1 \leq w(x) \leq t_2, \quad 1 \leq w(y) \leq t_2 - w(x) ,$$

and in each case the hypotheses imply that the sum vanishes. ∎

Example 10.6. It can be shown that the $OA(256, 16, 2, 5)$ constructed in Section 5.10 is a union of eight translates of an $OA(32, 16, 2, 3)$ Hadamard array. (This follows from the fact that the Nordstrom-Robinson code is a union of eight translates of the first-order Reed-Muller code of length 16 — see Chapter 15 of MacWilliams and Sloane, 1977; Hammons, Kumar, Calderbank, Sloane and Solé, 1994.) For A_3 and A_4 we take the trivial arrays $OA(2,4,2,1)$, $OA(16,4,2,4)$. From Theorem 10.5 we obtain an $OA(512,20,2,5)$. Similarly, using Kerdock codes, we obtain an $OA(2^{2m+1}, 2^m+m, 2, 5)$ for all even values of $m \geq 4$. ■

10.5 Orthogonal Arrays from Union of Translates of a Linear Code

The following construction is due to Srivastava and Chopra (1973) (see also Srivastava and Throop, 1990), and is a generalization of Theorem 4.6. It gives necessary and sufficient conditions for a union of translates of a linear code to form an orthogonal array.

Suppose C is a linear $(k, s^n, d)_s$ code defined by a $(k-n) \times k$ parity check matrix P (as in Section 4.2). Thus C consists of all vectors $u \in GF(s)^k$ such that $Pu^T = 0$. For $z \in GF(s)^{k-n}$ the set of vectors $u \in GF(s)^k$ such that $Pu^T = z^T$ is a translate, $C(z)$ say, of C, also containing s^n vectors.

In this construction we choose vectors $z_1, \ldots, z_m \in GF(s)^{k-n}$ and form an $ms^n \times k$ matrix A by juxtaposing the translates $C(z_1), \ldots, C(z_m)$. Let Z denote the $m \times (k-n)$ matrix whose rows are $z_1, \ldots, z_m$.

Theorem 10.7. *A is an $OA(ms^n, k, s, t)$ if and only if for every nonzero vector $y \in GF(s)^{k-n}$ with $w(yP) \leq t$ the $m \times 1$ array Zy^T has strength 1.*

10.6 Bounds on Large Orthogonal Arrays

It follows from Shannon's work (Shannon 1948, 1993) that the efficiency of a digital communications scheme can be increased by using longer and longer codes. Codes of length over 1000 are in common use. Fundamental theorems in coding theory give bounds on the error-correcting ability of long codes. The next pair of theorems give analogous bounds on the minimal size that is needed in orthogonal arrays with large numbers of factors when the ratio strength/number of factors is held fixed. These results are of primarily theoretical interest, and are unlikely to have any impact on current applications of orthogonal arrays.

Theorem 10.8. *Let s be a prime power. Given any τ, $0 < \tau < 0.5$, and any $\epsilon > 0$, there is a number k_0 depending only on τ and ϵ such that, for all $k \geq k_0$, there exists an $OA(\lambda s^t, k, s, t)$ with*

$$\frac{t}{k} \geq \tau$$

and

$$\frac{1}{k}\log_s \lambda \le H_s(\tau) - \tau + \epsilon \ ,$$

where

$$H_s(x) = -x \log_s x - (1-x)\log_s(1-x)$$

is the entropy function to base s.

Proof: The Gilbert-Varshamov bound (see for example Gilbert, 1952; Varshamov, 1957; MacWilliams and Sloane, 1977, Chapter 17; van Lint, 1982, Chapter 5) shows that linear codes exist over $GF(s)$ for which the ratios R = dimension / length and δ = minimal distance / length satisfy $\delta \ge H_s^{-1}(1-R) + o(1)$. By considering the dual codes and applying Theorem 4.9 we obtain the result stated in the theorem. ■

There is a similar bound in the other direction. For simplicity we only give the binary version.

Theorem 10.9. *Given any τ, $0 < \tau < 0.5$, and any $\epsilon > 0$, there is a number k_0 depending only on τ and ϵ such that, for all $k \ge k_0$, any $OA(\lambda s^t, k, 2, t)$ with*

$$\frac{t}{k} \ge \tau$$

must satisfy

$$\frac{1}{k}\log_2 \lambda \ge 1 - \tau - H_2\left(\frac{1}{2} - \sqrt{\tau(1-\tau)}\right) - \epsilon \ .$$

Proof: If such an orthogonal array exists then its dual distance is at least $t+1$. We may now apply the McEliece-Rodemich-Rumsey-Welch (1977) bound to the dual weight distribution (see MacWilliams and Sloane, 1977, Chapter 17). It does not matter that there is no "dual code" with this weight distribution: the McEliece et al. theorem only makes assertions about weight distributions. The stated result now follows easily from Theorem 35 of MacWilliams and Sloane (1977), Chapter 17. ■

Concerning the existence of large orthogonal arrays, besides the results mentioned in earlier chapters (some of which are summarized in Table 12.6), several asymptotic results are known:

(i) Ray-Chaudhuri and Singhi (1988) (see also Rosenberg, 1996) shows that for all values of $s \ge 2$ and $k \ge t \ge 2$, an $OA(\lambda s^t, k, s, t)$ always exists if λ is sufficiently large.

(ii) Blanchard (1994, 1995, 1996) shows that for any choice of $k \ge t \ge 1$, an $OA(s^t, k, s, t)$ of index 1 always exists if s is sufficiently large.

10.7 Compound Orthogonal Arrays

Compound orthogonal arrays were introduced by Rosenbaum (1994) for studying dispersion effects in factorial experiments (see also Section 11.8). Techniques for constructing such arrays were given by Rosenbaum (1994, 1996) and, the main source for this section, Hedayat and Stufken (1999).

Throughout this section orthogonal arrays may have mixed levels, and we will make use of the notation developed in Chapter 9.

Definition 10.10. Let

$$B = \begin{bmatrix} b_1 \\ \vdots \\ b_{N_1} \end{bmatrix}$$

be an $OA(N_1, s_1 s_2 \ldots s_{k_1}, t_1)$. For each $i \in \{1, \ldots, N_1\}$ let C_i be an array with parameters $OA(N_2, u_1 u_2 \ldots u_{k_2}, t_2)$, and form the $N_1 N_2 \times (k_1 + k_2)$ array

$$A = \begin{bmatrix} b_1 & \\ \vdots & C_1 \\ b_1 & \\ \vdots & \\ b_{N_1} & \\ \vdots & C_{N_1} \\ b_{N_1} & \end{bmatrix} .$$

Suppose A is itself an orthogonal array of strength t_3. Then an array with this structure is called a *compound orthogonal array of type* (t_1, t_2, t_3). We will denote such an array by the symbol $COA((N_1, N_2), (k_1, k_2), (s_1 s_2 \ldots s_{k_1}, u_1 u_2 \ldots u_{k_2}), (t_1, t_2, t_3))$.

Although the definition requires all the C_i to have strength t_2, we do not exclude the possibility that some C_i have greater strength. If C_i has strength $t_{2,i}$, we use t_2 to denote the minimum of the $t_{2,i}$'s.

We will consider only compound orthogonal arrays in which the parameters satisfy $N_1 \le s_1 s_2 \ldots s_{k_1}$ and $N_2 \le u_1 u_2 \ldots u_{k_2}$. To facilitate the discussion we will distinguish between four cases and use the letters C and F for complete and fractional, respectively.

The case CC: $N_1 = s_1 s_2 \ldots s_{k_1}$ and $N_2 = u_1 u_2 \ldots u_{k_2}$.

Using the notation of Definition 10.10, we use each possible level combination of the first k_1 factors to form B, and take the array consisting of all possible level combinations for the last k_2 factors for each of the C_i's. Then A contains every level combination of all $k_1 + k_2$ factors precisely once, so $t_1 = k_1$, $t_2 = k_2$ and $t_3 = k_1 + k_2$, each attaining its maximal possible value.

The case FC: $N_1 < s_1 s_2 \ldots s_{k_1}$ and $N_2 = u_1 u_2 \ldots u_{k_2}$.

For each of the C_i's we again take all possible level combinations once. Then $t_2 = k_2$ and $t_3 = t_1$; interest centers on finding an array B with maximal possible strength.

The case CF: $N_1 = s_1 s_2 \ldots s_{k_1}$ and $N_2 < u_1 u_2 \ldots u_{k_2}$.

Now B consists of all possible level combinations of the first k_1 factors, and each C_i has strength at least t_2 (but not all have strength t_2+1). Then $t_1 = k_1$ and $t_2 \le t_3 \le k_1 + t_2$. If possible we would like A to have strength $t_3 = k_1 + t_2$.

The case FF: $N_1 < s_1 s_2 \ldots s_{k_1}$ and $N_2 < u_1 u_2 \ldots u_{k_2}$.

With t_1 and t_2 as the strengths of B and the C_i's, respectively, it is not too hard to see that

$$\min(t_1, t_2) \le t_3 \le t_1 \ .$$

If possible we would like to achieve $t_3 = t_1$.

The acronym CC might as well stand for crystal clear, and this case needs no further attention. The case FC, while generally not an easy problem, is of the type encountered in the previous chapters. It remains to address the cases CF and FF.

For case CF we will attempt to construct a compound orthogonal array by appropriately grouping the rows and columns of an orthogonal array of one of the types considered in the previous chapters. Indeed, any orthogonal array of strength $t \ge 1$ is a compound orthogonal array for some set of parameters. More precisely, for $1 \le k_1 \le t$, an $OA(N, s_1 s_2 \ldots s_k, t)$ is also a $COA((s_1 s_2 \ldots s_{k_1}, N/(s_1 s_2 \ldots s_{k_1}))$, $(k_1, k - k_1)$, $(s_1 s_2 \ldots s_{k_1}, s_{k_1+1} s_{k_2+2} \cdots s_k)$, $(k_1, t_2, t))$, where $t_2 \ge t - k_1$. If t is the maximal possible strength for the original orthogonal array, then $t_3 = t$ is the maximal possible value for t_3. However, different orthogonal arrays with the same parameters can lead to different values of t_2, and a judicious selection is needed to obtain the maximal value for t_2. This is illustrated in the next example.

Example 10.11. Consider a 2-level orthogonal array $OA(2^6, 8, 2, 4)$. For $N = 64$, $k = 8$ and $s = 2$, a strength of $t = 4$ is indeed the maximum that can be achieved. Using the construction discussed above we can obtain a

$COA((8,8),(3,5),(2^3,2^5),(3,t_2,4))$ from such an orthogonal array, where $t_2 \geq 1$. And indeed, if we use the orthogonal array with parity check matrix (see Section 4.2)

$$P_1 = \begin{bmatrix} 1 & 1 & 1 & 1 & 1 & 0 & 0 & 0 \\ 0 & 0 & 0 & 1 & 1 & 1 & 1 & 1 \end{bmatrix},$$

and take the first three columns for the factors of B in Definition 10.10, we obtain a compound orthogonal array with the above parameters, with $t_2 = 1$. However, if instead we use the orthogonal array with parity check matrix

$$P_2 = \begin{bmatrix} 1 & 1 & 0 & 1 & 1 & 1 & 0 & 0 \\ 0 & 0 & 1 & 0 & 1 & 1 & 1 & 1 \end{bmatrix},$$

obtained by permuting the columns of P_1, we will obtain a $COA((8,8), (3,5), (2^3,2^5), (3,2,4))$ when taking the first three columns for the factors in B. In fact $t_2 = 2$ is the maximal possible value for t_2 in such a compound orthogonal array. ■

Table 10.12 gives the maximal possible value of t_2 in a 2-level $COA((2^{k_1},N_2), (k_1,k_2), (2^{k_1},2^{k_2}), (k_1,t_2,t_3))$, for $2 \leq k_1, k_2 \leq 5$, $N_2 = 2^{k_2-n_2}$, $1 \leq n_2 \leq k_2-1$, and where t_3 is the maximal possible value in an $OA(2^{k_1}N_2, k_1+k_2, 2, t_3)$.

Table 10.12. Maximal value for t_2 in any $COA((2^{k_1},N_2),(k_1,k_2),(2^{k_1},2^{k_2}), (k_1,t_2,t_3))$ for $2 \leq k_1, k_2 \leq 5$, $N_2 = 2^{k_2-n_2}$, $1 \leq n_2 \leq k_2 - 1$, where t_3 takes its maximal possible value. Here $*$ indicates that $t_2 > t_3 - k_1$ and # that t_2 can be increased if t_3 is not maximal.

N_2	k_1	k_2	t_3	t_2		N_2	k_1	k_2	t_3	t_2	
2	2	2	3	1		2	4	2	5	1	
2	2	3	2	1	*	2	4	3	3	1	*
4	2	3	4	2		4	4	3	6	2	
2	2	4	2	1	*	2	4	4	3	1	*
4	2	4	3	1		4	4	4	4	1	*
8	2	4	5	3		8	4	4	7	3	
2	2	5	2	0	#	2	4	5	3	1	*
4	2	5	3	1		4	4	5	3	1	*
8	2	5	3	2	*	8	4	5	5	2	*
16	2	5	6	4		16	4	5	8	4	
2	3	2	4	1		2	5	2	6	1	
2	3	3	3	1	*	2	5	3	4	1	*
4	3	3	5	2		4	5	3	7	2	
2	3	4	3	1	*	2	5	4	3	1	*
4	3	4	3	1	*	4	5	4	5	1	*
8	3	4	6	3		8	5	4	8	3	
2	3	5	3	0	#	2	5	5	3	1	*
4	3	5	3	1	*	4	5	5	4	1	*
8	3	5	4	2	*	8	5	5	5	2	*
16	3	5	7	4		16	5	5	9	4	

Those entries in Table 10.12 that are not marked with a $*$ satisfy $t_3 = k_1 + t_2$, and thus can be obtained from any orthogonal array $OA(2^{k_1}N_2, k_1 + k_2, 2, t_3)$ using any k_1 factors to form the array B. Those marked with a $*$ require more care, as demonstrated in Example 10.11. The construction of these arrays is addressed in Table 10.13. For each set of parameters this table gives $k_2 - n_2$ strings of numbers from 1 to $k_1 + k_2$ (with 0 being used for 10). Each string represents a row of a $(k_2 - n_2) \times (k_1 + k_2)$ parity check matrix with entries 0 and 1. The entry in position (i, j) is 1 if and only if j appears in the i-th string of numbers. For example, the first entry in Table 10.13 gives the strings 134 and 235 and corresponds to the parity check matrix

$$\begin{bmatrix} 1 & 0 & 1 & 1 & 0 \\ 0 & 1 & 1 & 0 & 1 \end{bmatrix} .$$

The strings in Table 10.13 are presented in such a way that the first k_1 factors can be used to form the array B. For those familiar with the terminology, the strings in Table 10.13 may also be thought of as representing defining contrasts for a regular fractional factorial (see also Section 11.5).

Table 10.13. Compound orthogonal arrays for the case CF with $t_2 > t_3 - k_1$.

N_2	k_1	k_2	t_3	t_2	Parity check matrix
2	2	3	2	1	134, 235
2	2	4	2	1	134, 235, 156
8	2	5	3	2	1345, 2367
2	3	3	3	1	1245, 1356
2	3	4	3	1	1245, 1267, 1356
4	3	4	3	1	12456, 1367
4	3	5	3	1	1245, 1356, 2478
8	3	5	4	2	12456, 13678
2	4	3	3	1	1256, 23457
2	4	4	3	1	1256, 1357, 1458
4	4	4	4	1	12356, 14578
2	4	5	3	1	1256, 1357, 1458, 23589
4	4	5	3	1	123456, 12578, 13579
8	4	5	5	2	123567, 145689
2	5	3	4	1	12367, 14568
2	5	4	3	1	123467, 12578, 14579
4	5	4	5	1	123467, 125689
2	5	5	3	1	1267, 134568, 146789, 156780
4	5	5	4	1	12678, 13690, 145670
8	5	5	5	2	1234678, 125690

There are just two cases in Table 10.12 where we can obtain a compound orthogonal array with a larger value of t_2 if the requirement that t_3 attains its maximal possible value is dropped. These are indicated by the symbol ♯. For the first of these, a compound orthogonal array with $t_2 = t_3 = 1$ exists; for the second, an array with $t_2 = 1$ and $t_3 = 2$ can be obtained.

A table similar to Table 10.12, but now for 3-level compound orthogonal arrays with $N_1 = 3^{k_1}$, $t_1 = k_1$, can be obtained with the help of Table 12.2. This is left as an exercise for the reader.

We now turn our attention to the case FF. For this case, if the maximal possible value of t_2 in an $OA(N_2, u_1u_2\dots u_{k_2}, t_2)$ is greater than or equal to the maximum of t_1 in an $OA(N_1, s_1s_2\dots s_{k_1}, t_1)$, then the maximum for t_3 in a compound orthogonal array that uses orthogonal arrays as above for the C_i's and B is equal to the maximum for t_1. Moreover, in building such a compound orthogonal array all the C_i's can be taken to be the same array. However, if the above maximum for t_2 is less than that for t_1, a larger value of t_3 may be obtained by not taking all the C_i's to be the same; if all the C_i's are the same, $t_3 = t_2 = \min(t_1, t_2)$.

Table 10.14. The maximal possible value for t_2 in a compound array $COA((N_1, N_2), (k_1, k_2), (2^{k_1}, 2^{k_2}), (t_1, t_2, t_3))$ for $3 \le k_1 \le 5$, $2 \le k_2 \le 5$, $N_1 = 2^{k_1 - n_1}$, $1 \le n_1 \le k_1 - 2$, $N_2 = 2^{k_2 - n_2}$, $1 \le n_2 \le k_2 - 1$, in which t_1 is the maximal possible strength in an $OA(N_1, k_1, 2, t_1)$, and t_3 is the smaller of t_1 and the maximal possible strength in an $OA(N_1N_2, k_1 + k_2, 2, t_3)$. Nos. 7, 17, 22, 24, 28, 30, 34 and 36 satisfy $t_1 > t_3$; in all other cases $t_1 = t_3$.

No.	N_1	N_2	k_1	k_2	t_1	t_3	t_2	No.	N_1	N_2	k_1	k_2	t_1	t_3	t_2
1	4	2	3	2	2	2	1	21	8	2	5	2	2	2	1
2	4	2	3	3	2	2	1	22	16	2	5	2	4	3	1
3	4	4	3	3	2	2	2	23	8	2	5	3	2	2	1
4	4	2	3	4	2	2	1	24	16	2	5	3	4	3	1
5	4	4	3	4	2	2	1	25	8	4	5	3	2	2	2
6	4	8	3	4	2	2	3	26	16	4	5	3	4	4	2
7	4	2	3	5	2	1	1	27	8	2	5	4	2	2	1
8	4	4	3	5	2	2	1	28	16	2	5	4	4	3	1
9	4	8	3	5	2	2	2	29	8	4	5	4	2	2	1
10	4	16	3	5	2	2	4	30	16	4	5	4	4	3	1
11	8	2	4	2	3	3	1	31	8	8	5	4	2	2	3
12	8	2	4	3	3	3	1	32	16	8	5	4	4	4	3
13	8	4	4	3	3	3	2	33	8	2	5	5	2	2	1
14	8	2	4	4	3	3	1	34	16	2	5	5	4	3	1
15	8	4	4	4	3	3	1	35	8	4	5	5	2	2	1
16	8	8	4	4	3	3	3	36	16	4	5	5	4	3	1
17	8	2	4	5	3	2	1	37	8	8	5	5	2	2	2
18	8	4	4	5	3	3	1	38	16	8	5	5	4	4	2
19	8	8	4	5	3	3	2	39	8	16	5	5	2	2	4
20	8	16	4	5	3	3	4	40	16	16	5	5	4	4	4

Table 10.15. Compound orthogonal arrays for the case FF.

No.	Parity check matrix	No.	Parity check matrix
1	123, 345	21	123, 145, 2467
2	123, 345, 1356	22	12345, 1267
3	123, 1456	23	123, 145, 2467, 2568
4	123, 1245, 1346, 2347	24	12345, 1267, 1368
5	123, 145, 2467	25	123, 145, 24678
6	123, 12456	26	12345, 12678
7	123, 145, 246, 347, 48	27	123, 145, 167, 268, 489
8	123, 1245, 13467, 23468	28	12345, 1267, 1368, 1469
9	123, 1456, 2478	29	123, 145, 1678, 2469
10	123, 145678	30	12345, 1267, 1368, 12489
11	1234, 1256	31	123, 145, 246789
12	1234, 1256, 1357	32	12345, 126789
13	1234, 12567	33	123, 145, 167, 189, 290, 460
14	1234, 1256, 1357, 1458	34	12345, 1267, 1368, 1469, 1560
15	1234, 12567, 1358	35	123, 145, 2467, 1689, 1780
16	1234, 125678	36	12345, 1267, 13689, 3590
17	1234, 1256, 1357, 1278, 12579	37	123, 145, 24678, 25690
18	1234, 1256, 1357, 14589	38	12345, 12678, 13690
19	1234, 12567, 13589	39	123, 145, 2467890
20	1234, 1256789	40	12345, 1267890

Example 10.16. As in Example 10.11, let $N = 64$, $k = 8$, with all factors at 2 levels. We also let $N_1 = N_2 = 8$ as before, but consider the case $k_1 = k_2 = 4$. Now both B and the C_i's of Definition 10.10 must be fractional factorials. Since the maximal strength in an $OA(8, 4, 2, t)$ is $t = 3$, it follows that $t_1 \leq 3$ and $t_2 \leq 3$ in a $COA((8, 8), (4, 4), (2^4, 2^4), (t_1, t_2, t_3))$. Considering the maximal strength in an $OA(64, 8, 2, t)$, it also follows that $t_3 \leq 4$. However, as seen above, we also know that $t_3 \leq t_1$. Hence the question becomes whether a compound orthogonal array of type $(3, 3, 3)$ exists for these parameters. The answer is affirmative, as is seen by taking the orthogonal array $OA(64, 8, 2, 3)$ with parity check matrix

$$P = \begin{bmatrix} 1 & 1 & 1 & 1 & 0 & 0 & 0 & 0 \\ 1 & 1 & 0 & 0 & 1 & 1 & 1 & 1 \end{bmatrix}$$

and using the first four columns for the factors of B in Definition 10.10. ∎

Table 10.14 gives the maximal possible value for t_2 in a $COA((N_1, N_2),$ (k_1, k_2), $(2^{k_1}, 2^{k_2})$, $(t_1, t_2, t_3))$ for $3 \leq k_1 \leq 5$, $2 \leq k_2 \leq 5$, $N_1 = 2^{k_1 - n_1}$, $1 \leq n_1 \leq k_1 - 2$, $N_2 = 2^{k_2 - n_2}$, $1 \leq n_2 \leq k_2 - 1$, in which t_1 is the maximal possible strength in an $OA(N_1, k_1, 2, t_1)$, and t_3 is the smaller of t_1 and the maximal possible strength in an $OA(N_1N_2, k_1 + k_2, 2, t_3)$. In all cases, the given value of t_2 is also its maximum if we do not insist that t_1 and t_3 attain their maximal values.

The case $N_1 = 2$ has not been included since it can only result in $t_1 = t_3 = 1$.

Table 10.15 presents parity check matrices for all the entries of Table 10.14, using the same notation as Table 10.13. Once again, factors 1 through k_1 can be used to obtain the level combinations for the array B of Definition 10.10.

10.8 Orthogonal Multi-Arrays

Balanced orthogonal multi-arrays, introduced by Sitter (1993), form a class of arrays that includes orthogonal arrays. While other members of the class have not nearly as many applications as orthogonal arrays at the present time, there is a bona fide application in sampling (Sitter, 1993) and the class is quite interesting from a combinatorial perspective.

In a multi-array each entry of the array is allowed to be a subset of a set of symbols. As in mixed orthogonal arrays the set of symbols may vary from one factor to another. (While the main application of these arrays is not in factorial experiments, it is convenient to continue to use the terminology and notation of the previous chapters, and, for example, to refer to the columns of the array as factors.) Following Mukerjee (1998) we will restrict attention to "proper" balanced orthogonal multi-arrays. We will also limit consideration to arrays of strength 2, even though an extension to higher strengths is immediately apparent.

Definition 10.17. An $N \times k$ array $A = (\alpha_{ij})$ in which α_{ij} is a subset of size n_j, $1 \le n_j \le s_j - 1$, of $S_j = \{0, 1, \ldots, s_j - 1\}$ is called a *proper balanced orthogonal multi-array* (of strength 2) if the following conditions are satisfied:

(i) for every $j \in \{1, 2, \ldots, k\}$, the values of $\sum_{i=1}^N \delta_{ij}^\ell$ and $\sum_{i=1}^N \delta_{ij}^\ell \delta_{ij}^{\ell'}$, $\ell \neq \ell'$, do not depend on the choice of ℓ, $\ell' \in S_j$; and

(ii) for every j, $j' \in \{1, 2, \ldots, k\}$, $j \neq j'$, the value of $\sum_{i=1}^N \delta_{ij}^\ell \delta_{ij'}^{\ell'}$ does not depend on the choice of $\ell \in S_j$ and $\ell' \in S_{j'}$,

where we define

$$\delta_{ij}^\ell = \begin{cases} 1 & \text{if } \ \ell \in \alpha_{ij} \ , \\ 0 & \text{otherwise,} \end{cases}$$

for $i = 1, 2, \ldots, N$, $j = 1, 2, \ldots, k$, $\ell = 0, 1, \ldots, s_j - 1$.

We will denote such an array by either of the symbols $BOMA(N, s_1s_2\dots s_k; n_1n_2\dots n_k)$ or $BOMA(N, s_1^{k_1}\dots s_u^{k_u}; n_1^{m_1}\dots n_v^{m_v})$, where $\sum_{h=1}^{u} k_h = \sum_{h=1}^{v} m_h = k$.

Clearly, if A is a $BOMA(N, s_1s_2\dots s_k; n_1n_2\dots n_k)$ then

$$\sum_{i=1}^{N} \delta_{ij}^{\ell} = Nn_j/s_j, \quad \text{for} \quad \ell \in S_j \ ,$$

$$\sum_{i=1}^{N} \delta_{ij}^{\ell}\delta_{ij}^{\ell'} = Nn_j(n_j-1)/(s_j(s_j-1)) \text{ for } \ell, \ell' \in S_j \ , \ \ell \neq \ell' \ ,$$

and

$$\sum_{i=1}^{N} \delta_{ij}^{\ell}\delta_{ij'}^{\ell'} = Nn_jn_{j'}/(s_js_{j'}) \text{ for } j \neq j', \ \ell \in S_j \ , \ell' \in S_{j'} \ .$$

Note that condition (i) in Definition 10.17 simply requires that for each factor the N subsets form a balanced incomplete block design (degenerate if $n_j = 1$). Condition (ii) pertains to the inter-factor structure of the array. The case when $n_j = 1$ for all $j \in \{1, 2, \dots, k\}$ corresponds to a (mixed level) orthogonal array of strength 2. The adjective "proper" in the definition refers to the requirement that, for each j, the sets $\alpha_{1j}, \alpha_{2j}, \dots, \alpha_{Nj}$ have the same size. Sitter's original definition did not require this.

Given a $BOMA(N, s_1s_2\dots s_k; n_1n_2\dots n_k)$, if the sets $\alpha_{11}, \dots, \alpha_{N1}$ are replaced by their complements in S_1, then it is easily seen that the resulting array is a $BOMA(N, s_1s_2\dots s_k; (s_1-n_1)n_2\dots n_k)$. This observation, which like most others in this section is due to Mukerjee (1998), implies that in constructing these arrays we may assume that $n_j \leq s_j/2$, for $j = 1, \dots, k$.

Example 10.18. Let $s_1 = s_2 = 4$ and $n_1 = n_2 = 2$. Conditions (i) and (ii) imply that $N \equiv 0 \pmod{12}$. A $BOMA(12, 4^2; 2^2)$ trivially exists. It is possible however to add more factors to this array without increasing the value of N. Table 10.19 gives a $BOMA(12, 4^3 3^1; 2^3 1^1)$. We write sets $\{i\}$ and $\{i, j\}$ simply as i and ij, respectively. ■

Table 10.19. A $BOMA(12, 4^3 3^1; 2^3 1^1)$.

01	01	01	0
01	23	23	0
02	02	02	1
02	13	13	1
03	03	03	2
03	12	12	2
12	03	12	2
12	12	03	2
13	02	13	1
13	13	02	1
23	01	23	0
23	23	01	0

The following result shows that no further factors can be adjoined to the *BOMA* in Table 10.19.

Theorem 10.20. *If a* BOMA$(N, s_1 s_2 \ldots s_k; n_1 n_2 \ldots n_k)$ *exists then*

$$N \geq 1 + \sum_{j=1}^{k} (s_j - 1) \ . \tag{10.4}$$

Proof: Let N_j denote the $s_j \times N$ incidence matrix for the balanced incomplete block design formed by the subsets for factor j in an array with parameters $BOMA(N, s_1 s_2 \ldots s_k; n_1 n_2 \ldots n_k)$. Using conditions (i) and (ii) in Definition 10.17, it follows easily that

$$N_j (I_N - P_{1_N}) N_{j'}^T = 0, \quad \text{if} \quad j \neq j' \ ,$$

and

$$N_j (I_N - P_{1_N}) N_j^T = \frac{N n_j (s_j - n_j)}{s_j (s_j - 1)} \left(I_{s_j} - \frac{1}{s_j} J_{s_j} \right) \ .$$

Note that this last matrix has rank $s_j - 1$. Defining the $N \times \left(1 + \sum_{j=1}^{k} s_j\right)$ matrix B to be

$$B = [1_N \ \ (I_N - P_{1_N}) N_1^T \ \cdots \ (I_N - P_{1_N}) N_k^T]$$

we see that $B^T B$ is the block-diagonal matrix

$$B^T B = \begin{bmatrix} N & 0_{s_1}^T & \cdots & 0_{s_k}^T \\ 0_{s_1} & N_1 (I_N - P_{1_N}) N_1^T & & \\ \vdots & & \ddots & \\ 0_{s_k} & & & N_k (I_N - P_{1_N}) N_k^T \end{bmatrix} ,$$

which has rank $1 + \sum_{j=1}^{k}(s_j - 1)$. Since this is also the rank of B, which has N rows, the result follows. ∎

Example 10.21. Equality in (10.4) holds for the array in Table 10.19, confirming that no factors can be added to that array without increasing N. While (10.4) is a necessary condition, it is of course not a sufficient condition for the existence of a $BOMA(N, s_1s_2 \dots s_k; n_1n_2 \dots n_k)$. Again taking $N = 12$, $s_1 = s_2 = 4$, $n_1 = n_2 = 2$, consider the cases

(i) $s_3 = 4$, $s_4 = s_5 = 2$, $n_3 = 2$, $n_4 = n_5 = 1$;

(ii) $s_3 = 3$, $s_4 = s_5 = s_6 = 2$, $n_3 = n_4 = n_5 = n_6 = 1$; and

(iii) $s_3 = s_4 = s_5 = s_6 = s_7 = 2$, $n_3 = n_4 = n_5 = n_6 = n_7 = 1$.

For each of these a *BOMA* would give equality in (10.4). However, none of these arrays exists. The nonexistence of a $BOMA(12, 4^3 2^2; 2^3 1^2)$ and a $BOMA(12, 4^2 2^5; 2^2 1^5)$ was shown in Mukerjee (1998). The nonexistence of a $BOMA(12, 4^2 3^1 2^3; 2^2 1^4)$ is left as an exercise for the reader. ∎

For many choices of the parameters s_j and n_j $(j = 1, \dots, k)$ that satisfy the divisibility conditions implied by (i) and (ii) of Definition 10.17 there is no *BOMA* that achieves equality in (10.4). Mukerjee (1998) provides some useful tools to address this problem, but it is clear that more work is needed.

We conclude this section with a construction that is essentially due to Sitter (1993), with a modification (to assure that the resulting arrays are proper) and extension due to Mukerjee (1998).

An α-resolvable $BIB(v, k, \lambda)$ design is a balanced incomplete block design for v varieties in blocks of size k such that each pair of varieties appears in λ blocks and such that the blocks can be partitioned into r/α sets of blocks, each set containing $\alpha v/k$ blocks, with each treatment appearing in α blocks of each set. The number of blocks that contain a given treatment is denoted by r; in a $BIB(v, k, \lambda)$ we have $r = \lambda(v - 1)/(k - 1)$.

To construct a *BOMA* with k factors, symbol sets of sizes $s_1, \dots, s_k$, and sets in the array of sizes $n_1, \dots, n_k$ by this method we need the following:

1. For each $j \in \{1, 2, \dots, k\}$, an α_j-resolvable $BIB(s_j, n_j, \lambda_j)$ for some positive integers α_j and λ_j. Denote the varieties in the *BIB* design by $0, 1, \dots, s_j - 1$, and let $r_j = \lambda_j(s_j - 1)/(n_j - 1)$.

2. An $OA(N, w_1w_2 \dots w_k, 2)$, where $w_j = \alpha_j s_j / n_j$.

From these ingredients we obtain a $BOMA(gN, s_1s_2\dots s_k; n_1n_2\dots n_k)$, where g is the least common multiple of $r_1/\alpha_1,\dots,r_k/\alpha_k$, by the following construction.

(a) Let $\mathcal{G}_{ji}$, $i = 1,\dots,r_j/\alpha_j$, denote the sets of blocks corresponding to the resolvable form of the $BIB(s_j, n_j, \lambda_j)$ in 1. Label the blocks in each set by $0, 1,\dots, w_j - 1$, where w_j is as defined in 2.

(b) Form an $OA(gN, w_1w_w\dots w_k, 2)$, say B, by the juxtaposition of g copies of the $OA(N, w_1w_2\dots w_k, 2)$ from 2, say A, where g is as defined above. Thus

$$B = 0_g \otimes A = \begin{bmatrix} A \\ \vdots \\ A \end{bmatrix} ,$$

where the entries for factor j are $0, 1,\dots, w_j - 1$.

(c) For $j = 1,\dots,k$, replace the entries of B in (b) by subsets of $\{0, 1,\dots, s_j - 1\}$ of size n_j as follows. For the first $gN/(r_j/\alpha_j) = \alpha_j gN/r_j$ entries for factor j, replace $0, 1,\dots, w_j - 1$ by the blocks with these labels (see (a)) in $\mathcal{G}_{j1}$; for the next $\alpha_j gN/r_j$ entries repeat this, but now using $\mathcal{G}_{j2}$; and so on.

The verification that the resulting array is a $BOMA(gN, s_1s_2\dots s_k; n_1n_2\dots n_k)$ is left as an exercise for the reader.

Mukerjee (1998) observes that it may be possible to add one or more factors to this array if the orthogonal array A in (b) can be written as the juxtaposition of several orthogonal arrays of strength 1. We omit the details here, but will illustrate the basic idea in the next example.

Table 10.22. An $OA(8, 7, 2, 2)$.

0	0	0	0	0	0	0
1	0	0	1	1	0	1
0	1	0	1	0	1	1
1	1	0	0	1	1	0
0	0	1	0	1	1	1
1	0	1	1	0	1	0
0	1	1	1	1	0	0
1	1	1	0	0	0	1

Example 10.23. We will use the above ideas to construct an array with parameters $BOMA(24, 4^7 3^1; 2^7 1^1)$. The first ingredient is a 1-resolvable $BIB(4, 2, 1)$. For $j = 1,\dots,7$ we will use

$$\mathcal{G}_{j1} = \{01, 23\},\quad \mathcal{G}_{j2} = \{02, 13\},\quad \mathcal{G}_{j3} = \{03, 12\} ,$$

where, for example, 01 denotes the block consisting of varieties 0 and 1. The next ingredient is the $OA(8,7,2,2)$ in Table 10.22. Since $r_j/\alpha_j = 3$, for $j = 1,\ldots,7$, we have $g = 3$, so we use three copies of this orthogonal array to obtain an $OA(24,7,2,2)$. Further, $\alpha_j gN/r_j = 8$, $j = 1,2,\ldots,7$; hence $\mathcal{G}_{j1}$ is used for the first 8 elements of factor j in the $OA(24,7,2,2)$, and so on. Using these blocks to replace entries of the orthogonal array produces the first seven columns of the array shown in Table 10.24. One additional factor can be added as shown in that table. Observe that the idea for adding one or more factors is similar to that used in Chapter 6 to add factors to an orthogonal array constructed from a difference scheme. ■

The *BOMA* in Table 10.24 achieves equality in (10.4).

Table 10.24. A $BOMA(24, 4^7 3^1; 2^7 1^1)$.

01	01	01	01	01	01	01	0
23	01	01	23	23	01	23	0
01	23	01	23	01	23	23	0
23	23	01	01	23	23	01	0
01	01	23	01	23	23	23	0
23	01	23	23	01	23	01	0
01	23	23	23	23	01	01	0
23	23	23	01	01	01	23	0
02	02	02	02	02	02	02	1
13	02	02	13	13	02	13	1
02	13	02	13	02	13	13	1
13	13	02	02	13	13	02	1
02	02	13	02	13	13	13	1
13	02	13	13	02	13	02	1
02	13	13	13	13	02	02	1
13	13	13	02	02	02	13	1
03	03	03	03	03	03	03	2
12	03	03	12	12	03	12	2
03	12	03	12	03	12	12	2
12	12	03	03	12	12	03	2
03	03	12	03	12	12	12	2
12	03	12	12	03	12	03	2
03	12	12	12	12	03	03	2
12	12	12	03	03	03	12	2

Research Problem 10.25. Develop tools for the construction of balanced orthogonal multi-arrays or for establishing their nonexistence.

10.9 Transversal Designs, Resilient Functions and Nets

In this section we briefly discuss connections between orthogonal arrays and some other combinatorial structures not discussed elsewhere in the book.

Transversal designs

A *transversal design* $TD_\lambda(k,s)$ is a triple $(\mathcal{V},\mathcal{G},\mathcal{B})$, where:

(i) $\mathcal{V}$ is a set of ks elements,

(ii) $\mathcal{G}$ is a partition of $\mathcal{V}$ into k disjoint sets called *groups*[1], each of size s,

(iii) $\mathcal{B}$ is a collection of k-subsets of $\mathcal{V}$ called *blocks*, such that

(iv) every unordered pair of elements from $\mathcal{V}$ is contained either in exactly one group or in exactly λ blocks.

A transversal design $TD_\lambda(k,s)$ is precisely the same as an $OA(\lambda s^2,k,s,2)$. To see this, let $A=(a_{ij})$ be the orthogonal array, and S the set of levels. We construct a transversal design by setting $\mathcal{V}=S\times\{1,2,\ldots,k\}$ and taking the groups to be the subsets $S\times\{j\}$, $1\le j\le k$. For each row of the array we define a block

$$B_i=\{(a_{ij},j):1\le j\le k\},\quad 1\le i\le\lambda s^2 .$$

It is easily checked that, because the orthogonal array has strength 2, we do indeed obtain a transversal design. The steps can be reversed to obtain an orthogonal array of strength 2 from a transversal design. Thus the two concepts describe the same combinatorial structure, but in different ways. The concept of an $(s,k;\lambda)$-*net* is also equivalent to an $OA(\lambda s^2,k,s,2)$, but we will not give the definition here (see Abel, Brouwer, Colbourn and Dinitz, 1996, and Colbourn, 1996b).

Some aspects of the theory of orthogonal arrays are more easily handled in the language of transversal designs. For example, it follows from the work of Hanani (see Hanani, 1975), Colbourn (1996a) and others that one can give an almost complete list of all cases in which an $OA(\lambda s^2,k,s,2)$ exists with $\lambda\ge 2$ and k at most 9. That is, for each $k\le 9$, there is a list stating for which values of $\lambda\ge 2$ and s there exists an $OA(\lambda s^2,k,s,2)$. For $k=8$ and 9 there is a quite short list of possible exceptions that have not yet been completely settled. (For $\lambda = 1$ these questions become questions about the existence of pairwise orthogonal Latin squares—see Theorem 8.28.) For further details the reader may consult the above references, and Beth, Jungnickel and Lenz (1986) and Colbourn (1996b).

[1] Also called groops, an ugly word.

The following is a construction of orthogonal arrays due to M. Grieg, which was originally stated in terms of transversal designs. We give only the most interesting special case.

Theorem 10.26. *If $s-1$ and $s+1$ are both prime powers then there exists an $OA(2s^2, s, s, 2)$.*

Proof:
CONSTRUCTION: Begin by constructing a linear $OA((s+1)^2, s+2, s+1, 2)$ over $GF(s+1)$, using the Rao-Hamming construction (Sections 3.4, 5.3). This contains runs in which no factors are at level 0, and so by Problem 3.8 may be assumed to contain the runs $00\ldots0$ and $11\ldots1$. In this array, replace every 1 by a 0, delete the two runs $00\ldots0$, and delete any two columns, obtaining a partial array L. Second, construct an $OA((s-1)^2, s, s-1, 2)$, M, say, on the $s-1$ symbols $2, \ldots, s$, and take the union of the runs in L and M, obtaining an array with symbols $0, 2, \ldots, s$. There are a total of $(s+1)^2 - 2 + (s-1)^2 = 2s^2$ runs.

VERIFICATION: Consider any two columns, and consider two symbols a and b. If $(a, b) = (0, 0)$, these two symbols are contained in the runs from L that formerly contained $(0, 1)$ and $(1, 0)$ in these two columns. If $a = 0$, $b \neq 0$, these two symbols are contained in the runs from L that formerly contained $(0, b)$ and $(1, b)$. Finally, if $a \neq 0$, $b \neq 0$, they are contained in one run from L and one from M. ■

This construction produces arrays with parameters $OA(72, 6, 6, 2)$, $OA(200, 10, 10, 2)$, $OA(288, 12, 12, 2)$, $OA(648, 18, 18, 2)$, among others. (Note however that we can obtain an $OA(72, 7, 6, 2)$ using a difference scheme—see Table 12.7.)

More general versions are given by Colbourn, 1996a and Buratti, 1998. For example, it is easy to modify the above proof to show that if there is an $OA((s-1)^2, k_1, s, 2)$ and an $OA((s+1)^2, k_2, s, 2)$ then there exists an $OA(2s^2, k, s, 2)$ where $k = min\{k_1, k_2\}$.

The difference schemes found by Abel and Cheng (1994) that are mentioned in Table 6.67 were also originally constructed via transversal designs.

Resilient functions

The motivation here is from cryptography and computer science. A function $f : \mathbb{Z}_2^k \rightarrow \mathbb{Z}_2^m$, with $k \geq m + t$, is called *t-resilient* if the vectors $f^{-1}(y_1, \ldots, y_m)$ form the rows of an $OA(2^{k-m}, k, 2, t)$ for every choice of $(y_1, \ldots, y_m) \in \mathbb{Z}_2^m$.

These orthogonal arrays are obviously disjoint and partition all of $\mathbb{Z}_2^k$ — such a partition of $\mathbb{Z}_2^k$ into disjoint orthogonal arrays is called a *large set* of orthogonal arrays.

Thus a t-resilient function is equivalent to a large set of $OA(2^{k-m}, k, 2, t)$'s. The translates of a *linear* orthogonal array clearly form a large set, and so provide a large number of examples of t-resilient functions. Many nonlinear orthogonal arrays also have the property that their translates form a large set: arrays constructed from the Nordstrom-Robinson, Kerdock, Preparata, etc. codes (see Chapter 5) are examples.

The application of t-resilient functions in cryptography arises because of the following property.

Lemma 10.27. *Let $f : \mathbb{Z}_2^k \to \mathbb{Z}_2^m$ be a t-resilient function.*

(i) Let $Pr(y_1, \ldots, y_m)$ denote the probability that $f(x_1, \ldots, x_k) = (y_1, \ldots, y_m)$ when the x_i's are chosen at random. Then $Pr(y_1, \ldots, y_m) = 2^{-m}$, independently of the choice of $(y_1, \ldots, y_m)$.

(ii) Let $Pr(y_1, \ldots, y_m \mid x_{i_1} = c_1, \ldots, x_{i_t} = c_t)$ denote the probability that the output of f is $(y_1, \ldots, y_m)$, given that t of the x_i are specified in advance. Then this probability is also equal to 2^{-m}, for all choices of the y_j's, the x_i's and the c_h's.

We omit the straightforward proof.

In words, the lemma states that the outputs of a t-resilient function are uniformly distributed, and remain so even if an adversary can gain control over up to t of the input variables.

For further information we refer the reader to the extensive literature on this subject: see for example Bennett, Brassard and Robert (1986), Bierbrauer (1995), Bierbrauer, Gopalakrishnan and Stinson (1996), Camion, Carlet, Charpin and Sendrier (1991), Camion and Canteaut (1996), Chor et al. (1985), Friedman (1992), Stinson (1993, 1997). The survey article by Colbourn, Dinitz and Stinson (1999) is especially recommended.

(t, s, m)-nets

A second kind of "net", unrelated to those mentioned at the beginning of this section, has been studied in connection with numerical integration and random number generation. These (t, s, m)*-nets* were introduced by Niederreiter (1987) and have since been studied by many authors. They are equivalent to a certain kind of generalized orthogonal array called an *ordered* orthogonal array. Since they are not directly related to the rest of the book, we shall say no more about them, but refer the reader to Clayman et al. (1999), pp. 478–480 of Colbourn and Dinitz (1996a), Martin and Stinson (1999a, 1999b), Niederreiter (1987, 1988, 1992a, 1992b).

10.10 Schematic Orthogonal Arrays

An orthogonal array $OA(N, k, s, t)$ is called *schematic* if its runs form an association scheme with respect to Hamming distance, that is if the following condition is satisfied. Let $R_1, \ldots, R_N$ be the runs. The condition is that, if $\text{dist}(R_a, R_b) = h$, then the number of $c \in \{1, \ldots, N\}$ such that

$$\text{dist}(R_a, R_c) = i \quad \text{and} \quad \text{dist}(R_b, R_c) = j$$

is independent of $a, b \in \{1, \ldots, N\}$ (but may depend on h, i, j), for all $h, i, j \in \{0, \ldots, k\}$. Association schemes are of great importance in combinatorics (see Delsarte, 1973; Sloane, 1975; Goethals, 1979; Bannai and Ito, 1984).

It would be of great interest to classify all schematic orthogonal arrays. The following sufficient condition, due to Delsarte (1973), provides many examples.

Theorem 10.28. *Let σ be the number of distinct nonzero Hamming distances between the runs of an $OA(N, k, s, t)$. If $t \geq 2\sigma - 2$ then the orthogonal array is schematic.*

Example 10.29. The Rao-Hamming arrays of Theorem 3.20 have $t = 2$ and $\sigma = 1$ and are therefore schematic. ■

The condition of Theorem 10.28 is certainly not necessary, since Calderbank and Goethals (1984) have found codes such that the corresponding orthogonal arrays are schematic with $t = \sigma = 3$.

Some preliminary results on the classification of schematic orthogonal arrays have been obtained by Atsumi (1983) and Yoshizawa (1987). For example, Atsumi (1983) has established the following result.

Theorem 10.30. *For any given values of k and λ there are only finitely many values of s for which there exists a schematic $OA(\lambda s^t, k, s, t)$ with $2 \leq t \leq k$.*

This problem is a long way from being solved.

Research Problem 10.31. Classify all schematic orthogonal arrays.

10.11 Problems

10.1. In reference to Theorem 10.1 show that if the orthogonal array A_i has generator matrix of the form $[I\ M_i]$ $(i = 1, 2)$ then

$$\begin{bmatrix} I & M_1 & 0 & 0 \\ 0 & 0 & I & M_2 \\ 0 & I & 0 & -I \end{bmatrix}$$

is a generator matrix for the orthogonal array A_3.

10.2. (i) Let A_i be an $OA(N_i, k_i, s, t_i)$ for $i = 1, 2$ with $k_1 < k_2$. For a row u of A_2, let u^* denote the first k_1 elements of u. Show that the set of all vectors $(u, u^* + v)$, where u is a row of A_2 and v is a row of A_1, forms an $OA(N_1N_2, k_1 + k_2, s, t_3)$ with $t_3 = \min\{2t_1 + 1, t_2\}$.

(ii) Let A_i be an $OA(N_i, k_i, s, t_i)$ for $i = 1, 2$ with $k_1 > k_2$. For a row u of A_2, let $\bar{u} = (u\ 0 \cdots 0)$, i.e. $\bar{u}$ is u padded with $k_1 - k_2$ zeros. Show that the set of all vectors $(u, \bar{u} + v)$, where u is a row of A_2 and v is a row of A_1, forms an $OA(N_1N_2, k_1 + k_2, s, t_3)$ with $t_3 = \min\{2t_1 + 1 - \min\{k_1 - k_2, t_1 + 1\}, t_2\}$.

10.3. Provide compound orthogonal arrays with all factors at 2 levels for the following parameters.

(i) $N_1 = 8$, $N_2 = 2$, $k_1 = 3$, $k_2 = 5$, $t_1 = 3$, $t_2 = 1$, $t_3 = 2$.

(ii) $N_1 = 8$, $N_2 = 4$, $k_1 = 3$, $k_2 = 3$, $t_1 = 3$, $t_2 = 2$, $t_3 = 5$.

(iii) $N_1 = 16$, $N_2 = 4$, $k_1 = 6$, $k_2 = 3$, and t_1, t_2 and t_3 as large as possible.

10.4. Find all possible parameters for which a $COA((N_1, N_2), (k_1, k_2), (2^{k_1}, 2^{k_2}), (t_1, t_2, t_3))$ with $N_1N_2 = 64$ exists. You may restrict your answer to $N_1 = 2^{k_1 - n_1}, 0 \leq n_1 \leq k_1 - 2$, and $N_2 = 2^{k_2 - n_2}, 0 \leq n_2 \leq k_2 - 1$.

10.5. For $2 \leq k_1, k_2, \leq 5$, study the existence of 3-level compound orthogonal arrays.

10.6. Show that a $BOMA(12, 4^2 3^1 2^3; 2^2 1^4)$ does not exist.

10.7. Verify that the construction in Section 10.8 does indeed yield an array with parameters $BOMA(gN, s_1 s_2 \ldots s_k; n_1 n_2 \ldots n_k)$.

Chapter 11

Statistical Application of Orthogonal Arrays

Rao (1947) introduced orthogonal arrays because of their desirable statistical properties when used in "fractional factorial" experiments. Nowadays the main statistical application of orthogonal arrays, with mixed levels or otherwise, is still as fractional factorials, although other applications have been discovered. We will present the main application in considerable detail, while only giving key references for the other applications. Unless stated otherwise, throughout this chapter the term orthogonal array is to be interpreted as including mixed level arrays.

11.1 Factorial Experiments

In most areas of scientific research, experimentation is a major tool for acquiring new knowledge or a better understanding of what is happening. Experimentation may for example help in finding ways to increase the yield of a chemical reaction; it may help in understanding how to improve or maintain the quality of a manufactured product; or it may help to identify varieties of a crop that provide a higher yield. Discussions of the basic principles of experimental design and many examples of designed experiments can be found in the various text books on the design of experiments, such as Box and Draper (1987), Cochran and Cox (1957), Cox (1958), Hinkelmann and Kempthorne (1994), Mead (1988), Montgomery (1997) and Myers and Montgomery (1995). An indispensable reference for the class of experiments we will focus on, factorial experiments, is Raktoe, Hedayat and Federer (1981).

Factorial experiments usually aim at studying how changes in the levels of various factors affect a response variable of interest. As a first step, typically in consultation with a technician, engineer or scientist and possibly based on experience with or knowledge of the process under study, various factors that might affect the response variable are identified. Some of these factors may be of a *quantitative* nature, such as the temperature at which a chemical reaction takes place or the amount of potassium that is added to the soil in an agricultural experiment, while others may be of a *qualitative* nature, such as the type of catalyst used in a chemical reaction or the brand of fertilizer used in an agricultural experiment.

For each of the factors that are identified two or more levels are selected for inclusion in the experiment. For a qualitative factor the choice of the levels is often clear: only the two types of catalyst or three brands of fertilizer that are available can be used. For a quantitative factor an appropriate choice of levels can be more involved. The number of temperature levels that can be selected is, in principle, only limited by the precision of the instruments used in the experiment. A choice of levels for a quantitative factor can be influenced by, among other things, the precise objectives of the experiment (at an exploratory stage of experimentation, for example, it is rather common to use only a small number of levels, often two, for each of the factors), prior knowledge or belief concerning the importance of the factor and an "optimal" level for it, and an understanding of the minimal change in level required to produce a change in the response. These are, indeed, very subjective considerations, and experts, working independently, are unlikely to reach exactly the same recommendations for the same problem.

Once the factors and levels have been selected we may form the collection of all possible level combinations. (Other names for the level combinations are treatments, treatment combinations, or assemblies, cf. Chapter 1.) A "complete factorial" experiment would make measurements at each of the possible level combinations. However, the number of level combinations is often so large that this is impracticable, and a subset of level combinations must be judiciously selected to be used, resulting in a "fractional factorial" experiment. It is in this selection process that orthogonal arrays can play an important role.

Analysis of data from factorial experiments has traditionally concentrated on measuring the impact of factors or their interactions on variability in the average value of a response of interest. Henceforth, when we refer to variability, it is this type of variability that we will mean. More recently, spurred on by the contributions of Taguchi (1986, 1987), the possible effects of factors or their interactions on the *variance* of a response variable has also received the attention that it deserves. For example, reducing the variance in the quality of a product in a manufacturing process is an important consideration; the levels for factors that affect the quality of the product should not only assure that the *average* quality of items produced meets the desired specifications, but also that the variance in quality is small and that *each* individual meets the specifications.

This insistence on a small variance, even when some uncontrollable factors change (i.e. on product *robustness*), is perhaps one of the major contributions to statistics made by Taguchi, and in a broader sense by the early *total quality management* (or *TQM*) advocates, especially W. E. Deming and J. M. Juran.

11.2 Notation and Terminology

Let $A_1, A_2, \ldots, A_k$ denote the k factors to be included in the experiment. The number of levels for factor A_l will be denoted by s_l, $l = 1, 2, \ldots, k$. The levels of factor A_l can be coded as $0, 1, \ldots, s_l - 1$. This coding may be quite arbitrary for a qualitative factor, but is usually according to rank for a quantitative factor. The level combinations can then be represented by the k-tuples $(j_1, j_2, \ldots, j_k)$, where $0 \leq j_l \leq s_l - 1$, $l = 1, 2, \ldots, k$. We will denote the set of all level combinations by L and let $M = s_1 s_2 \cdots s_k$ be the number of elements of L.

It is sometimes desirable to specify an ordering for the elements of L. Statisticians commonly use what is called the *standard order* of the level combinations, or reverse lexicographic order. Precedence of one level combination over another is decided on a lexicographic basis with the primary criterion being the level of the last factor, with ties in the level of the last factor being decided by the next-to-last factor, and so on. For example, if $k = 3, s_1 = s_2 = 2$ and $s_3 = 3$, the standard order of the level combinations is 000, 100, 010, 110, 001, 101, 011, 111, 002, 102, 012, 112, whereas the usual lexicographic order would be 000, 001, 002, 010, 011, 012, 100, 101, 102, 110, 111, 112. In the standard order the levels of the earliest factors change the fastest.

Let N denote the number of experimental units that are made available for the experiment. Each unit will be assigned to a particular level combination, often via some method of randomization. By r_τ, $\tau \in L$, we denote the number of units, possibly zero, that are assigned to level combination τ. If $r_\tau > 0$, then we will use $Y_{\tau j}$, $j = 1, 2, \ldots, r_\tau$, to denote the random variable corresponding to the response of the j-th unit that is assigned to level combination τ. A commonly used statistical model for a continuous response variable may be written as

$$Y_{\tau j} = \mu_\tau + \epsilon_{\tau j} \ , \tag{11.1}$$

where μ_τ denotes a nonrandom population mean for possible observations with level combination τ, and $\epsilon_{\tau j}$ denotes a nonobservable random deviation from the population mean μ_τ for the j-th unit that receives τ. We will assume that the $\epsilon_{\tau j}$'s, which are also referred to as random errors, have mean 0 and constant variance σ^2. We will briefly return to the case where the variance may vary from one level combination to another in Section 11.8.

The model (11.1) specifies that the expected value and variance of $Y_{\tau j}$ are

$$E(Y_{\tau j}) = \mu_\tau, \quad \text{Var } (Y_{\tau j}) = \sigma^2 .$$

We will henceforth also assume that the $Y_{\tau j}$'s are uncorrelated. The reasonableness of these and other model assumptions should be studied carefully when planning any experiment. Considerations affecting their validity include, among other things, the nature of the experimental or observational units and the type of measurement being made. Diagnostic plots are also a helpful tool to detect possible violations of the assumptions after data have been collected (see for example Montgomery, 1997).

If systematic differences among the experimental units are known or suspected, it might be prudent to group the units into blocks of similar units. This would affect the commonly used method of randomization for assigning units to level combinations. It would also result in an additional term in model (11.1) to account for possible differences between blocks. The main reason for blocking is not to study differences between the blocks, but to increase the sensitivity of the experiment for inferences concerning the population means μ_τ by adequately accounting for variability among the $Y_{\tau j}$'s that is caused by systematic differences between the experimental units. We will discuss blocking in more detail in Section 11.6. For now we will assume that the experimental units are homogeneous, or at least that no major systematic differences between the units are believed to be present.

It will often be convenient to represent the model (11.1) in matrix form. We will use μ to denote the $M \times 1$ column vector of population means μ_τ, where the means are ordered according to the standard order of the corresponding level combinations. (Note that in this chapter "vector" usually means "column vector", except when we are discussing codes.) By Y we will denote the $N \times 1$ vector of random variables $Y_{\tau j}$, also in standard order based on the subscript τ. Similarly, ϵ will be used to denote the corresponding $N \times 1$ vector of random errors. The model (11.1) may then be written as

$$Y = X\mu + \epsilon , \tag{11.2}$$

where X is an $N \times M$ matrix of zeros and ones whose entries are obtained as follows. Label the columns by the level combinations, in standard order, and the rows by the subscripts of corresponding entries in Y; the entry in position $(\tau j, \tau')$ is equal to 1 if $\tau = \tau'$ and is otherwise 0.

Observe that each row of X contains exactly one 1. If $r_\tau = 0$ for $\tau \in L$, then the column of X labeled by τ will have all entries equal to 0. If $r_\tau = 1$ for all $\tau \in L$, X is simply the identity matrix of order M.

Comparisons of population means for different level combinations are made through treatment contrasts. A *treatment contrast* (or simply *contrast*) is a linear combination of the population means with coefficients that add to zero. Thus, for a known $M \times 1$ vector c, we say that $c^T\mu$ is a treatment contrast if $c^T 1_M = 0$. Two treatment contrasts $c_1^T\mu$ and $c_2^T\mu$ are said to be *orthogonal* if $c_1^T c_2 = 0$.

A treatment contrast $c^T\mu$ is said to be *estimable*, under a particular model and a particular selection of N level combinations, if there is a linear combination of the $Y_{\tau j}$'s, say h^TY, such that the expected value $E(h^TY)$ is equal to $c^T\mu$. Any such function h^TY is said to be an *unbiased estimator* of $c^T\mu$.

Under model (11.2) a treatment contrast $c^T\mu = \sum_{\tau\in L} c_\tau\mu_\tau$ is estimable if and only if $c_\tau = 0$ for all τ with $r_\tau = 0$. However, if we consider models that are derived from model (11.2) by making additional assumptions, other treatment contrasts may also become estimable. We will return to this issue when additional model assumptions are studied.

An important tool for the analysis of data from factorial and other experiments is the analysis of variance (ANOVA). Details will be given in Section 11.4. The analysis of variance attempts to quantify the contributions from different sources of variability to the overall variability in the observations. An ANOVA table is often a useful starting point for identifying the most important factors causing the variability in the observations and acquiring an understanding of how these factors affect the response variable. A further analysis may then be based on treatment contrasts.

11.3 Factorial Effects

With this section we begin a more rigorous and detailed explanation of some of the concepts introduced in Section 11.2. In particular, in this section and Section 11.4 we will define main-effects and interactions of factors, and discuss the use of treatment contrasts and ANOVA in data analysis. The following basic example will serve to introduce some of these concepts.

Example 11.1. In a $2^2 (= 2 \times 2)$ factorial experiment, two levels are selected for each of two factors. The level combinations are 00, 10, 01 and 11, where (j_1, j_2) has been abbreviated to j_1j_2. The 4×1 vector of population means is

$$\mu = (\mu_{00}, \mu_{10}, \mu_{01}, \mu_{11})^T .$$

A treatment contrast that is frequently of interest in a 2^2 factorial experiment is the contrast

$$\frac{1}{2}(\mu_{11} - \mu_{01} + \mu_{10} - \mu_{00}) . \tag{11.3}$$

Both $\mu_{11} - \mu_{01}$ and $\mu_{10} - \mu_{00}$ can be interpreted as measuring the effect of changing the level of the first factor, A_1, while keeping the level of A_2 fixed.

The contrast in (11.3) therefore represents the effect of A_1, averaged over both levels of A_2, and is referred to as the *main-effect* of factor A_1.

The divisor of 2 in the contrast in (11.3) is fairly common, although not used by all authors (compare, for example, John, 1971, who uses a divisor of 4). The general definition given later in this section produces a divisor of 2 in (11.3), but gives a different divisor than those commonly used in 2^k factorial experiments when k is greater than 2.

The main-effect of factor A_2 is similarly defined to be

$$\frac{1}{2}(\mu_{11} - \mu_{10} + \mu_{01} - \mu_{00}) \,.$$

A third contrast that is often of interest is

$$\frac{1}{2}(\mu_{11} - \mu_{01} - \mu_{10} + \mu_{00}) \,.$$

This compares the effect of factor A_1 when A_2 is at level 1 ($\mu_{11} - \mu_{01}$) to the effect of A_1 when A_2 is at level 0 ($\mu_{10} - \mu_{00}$). It studies, therefore, whether the effect of A_1 depends on the level of A_2, or equally, whether the effect of A_2 depends on the level of A_1. This contrast is referred to as the *interaction effect* of factors A_1 and A_2.

Observe that the two main-effects and the interaction effect are three pairwise orthogonal treatment contrasts. An analysis of data from a factorial experiment, and an explanation of how factors affect the response variable, would typically be based on studying these main-effects and interactions. It should however be stressed that it is not necessary to use these particular contrasts. If an investigator is more interested in, for example, $\mu_{11} - \mu_{10}$, $\mu_{01} - \mu_{00}$ and $\mu_{11} - \mu_{01} + \mu_{10} - \mu_{00}$, then these pairwise orthogonal contrasts could be used instead of the earlier set. Other choices, not necessarily consisting of three pairwise orthogonal contrasts, could also be considered.

An alternative way of obtaining the main-effects and interaction in this example is by defining the orthogonal matrices Z_1 and Z_2 to be

$$Z_1 = Z_2 = \frac{1}{\sqrt{2}} \begin{bmatrix} 1 & 1 \\ -1 & 1 \end{bmatrix} ,$$

and by considering the vector

$$(Z_2 \otimes Z_1)\mu = \frac{1}{2} \begin{bmatrix} 1 & 1 & 1 & 1 \\ -1 & 1 & -1 & 1 \\ -1 & -1 & 1 & 1 \\ 1 & -1 & -1 & 1 \end{bmatrix} \begin{bmatrix} \mu_{00} \\ \mu_{10} \\ \mu_{01} \\ \mu_{11} \end{bmatrix} = \frac{1}{2} \begin{bmatrix} \mu_{00} + \mu_{10} + \mu_{01} + \mu_{11} \\ -\mu_{00} + \mu_{10} - \mu_{01} + \mu_{11} \\ -\mu_{00} - \mu_{10} + \mu_{01} + \mu_{11} \\ \mu_{00} - \mu_{10} - \mu_{01} + \mu_{11} \end{bmatrix}$$

The last three elements of this vector correspond to the main-effect of A_1, the main-effect of A_2, and the interaction effect of A_1 and A_2, respectively. ■

In general, to each factor A_l we associate an $(s_l - 1) \times s_l$ matrix B_l (where s_l is the number of levels of A_l) in such a way that

$$Z_l = \begin{bmatrix} 1/\sqrt{s_l}\, 1_{s_l}^T \\ B_l \end{bmatrix} = (b^{(l)}(i,j)) \ , \ i,j \ = \ 0,\ldots,s_l - 1 \ , \tag{11.4}$$

is an orthogonal matrix, for $l = 1, 2, \ldots, k$. Define the $M \times 1$ vector β to be

$$\beta = (Z_k \otimes Z_{k-1} \otimes \cdots \otimes Z_1)\mu \ , \tag{11.5}$$

and denote the elements of β by $\beta_{i_1 i_2 \cdots i_k}$, for $0 \le i_l \le s_l - 1$, $l = 1, 2, \ldots, k$, with the subscripts arranged in standard order. For example, if $k = 2$, $s_1 = 2$ and $s_2 = 3$, then

$$\beta = (\beta_{00}, \beta_{10}, \beta_{01}, \beta_{11}, \beta_{02}, \beta_{12})^T \ .$$

Since $Z_k \otimes Z_{k-1} \otimes \cdots \otimes Z_1$ is an orthogonal matrix, (11.5) implies that

$$\mu = (Z_k^T \otimes Z_{k-1}^T \otimes \cdots \otimes Z_1^T)\beta \ .$$

Note that the last $M - 1$ elements of β are pairwise orthogonal treatment contrasts. These treatment contrasts are normalized in the sense that if $\beta_{i_1 i_2 \cdots i_k} = c^T \mu$, then $c^T c = 1$.

The relationship between μ and β allows us to rewrite model (11.2) as

$$Y = X(Z_k^T \otimes Z_{k-1}^T \otimes \cdots \otimes Z_1^T)\beta + \epsilon \ . \tag{11.6}$$

This reparameterization of the original model is essential for our discussion of the analysis of data from fractional factorial experiments.

Let $I(i_1, i_2, \ldots, i_k)$ denote the index set corresponding to the coordinates of the nonzero elements in $(i_1, i_2, \ldots, i_k)$:

$$I(i_1, i_2, \ldots, i_k) = \{l \in \{1, 2, \ldots, k\} : i_l > 0\} \ .$$

If $I(i_1, i_2, \ldots, i_k) = \{l_1\}$, then we will say that $\beta_{i_1 i_2 \cdots i_k}$ is a *component of the main-effect* of factor A_{l_1}. If $I(i_1, i_2, \ldots, i_k)$ has cardinality $m \ge 2$, say $I(i_1, i_2, \ldots, i_k) = \{l_1, l_2, \ldots, l_m\}$, then we will say that $\beta_{i_1 i_2 \ldots i_k}$ is a *component of the m-factor interaction* of factors $A_{l_1}, A_{l_2}, \ldots, A_{l_m}$. If

$$\prod_{l \in I(i_1, i_2, \ldots, i_k)} (s_l - 1) = 1 \ ,$$

that is, if $s_l = 2$ for all $l \in I(i_1, i_2, \ldots, i_k)$, we will sometimes omit the phrase "a component of" in the above definitions. Finally, we will refer to $\beta_{00\cdots 0}$ as the *intercept parameter*.

The elements of β depend of course on the choice of the matrices B_l, $l = 1, 2, \ldots, k$. The selection of these matrices might involve consideration of treatment contrasts that are of special interest. A common choice for these matrices with quantitative factors corresponds to fitting an orthogonal polynomial model. If $x_1 < x_2 < \cdots < x_{s_l}$ denote the s_l numerical, noncoded levels of a quantitative factor A_l, then we define polynomials $b_0^{(l)}, b_1^{(l)}, \ldots, b_{s_l-1}^{(l)}$ by

(i) the degree of $b_i^{(l)}$ is i, for $i = 0, 1, \ldots, s_l - 1$, and

(ii) $\sum_{j=1}^{s_l} b_i^{(l)}(x_j) b_{i'}^{(l)}(x_j) = 1$ if $i = i'$, and 0 if $i \neq i'$, $i, i' = 0, 1, \ldots, s_l - 1$.

The matrix B_l is then defined to be

$$B_l = \begin{bmatrix} b_1^{(l)}(x_1) & \cdots & b_1^{(l)}(x_{s_l}) \\ \cdots & \cdots & \cdots \\ b_{s_l-1}^{(l)}(x_1) & \cdots & b_{s_l-1}^{(l)}(x_{s_l}) \end{bmatrix}.$$

Use of these matrices is especially common if the levels of the factor are equally spaced, that is, if $x_j - x_{j-1}$ is independent of j, for $j = 2, 3, \ldots, s_l$. In this case the matrix B_l may, after a suitable normalization, be obtained from the standard tables in (for example) Fisher and Yates (1963), Table XXIII; Pearson and Hartley (1966), Table 47; or Raktoe, Hedayat and Federer (1981), p. 49. In this case the polynomials $b_i^{(l)}(x)$ are sometimes called *discrete Chebyshev polynomials* — see Abramowitz and Stegun (1964), §22.17; Erdélyi (1953), §10.23; Szegö (1967), §2.8.

With the above definitions, the main-effect of factor A_l consists of $s_l - 1$ components. The interaction effect of factors $A_{l_1}, A_{l_2}, \ldots, A_{l_m}$ consists of $(s_{l_1} - 1)(s_{l_2} - 1) \cdots (s_{l_m} - 1)$ components. A set of such components is referred to as the main-effect or interaction effect, respectively.

While all factors that are included in an experiment are initially thought of as potentially important for explaining variability in the response variable, in most experiments only a few factors will eventually prove to be truly important. Moreover, usually only a small subset of the effects based on those few factors will explain most of the variability. Often some main-effects and interactions of at most three factors suffice to understand how the important factors affect the response. This *effects sparsity* assumption is essential in justifying the use of fractional factorials.

The value of M, the number of possible level combinations, can be enormous if there are a large number of factors, even if all factors are only at two levels. Because of budget or time constraints, it may not be possible or practical to obtain a response at each of the possible level combinations. In such cases data will only be collected for some of the level combinations, and a judicious selection of these level combinations is crucial. Any collection of the level combinations that does not use all possible combinations is called a *fractional factorial*, or simply a *fraction*.

The use of fractional factorials, however, requires some additional assumptions about the model. For any level combination $\tau \in L$ and any main-effect or interaction effect, there is at least one component of the effect in which the coefficient of μ_τ is nonzero. If $r_\tau = 0$ this component will not be estimable under model (11.6). Hence, for any fractional factorial and any effect, we are faced with the highly undesirable situation that, under model (11.6), at least one component of the effect is not estimable.

This difficulty can be overcome if we are willing to make some simplifying assumptions about the model. By assuming that some of the elements in the parameter vector β are negligible, and so can be set equal to zero, we obtain a simpler, reduced model from model (11.6). An assumption of this type is often not unreasonable, as explained earlier, since only a few effects can usually explain most of the variability in the response variable. Thus, in some applications, the model (11.6) may be reduced by leaving only the intercept parameter and all components of main-effects in the model. In other applications some or all components of two- or three-factor interactions may also be left in the model. After a reduction of the model in this way, some components that were previously not estimable may now become estimable.

It should however also be stressed that, while a long history of empirical evidence provides strong support for the assumption that higher order interactions are usually negligible, there is no law that states that higher order interactions are never important for explaining variability in a response. The general ideas discussed here should therefore always be used judiciously. For a particular application, knowledge of the subject matter should play a role in judging whether certain assumptions are reasonable or not.

The reduced version of model (11.6) may be written as

$$Y = XU\gamma + \epsilon \,, \tag{11.7}$$

where γ is an $R\times 1$ vector consisting of the R elements of β that are not set equal to zero, and U is the $M \times R$ submatrix of the $M \times M$ matrix $Z_k^T \otimes Z_{k-1}^T \otimes \cdots \otimes Z_1^T$ consisting of the R columns in the latter matrix that correspond to the R elements of β that remain in γ.

Example 11.2. Let $k = 3$, $s_1 = s_2 = s_3 = 2$, $B_1 = B_2 = B_3 = \frac{1}{\sqrt{2}}[-1\ 1]$, and $N = 4$. If we choose to run the experiment at the four level combinations 000, 110, 101 and 011, then the 4×8 matrix X in (11.7), shown with the row and column labels described earlier, is given by

$$\begin{array}{c} \\ 000 \\ 110 \\ 101 \\ 011 \end{array} \begin{array}{c} \begin{array}{cccccccc} 000 & 100 & 010 & 110 & 001 & 101 & 011 & 111 \end{array} \\ \begin{bmatrix} 1 & 0 & 0 & 0 & 0 & 0 & 0 & 0 \\ 0 & 0 & 0 & 1 & 0 & 0 & 0 & 0 \\ 0 & 0 & 0 & 0 & 0 & 1 & 0 & 0 \\ 0 & 0 & 0 & 0 & 0 & 0 & 1 & 0 \end{bmatrix} \end{array}$$

(Since $r_\tau = 1$ for all $\tau \in L$ with $r_\tau > 0$, the labels for the rows have been simplified from τj to τ.) The parameter vectors μ and β are given by

$$\mu = (\mu_{000}, \mu_{100}, \mu_{010}, \mu_{110}, \mu_{001}, \mu_{101}, \mu_{011}, \mu_{111})^T$$

and

$$\beta = (\beta_{000}, \beta_{100}, \beta_{010}, \beta_{110}, \beta_{001}, \beta_{101}, \beta_{011}, \beta_{111})^T ,$$

where in both cases the elements are arranged according to standard order of the subscripts. With

$$Z_l = \begin{bmatrix} 1/\sqrt{2}\, 1_2^T \\ B_l \end{bmatrix}, \quad l = 1, 2, 3 ,$$

we find that

$$Z_3 \otimes Z_2 \otimes Z_1 = \frac{1}{2\sqrt{2}} \begin{bmatrix} 1 & 1 & 1 & 1 & 1 & 1 & 1 & 1 \\ -1 & 1 & -1 & 1 & -1 & 1 & -1 & 1 \\ -1 & -1 & 1 & 1 & -1 & -1 & 1 & 1 \\ 1 & -1 & -1 & 1 & 1 & -1 & -1 & 1 \\ -1 & -1 & -1 & -1 & 1 & 1 & 1 & 1 \\ 1 & -1 & 1 & -1 & -1 & 1 & -1 & 1 \\ 1 & 1 & -1 & -1 & -1 & -1 & 1 & 1 \\ -1 & 1 & 1 & -1 & 1 & -1 & -1 & 1 \end{bmatrix} .$$

The parameter vectors β and μ are related by $\beta = (Z_3 \otimes Z_2 \otimes Z_1)\mu$, or equivalently $\mu = (Z_3^T \otimes Z_2^T \otimes Z_1^T)\beta$. For a model that contains only the intercept parameter and the three main-effects we have

$$\gamma = (\beta_{000}, \beta_{100}, \beta_{010}, \beta_{001})^T$$

and

$$U^T = \frac{1}{2\sqrt{2}} \begin{bmatrix} 1 & 1 & 1 & 1 & 1 & 1 & 1 & 1 \\ -1 & 1 & -1 & 1 & -1 & 1 & -1 & 1 \\ -1 & -1 & 1 & 1 & -1 & -1 & 1 & 1 \\ -1 & -1 & -1 & -1 & 1 & 1 & 1 & 1 \end{bmatrix} .$$

Thus U^T contains only the four rows of $Z_3 \otimes Z_2 \otimes Z_1$ that define $\beta_{000}, \beta_{100}, \beta_{010}$ and β_{001}. Further,

$$XU = \frac{1}{2\sqrt{2}} \begin{bmatrix} 1 & -1 & -1 & -1 \\ 1 & 1 & 1 & -1 \\ 1 & 1 & -1 & 1 \\ 1 & -1 & 1 & 1 \end{bmatrix} .$$

Observe that $2\sqrt{2}XU$ is a Hadamard matrix of order 4, and so XU has rank 4. As we will see, it is the rank of XU that determines whether all elements of γ are estimable under the reduced model. ■

At times it is convenient to partition the model (11.7) into two parts:

$$Y = XU_1\gamma_1 + XU_2\gamma_2 + \epsilon \ , \tag{11.8}$$

where the entries of γ_1 and γ_2 partition those of γ and where the columns of U_1 and U_2 are the corresponding partition of the columns of U. The motivation for this partitioning of the model is that we may not always need to estimate all the parameters that occur in the model. The parameters that we want to estimate are part of γ_1, while γ_2 contains any remaining model parameters. For example, for a model that contains the intercept parameter, all components of the main-effects, and all components of the two-factor interactions, we may at times be interested in estimating only the components of the main-effects (in the presence of the other parameters). The components of the main-effects would then belong to γ_1, while the intercept parameter and the components of the two-factor interactions would belong to γ_2.

The matrix C defined by

$$C = U_1^T X^T (I_N - XU_2(U_2^T X^T XU_2)^- U_2^T X^T) XU_1 \ , \tag{11.9}$$

where the superscript $-$ denotes a generalized inverse, is known as the *information matrix* for γ_1 under model (11.8). Here and elsewhere, if γ_2 does not contain any parameters, we should interpret the matrix

$$I_N - XU_2(U_2^T X^T XU_2)^- U_2^T X^T$$

as I_N; then C reduces to

$$C = U_1^T X^T XU_1 = U^T X^T XU \ , \tag{11.10}$$

which is the information matrix for γ under model (11.7). It is well known that all elements of γ_1 are estimable under model (11.8) if and only if C is nonsingular. The vector of ordinary least squares estimators of the elements of γ_1 is in that case given by

$$\widehat{\gamma}_1 = C^{-1} U_1^T X^T (I_N - XU_2(U_2^T X^T XU_2)^- U_2^T X^T) Y \ .$$

The variance-covariance matrix for this vector of estimators is given by

$$\text{Var}\ (\widehat{\gamma}_1) = \sigma^2 C^{-1} \ .$$

To obtain an unbiased estimator of the error variance σ^2 we consider the quadratic form in Y

$$Y^T (I_N - P_{[XU_1 \ XU_2]}) Y \ ,$$

where P_W denotes the orthogonal projection matrix $W(W^TW)^-W^T$ onto the column space of the matrix W. Under model (11.8), the expected value of this quadratic form is equal to $(N-R_0)\sigma^2$, where R_0 denotes the rank of the matrix $[XU_1\ XU_2]$. Hence

$$\widehat{\sigma}^2 = Y^T(I_N - P_{[XU_1\ XU_2]})Y/(N-R_0)$$

is an unbiased estimator of σ^2.

Under the additional assumption that ϵ has a multivariate normal distribution, we find that lower and upper limits for a $100(1-\alpha)\%$ confidence interval for $c^T\gamma$, where c is a known vector, are given by

$$c^T\widehat{\gamma} \pm t_{N-R_0;\alpha/2}\,\widehat{\sigma}\sqrt{c^TC^{-1}c}\ ,$$

where $t_{N-R_0;\alpha/2}$ denotes the $100(1-\alpha/2)$ percentile of the central t-distribution with $N-R_0$ degrees of freedom.

The precise appearance of an ANOVA table depends on the assumed model, on the objectives of the experiment, and also on the level combinations that are used in the experiment. We will postpone the ANOVA discussion until the next section when particular selections of level combinations based on orthogonal arrays are considered.

11.4 Analysis of Experiments Based on Orthogonal Arrays

It is probably already obvious to the reader how an orthogonal array can be used to specify the level combinations in a factorial experiment. Using the notation for mixed orthogonal arrays, the rows of an $OA(N, s_1s_2\cdots s_k, t)$ specify N level combinations for an $s_1s_2\cdots s_k$ factorial experiment. For the remainder of this chapter we will use the term orthogonal array not only to denote the array but also to represent the N level combinations formed by the rows of the array. Thus an orthogonal array specifies a (fractional) factorial.

Part of the appeal that orthogonal arrays have in factorial experiments will become clear from the following result. The statistical relevance of the strength of an orthogonal array will also become apparent.

Theorem 11.3. *If an orthogonal array $OA(N, s_1s_2\cdots s_k, t)$, $t \geq 2$, is used in a factorial experiment then*

(i) if t is even, γ_2 is absent, and $\gamma_1 = \gamma$ consists of the intercept parameter, all components of main-effects, and all components of interactions of at most $t/2$ factors, then all elements of γ_1 are estimable under model (11.7);

(ii) if t is odd and γ_1 consists of the intercept parameter, all components of main-effects, and all components of interactions of at most $(t-1)/2$

factors, while γ_2 consists of all components of interactions of precisely $(t+1)/2$ factors, then all elements of γ_1 are estimable under model (11.8).

Proof: We must show that the matrix C defined in (11.9) is nonsingular. If t is even this will follow if we can show that the columns of the matrix XU_1 are orthogonal when the level combinations form an orthogonal array of strength t. If t is odd we will show in addition that each column of XU_1 is orthogonal to each column of XU_2.

To accomplish this, we label the rows and columns of XU by the level combinations in the orthogonal array and the model parameters, respectively. If a row is labeled by $(j_1, j_2, \ldots, j_k)$ and a column by $\beta_{i_1 i_2 \cdots i_k}$, the corresponding entry in XU is

$$\prod_{l=1}^{k} b^{(l)}(i_l, j_l) ,$$

where the elements $b^{(l)}(i_l, j_l)$ are defined as in equation (11.4). Thus the inner product of the columns corresponding to $\beta_{i_1 i_2 \cdots i_k}$ and $\beta_{f_1 f_2 \cdots f_k}$ in XU is

$$\sum_{(j_1, j_2, \ldots, j_k)} \prod_{l=1}^{k} b^{(l)}(i_l, j_l) b^{(l)}(f_l, j_l) , \qquad (11.11)$$

where the summation is over all level combinations of the orthogonal array. From the definitions of γ_1 and γ_2 it follows that, in order to show the stated orthogonality of columns in XU, we need only be concerned with expressions in (11.11) for which the number of nonzero elements in

$$\{i_1, i_2, \ldots, i_k, f_1, f_2, \ldots, f_k\}$$

is at most t, irrespective of whether t is even or odd. Furthermore, $b^{(l)}(0, j_l) = 1/\sqrt{s_l}$ for all $l = 1, 2, \ldots, k$ and all $j_l = 0, 1, \ldots, s_l - 1$. The proof that the expression in (11.11) vanishes is now entirely analogous to the argument used in the proof of Theorem 2.1. ■

Of course, the conclusions of Theorem 11.3 remain valid if γ_1 or γ_2 contains only some of the components specified in the theorem.

The argument used in the above proof shows that when an orthogonal array of strength t is used for the level combinations, with a model of the specified type, the information matrix C for γ_1 is a nonsingular diagonal matrix. In fact, since the matrices Z_l are orthogonal, it is easy to see that

$$C = (N/M) I_{R_1} ,$$

where R_1 denotes the number of parameters that belong to γ_1. (Thus $R_1 = R_0 = R$ if $\gamma_1 = \gamma$.) Hence, for an orthogonal array and a model as in Theorem 11.3, we have

$$\text{Var}\ (\widehat{\gamma}_1) = \sigma^2 (M/N) I_{R_1} . \qquad (11.12)$$

Equation 11.12 implies two other important properties of a fractional factorial experiment based on an orthogonal array, for a model as in Theorem 11.3. First, it provides meaning to the word *orthogonal* in orthogonal arrays: least squares estimators for different elements in γ_1 are orthogonal in the sense that they are uncorrelated. This is an important property when trying to interpret the results from a factorial experiment; if estimators of some elements of γ_1 were highly correlated it could be very difficult to understand how factors affect the response variable.

To describe the second property, we first need to define the efficiency of a fractional factorial experiment for a particular parameter relative to a complete factorial with one replicate of each possible level combination. If $r_\tau = 1$ for all $\tau \in L$, then, from (11.12), Var $(\widehat{\beta}_{i_1 i_2 \cdots i_k}) = \sigma^2$ for any element $\beta_{i_1 i_2 \cdots i_k}$ of γ_1. Under a model as in Theorem 11.3, we define the *relative efficiency* of a fraction for $\beta_{i_1 i_2 \cdots i_k}$ to be the normalized ratio of the variance of the least squares estimator of $\beta_{i_1 i_2 \cdots i_k}$ under the complete factorial with $r_\tau = 1$ for all $\tau \in L$ and the variance of the least squares estimator of this parameter under this fraction. The normalization is performed by multiplying the ratio by M/N, to adjust for the difference in the number of observations in the two factorials. The ratio is computed under the assumption that the error variance σ^2 is the same in numerator and denominator. It is easily seen from (11.12) that this normalized ratio is

$$(M/N)\frac{\sigma^2}{\sigma^2(M/N)} = 1$$

for any parameter in γ_1 when an orthogonal array of strength t is used. If other fractions are used, the relative efficiency may be less than 1 for some parameters, but greater than 1 for others.

With an orthogonal array and a model as described in Theorem 11.3 it is easy to set up an ANOVA table. All main-effects and interactions of at most $\lfloor t/2 \rfloor$ factors are used as sources of variability, and if t is odd interactions of $(t+1)/2$ factors are combined to form one source.

For the computation of sums of squares we require some additional terminology. Unfortunately, the notation needed to handle the general case is fairly complicated. In particular cases, as we shall see in the examples, a simpler notation can be used.

For a subset $\mathcal{G}$ of $\{1, 2, \ldots, k\}$ with cardinality g, say $\mathcal{G} = \{l_1, l_2, \ldots, l_g\}$, and for a $1 \times g$ vector $\alpha = (j_{l_1}, j_{l_2}, \ldots, j_{l_g})$ with $0 \le j_{l_u} \le s_{l_u} - 1$ for $u = 1, 2, \ldots, g$, we define $A(\mathcal{G}; \alpha)$ to consist of all distinct level combinations in the orthogonal array with factor A_{l_u} at level j_{l_u} for $u = 1, 2, \ldots, g$. We also define

$$r(\mathcal{G}; \alpha) = \sum_{\tau \in A(\mathcal{G}; \alpha)} r_\tau \, .$$

If $g \le t$, then for an orthogonal array of strength t, $r(\mathcal{G}; \alpha) = N/(s_{l_1} s_{l_2} \cdots s_{l_g})$, and we will simply write $r(\mathcal{G})$ for this quantity. Further, we define

$$T(\mathcal{G};\alpha) = \sum_{\tau \in A(\mathcal{G};\alpha)} \sum_{j=1}^{r_\tau} Y_{\tau j} ,$$

the total response over all level combinations in $A(\mathcal{G};\alpha)$.

Now let $\mathcal{H} \subset \{1,2,\ldots,k\}$ be a set of at most $\lfloor t/2 \rfloor$ elements. For an orthogonal array $OA(N, s_1 s_2 \cdots s_k, t)$ the *sum of squares* corresponding to $\mathcal{H}$ may be defined to be

$$\text{SS } \mathcal{H} = \sum_{\mathcal{G} \subset \mathcal{H}} (-1)^{|\mathcal{H}|-|\mathcal{G}|} \sum_{\alpha} \frac{T(\mathcal{G};\alpha)^2}{r(\mathcal{G})} , \tag{11.13}$$

where $|\mathcal{H}|$ and $|\mathcal{G}|$ stand for the cardinalities of the sets $\mathcal{H}$ and $\mathcal{G}$, and where the summation over α is over all possible level combinations for the factors corresponding to $\mathcal{G}$.

Example 11.4. Let $k \geq 3$, $s_1 = s_2 = 2$, $s_3 = 3$. For an orthogonal array of strength 4, let $T_{j_1\cdot\cdot}$ stand for $T(\{1\},(j_1))$, $T_{j_1\cdot j_3}$ for $T(\{1,3\},(j_1,j_3))$, and so on, where $j_1 \in \{0,1\}$, $j_3 \in \{0,1,2\}$. Setting $\mathcal{H} = \phi$ we obtain

$$\text{SS } \mathcal{H} = \sum_{\alpha} \frac{T(\phi;\alpha)^2}{r(\phi)} = \frac{T^2_{\cdots}}{N} ,$$

where we have used the conventions that, for "a vector α of length 0",

$$\begin{aligned} A(\phi;\alpha) &= \text{all level combinations in the orthogonal array} , \\ T(\phi;\alpha) &= \sum_{\tau} \sum_{j=1}^{r_\tau} Y_{\tau j} \text{ is the sum of all observations, and} \\ r(\phi) &= N . \end{aligned}$$

If we take $\mathcal{H} = \{1\}$, we find from (11.13) that

$$\text{SS } \mathcal{H} = \frac{T^2_{0\cdot\cdot}}{N/2} + \frac{T^2_{1\cdot\cdot}}{N/2} - \frac{T^2_{\cdots}}{N} .$$

This expression is equal to

$$\frac{(T_{0\cdot\cdot} - T_{\cdots}/2)^2}{N/2} + \frac{(T_{1\cdot\cdot} - T_{\cdots}/2)^2}{N/2} ,$$

so that SS $\mathcal{H}$ is a measure of the difference in response caused by the factor A_1.

By taking $\mathcal{H} = \{1,3\}$, we find

$$\begin{aligned} \text{SS } \mathcal{H} &= \frac{T^2_{0\cdot 0}}{N/6} + \frac{T^2_{1\cdot 0}}{N/6} + \frac{T^2_{0\cdot 1}}{N/6} + \frac{T^2_{1\cdot 1}}{N/6} + \frac{T^2_{0\cdot 2}}{N/6} + \frac{T^2_{1\cdot 2}}{N/6} \\ &- \frac{T^2_{0\cdot\cdot}}{N/2} - \frac{T^2_{1\cdot\cdot}}{N/2} - \frac{T^2_{\cdot\cdot 0}}{N/3} - \frac{T^2_{\cdot\cdot 1}}{N/3} - \frac{T^2_{\cdot\cdot 2}}{N/3} + \frac{T^2_{\cdots}}{N} , \end{aligned}$$

an expression that can be rewritten as

$$\sum_{i=0}^{1}\sum_{j=0}^{2}\frac{(T_{i\cdot j}-T_{i\cdot\cdot}/3-T_{\cdot\cdot j}/2+T_{\cdot\cdot\cdot}/6)^2}{N/6}\ .$$

It is a measure of the difference in response due to factors A_1 and A_3 beyond what can be explained by the main-effects of A_1 and A_3. ∎

In general, when $\mathcal{H}=\phi$ we refer to the corresponding sum of squares as the sum of squares for the intercept parameter or for the mean. If $\mathcal{H}=\{l_1\}$, we call SS $\mathcal{H}$ the sum of squares for the main-effect of factor A_{l_1}. If $\mathcal{H}=\{l_1,l_2\}$, reference is to the sum of squares for the interaction of factors A_{l_1} and A_{l_2}. We will also write SS A_{l_1} and SS $A_{l_1}A_{l_2}$ instead of SS $\mathcal{H}$. The extension to interactions of more than two factors is clear. Furthermore, when we refer to a set $\mathcal{H}=\{l_1,l_2\}$, for example, as a factorial effect, we will simply mean the two-factor interaction $A_{l_1}A_{l_2}$.

Using the notation of (11.7), we define the *model sum of squares* to be

$$Y^T P_{XU} Y\ .$$

With an orthogonal array of strength t and a model as in Theorem 11.3, it is immediately seen that the model sum of squares is equal to

$$\frac{N}{M}\widehat{\gamma}^T\widehat{\gamma},\quad \text{if}\quad t \text{ is even}\ , \tag{11.14}$$

or

$$\frac{N}{M}\widehat{\gamma}_1^T\widehat{\gamma}_1+Y^TP_{XU_2}Y,\quad \text{if}\quad t \text{ is odd}\ . \tag{11.15}$$

We may think of $Y^TP_{XU_2}Y$ as that part of the model sum of squares due to the presence of the interactions of precisely $(t+1)/2$ factors in the model.

Finally, again for an orthogonal array of strength t and a model as in Theorem 11.3, we define the sum of squares for a single component of a main-effect or interaction effect of at most $\lfloor t/2\rfloor$ factors, say $\beta_{i_1i_2\cdots i_k}$, to be

$$\frac{N}{M}\widehat{\beta}^2_{i_1i_2\cdots i_k}\ . \tag{11.16}$$

This would be precisely the model sum of squares if the model contained only the parameter $\beta_{i_1i_2\cdots i_k}$.

As mentioned earlier, the main-effects and interactions of at most $\lfloor t/2\rfloor$ factors are to be used as sources in an ANOVA table. For an effect that is found to be important, in order to learn more about how it affects the response we would like to use the components corresponding to this effect (if there is more than one) to partition the sum of squares for this effect into sums of squares for these components. The following result gives relations between the various sums of squares that support this idea.

Theorem 11.5. *If an orthogonal array $OA(N, s_1 s_2 \cdots s_k, t)$, $t \geq 2$, is used in a factorial experiment, and if the model is as specified in Theorem 11.3, then*

(i) for any main-effect or interaction effect of at most $\lfloor t/2 \rfloor$ factors, the sum of squares for that effect is equal to the total of the sums of squares for the components of the effect; and

(ii) the model sum of squares is equal to the total of the sum of squares for the mean, the sums of squares for the main-effects, the sums of squares for all interactions of at most $\lfloor t/2 \rfloor$ factors, and, if t is odd, the remainder sum of squares caused by the presence of interactions of $(t+1)/2$ factors.

Proof: For (i), first observe that, after some algebra, (11.13) can be rewritten as

$$\mathrm{SS}\ \mathcal{H} = \frac{M}{N} Y^T X \left[Q_k(\mathcal{H}) \otimes Q_{k-1}(\mathcal{H}) \otimes \cdots \otimes Q_1(\mathcal{H})\right] X^T Y \ , \tag{11.17}$$

where, for $l = 1, 2, \ldots, k$,

$$Q_l(\mathcal{H}) = \begin{cases} I_{s_l} - \frac{1}{s_l} J_{s_l}, & \text{if } l \in \mathcal{H} \ , \\ \frac{1}{s_l}\ J_{s_l}, & \text{otherwise} \ . \end{cases}$$

Writing

$$Z_l = \begin{bmatrix} b_0^{(l)^T} \\ \vdots \\ b_{s_l-1}^{(l)^T} \end{bmatrix}$$

for the matrix Z_l in (11.4), we see that, for $l \in \mathcal{H}$,

$$Q_l(\mathcal{H}) = b_1^{(l)} b_1^{(l)^T} + b_2^{(l)} b_2^{(l)^T} + \cdots + b_{s_l-1}^{(l)} b_{s_l-1}^{(l)^T} \ .$$

As a result, we can write the expression in (11.17) as a sum of $\prod_{l \in \mathcal{H}} (s_l - 1)$ terms, each term having the form

$$\frac{M}{N} Y^T X \left[b_{i_k}^{(k)} b_{i_k}^{(k)^T} \otimes b_{i_{k-1}}^{(k-1)} b_{i_{k-1}}^{(k-1)^T} \otimes \cdots \otimes b_{i_1}^{(1)} b_{i_1}^{(1)^T} \right] X^T Y \ ,$$

where $i_l = 0$ if $l \notin \mathcal{H}$ and $i_l \in \{1, 2, \ldots, s_l - 1\}$ if $l \in \mathcal{H}$. Since

$$\widehat{\beta}_{i_1 i_2 \cdots i_k} = \frac{M}{N} (b_{i_k}^{(k)^T} \otimes b_{i_{k-1}}^{(k-1)^T} \otimes \cdots \otimes b_{i_1}^{(1)^T}) X^T Y \ ,$$

we see that

$$\mathrm{SS}\ \mathcal{H} = \sum_{(i_1, i_2, \ldots, i_k)} \frac{N}{M} \widehat{\beta}^2_{i_1 i_2 \cdots i_k} \ ,$$

where the summation is over all $(i_1, i_2, \ldots, i_k)$ with $i_l \neq 0$ if and only if $l \in \mathcal{H}$. This establishes (i).

The result in (ii) follows immediately from part (i) and the expressions for the model sum of squares in (11.14) and (11.15). ∎

It is worth pointing out the following more direct verification of (ii). Clearly

$$\begin{aligned}
&(Z_k^T \otimes Z_{k-1}^T \otimes \cdots \otimes Z_1^T)(Z_k \otimes Z_{k-1} \otimes \cdots \otimes Z_1) = I_{s_k} \otimes I_{s_{k-1}} \otimes \cdots \otimes I_{s_1} \\
= &\left(I_{s_k} - \frac{1}{s_k}J_{s_k} + \frac{1}{s_k}J_{s_k}\right) \otimes \cdots \otimes \left(I_{s_1} - \frac{1}{s_1}J_{s_1} + \frac{1}{s_1}J_{s_1}\right) \\
= &\sum_{\mathcal{H} \subset \{1,2,\ldots,k\}} Q_k(\mathcal{H}) \otimes Q_{k-1}(\mathcal{H}) \otimes \cdots \otimes Q_1(\mathcal{H}) ,
\end{aligned}$$

where the summation is over all subsets $\mathcal{H}$ of $\{1, 2, \ldots, k\}$ and where the matrices $Q_l(\mathcal{H})$ are as in the proof of Theorem 11.5. With U_1 as the submatrix of $Z_k^T \otimes Z_{k-1}^T \otimes \cdots \otimes Z_1^T$ consisting of all columns corresponding to the intercept parameter, all components of main-effects and all components of interactions of at most $\lfloor t/2 \rfloor$ factors, it is then not hard to show that

$$U_1 U_1^T = \sum_{\substack{\mathcal{H} \subset \{1,2,\ldots,k\} \\ |\mathcal{H}| \le \lfloor t/2 \rfloor}} Q_k(\mathcal{H}) \otimes Q_{k-1}(\mathcal{H}) \otimes \cdots \otimes Q_1(\mathcal{H}) ,$$

where the summation is now over only those sets $\mathcal{H}$ with cardinality at most $\lfloor t/2 \rfloor$. Hence

$$\begin{aligned}
Y^T P_{XU_1} Y &= \frac{M}{N} Y^T X U_1 U_1^T X^T Y \\
&= \sum_{\substack{\mathcal{H} \subset \{1,2,\ldots,k\} \\ |\mathcal{H}| \le \lfloor t/2 \rfloor}} \frac{M}{N} Y^T X (Q_k(\mathcal{H}) \otimes Q_{k-1}(\mathcal{H}) \otimes \cdots \otimes Q_1(\mathcal{H})) X^T Y \\
&= \sum_{\substack{\mathcal{H} \subset \{1,2,\ldots,k\} \\ |\mathcal{H}| \le \lfloor t/2 \rfloor}} \mathrm{SS}\ \mathcal{H} ,
\end{aligned}$$

establishing the claim in (ii).

Now that we are aware of relationships between the various sums of squares when using an orthogonal array of strength t, we are sufficiently prepared to discuss the use of ANOVA tables. We stress again that these relationships and the following discussions are valid only under the assumption that the fractional factorial experiment is an orthogonal array.

All the models that we are concerned with will include the intercept parameter as one of the model parameters. In such cases, the ANOVA table commonly uses the corrected total sum of squares

$$Y^T Y - Y^T P_{1_N} Y = Y^T Y - \frac{1}{N} Y^T J_N Y$$

(instead of $Y^T Y$), which subtracts N times the square of the mean from the (uncorrected) total sum of squares. As a result of this correction, the mean or intercept parameter will not be used as a source in the ANOVA table. The corrected model sum of squares, using the notation of model (11.7), is now given by

$$Y^T P_{XU} Y - Y^T P_{1_N} Y \ .$$

With an orthogonal array of strength t and the model as specified in Theorem 11.3, the corrected model sum of squares can be partitioned into sums of squares for each of the main-effects, for each of the interactions of at most $\lfloor t/2 \rfloor$ factors, and, when t is odd, a remainder for the presence of the interactions of $(t+1)/2$ factors.

Each sum of squares is a quadratic form in Y, say $Y^T A Y$, where A is an orthogonal projection matrix. The "degrees of freedom" for a sum of squares is defined to be the rank of A. There are $N-1$ degrees of freedom for the corrected total sum of squares. For a factorial effect, say the interaction of $A_{l_1}, A_{l_2}, \ldots, A_{l_m}$, $1 \le m \le \lfloor t/2 \rfloor$, which is a main-effect if $m = 1$, there are $(s_{l_1} - 1)(s_{l_2} - 1) \cdots (s_{l_m} - 1)$ degrees of freedom. If t is odd, we let γ_2 in (11.8) consist of all components of the interactions of precisely $(t+1)/2$ factors. The degrees of freedom for the sum of squares due to the presence of γ_2 in the model is then equal to the rank of XU_2. Finally, by subtraction we find that there are $N - \text{rank}(XU) = N - R_0$ degrees of freedom for the error sum of squares (in the notation of Section 11.3).

An ANOVA table consists of various rows and columns. The first column names the sources of variability. These sources include all main-effects, all interactions of at most $\lfloor t/2 \rfloor$ factors, and error. If t is odd, another source of variability is the presence of γ_2 in the model. The second and third columns contain the degrees of freedom and sums of squares for the various sources of variability. The totals of these two columns should be the total degrees of freedom $N-1$ and the sum of squares for the corrected total, respectively. The next column provides the mean square for each source, which is simply the sum of squares divided by the corresponding degrees of freedom. A final column, with the entry for the error row left blank, contains for each source the ratio of its mean square to the mean square for error.

For a factorial effect $\mathcal{H}$ that is listed as a source of variability, the ratio in the last column of the ANOVA table,

$$F(\mathcal{H}) = \frac{\text{Mean square } \mathcal{H}}{\text{Mean square error}} \ ,$$

follows an F-distribution if the random error vector ϵ in (11.8) follows a multivariate normal distribution $N(0_N, \sigma^2 I_N)$ (cf. Hinkelmann and Kempthorne, 1994, §4.17.3). The degrees of freedom of this F-distribution are the degrees of freedom for $\mathcal{H}$ and the degrees of freedom for error. The distribution is a central F-distribution if and only if all components of the effect $\mathcal{H}$ are 0. The null hypothesis that all components of $\mathcal{H}$ are 0 is, at significance level α, rejected in favor of the alternative hypothesis that some components are not 0 if

$$F(\mathcal{H}) \geq F_{df\ \mathcal{H},\ df\ \text{error};\ \alpha}\ ,$$

where $F_{df\ \mathcal{H},\ df\ \text{error};\ \alpha}$ denotes the $100(1-\alpha)$ percentile of the central F-distribution with the indicated degrees of freedom.

If t is odd, the ratio

$$F(\gamma_2) = \frac{\text{Mean square } \gamma_2}{\text{Mean square error}}$$

can be used in a similar way, though the conclusion is slightly different. Since not all (and possibly none) of the components of γ_2 are estimable, if $F(\gamma_2)$ does not exceed the $100(1-\alpha)$ percentile of the relevant F-distribution we can only conclude that in the subspace of dimension $\text{rank}(XU_2)$ of estimable linear functions of components of interactions of $(t+1)/2$ factors there is not enough evidence to conclude that any of these linear functions differ significantly from 0. On the other hand, if $F(\gamma_2)$ exceeds the relevant quantile, then there is evidence to conclude that at least some of the components of interactions of $(t+1)/2$ factors are important and cannot be neglected. However, it may not be possible without obtaining further data to decide which of those components are important.

If $\mathcal{H}$ is a significant effect, there is usually a need to study more carefully how $\mathcal{H}$ affects the response. Tables of mean or total responses for all possible level combinations of the factors in $\mathcal{H}$ are often very useful in such situations. (See, for example, Snedecor and Cochran, 1989, §16.13.) Studying the components of $\mathcal{H}$ can also be illuminating, provided these are chosen in a meaningful way.

The following example illustrates some of the above ideas. Other examples may be found in the design of experiments text books referred to in Section 11.1.

Example 11.6. Consider the (artificial) data presented in Table 11.7. The design is a fractional factorial for a $2^4 \times 3$ factorial experiment, with A_1, A_2, A_3, A_4 denoting the two-level factors and A_5 the three-level factor. This is an orthogonal array $OA(24, 2^4 \times 3, 3)$, with level combinations written in standard order.

Table 11.7. Data from a $2^4 \times 3$ fractional factorial experiment.

A_1	A_2	A_3	A_4	A_5	Response
0	0	0	0	0	7.6
1	1	0	0	0	16.6
1	0	1	0	0	12.2
0	1	1	0	0	17.8
1	0	0	1	0	10.7
0	1	0	1	0	20.5
0	0	1	1	0	10.3
1	1	1	1	0	17.7
0	0	0	0	1	16.4
1	1	0	0	1	24.1
1	0	1	0	1	18.1
0	1	1	0	1	25.7
1	0	0	1	1	20.1
0	1	0	1	1	24.7
0	0	1	1	1	17.7
1	1	1	1	1	24.3
0	0	0	0	2	23.9
1	1	0	0	2	29.4
1	0	1	0	2	25.4
0	1	1	0	2	32.2
1	0	0	1	2	20.8
0	1	0	1	2	28.0
0	0	1	1	2	24.0
1	1	1	1	2	31.4

Taking the matrices Z_l in (11.4) to be

$$Z_1 = Z_2 = Z_3 = Z_4 = \frac{1}{\sqrt{2}} \begin{bmatrix} 1 & 1 \\ -1 & 1 \end{bmatrix}, \quad Z_5 = \frac{1}{\sqrt{3}} \begin{bmatrix} 1 & 1 & 1 \\ -\sqrt{6}/2 & 0 & \sqrt{6}/2 \\ 1/\sqrt{2} & -\sqrt{2} & 1/\sqrt{2} \end{bmatrix},$$

we compute the 24×48 matrix $X(Z_5^T \otimes Z_4^T \otimes Z_3^T \otimes Z_2^T \otimes Z_1^T)$. For the model we let γ_1 consist of the intercept parameter and the components of the main-effects, and let γ_2 consist of all components of the two-factor interactions. The 7×1 vector γ_1 is then given by

$$\gamma_1 = (\beta_{00000}, \beta_{10000}, \beta_{01000}, \beta_{00100}, \beta_{00010}, \beta_{00001}, \beta_{00002})^T,$$

with the components in standard order according to their subscripts. The vector γ_2 consists of the 14 components of the 10 two-factor interaction effects. The matrices XU_1 and XU_2 are therefore 24×7 and 24×14 submatrices of $X(Z_5^T \otimes Z_4^T \otimes Z_3^T \otimes Z_2^T \otimes Z_1^T)$, respectively. It is easily verified via computation that $U_1^T X^T X U_1 = \frac{1}{2} I_7$ and $U_1^T X^T X U_2 = 0_{7\times 14}$, as claimed in and after the proof of Theorem 11.3. Hence, referring back to (11.10), $C = I_7/2$ and

$$\widehat{\gamma}_1 = C^{-1} U_1^T X^T Y = (144.22, 0.58, 24.60, 4.04, 0.23, 35.96, -2.80)^T \ .$$

The rank of XU_2 is equal to 11, so there are 11 degrees of freedom for the sum of squares due to the presence of two-factor interactions. This sum of squares is

$$Y^T P_{XU_2} Y = 26.98 \ .$$

The sums of squares for the various main-effects are perhaps most easily computed as suggested by the expression for the model SS in (11.16). For example,

$$\text{SS } A_1 = \frac{24}{48}(0.58)^2 = 0.17 \ .$$

As a first alternative to this computation we can use (11.13) to obtain

$$\text{SS } A_1 = \frac{(248.8)^2}{12} + \frac{(250.8)^2}{12} - \frac{(499.6)^2}{24} = 0.17 \ ,$$

where 248.8, 250.8 and 499.6 are the total of all observations with A_1 at level 0, the total of all observations with A_1 at level 1, and the total of all observations, respectively. As a second alternative, following (11.17) this sum of squares can be computed as

$$2Y^T X (Q_5 \otimes Q_4 \otimes Q_3 \otimes Q_2 \otimes Q_1) X^T Y \ ,$$

where

$$Q_1 = \frac{1}{2}\begin{bmatrix} 1 & -1 \\ -1 & 1 \end{bmatrix}, \quad Q_2 = Q_3 = Q_4 = \frac{1}{2}\begin{bmatrix} 1 & 1 \\ 1 & 1 \end{bmatrix}, \quad Q_5 = \frac{1}{3}\begin{bmatrix} 1 & 1 & 1 \\ 1 & 1 & 1 \\ 1 & 1 & 1 \end{bmatrix} .$$

The results of similar computations for other sums of squares are given in Table 11.8.

Table 11.8. ANOVA table for Example 11.6.

Source	*df*	SS	MS	*F*
A_1	1	0.17	0.17	0.06
A_2	1	302.46	302.46	107.45 *
A_3	1	8.17	8.17	2.90
A_4	1	0.03	0.03	0.01
A_5	2	650.34	325.17	115.52 *
2-factor interactions	11	26.98	2.45	0.87
Error	6	16.89	2.81	
Total (corrected)	23	1005.03		

*F-values with a p-value less than 0.10

The ANOVA table suggests that only the main-effects of A_2 and A_5 are important for explaining the variability in the data. Based on the estimate $\widehat{\beta}_{01000} = 24.60$ and the estimate of Var $(\widehat{\beta}_{01000})$, which is $\frac{M}{N}$ MS Error $= 5.63$, we obtain a 95% confidence interval for β_{01000} as

$$\widehat{\beta}_{01000} \pm t_{6;.025}\sqrt{2 \text{ MS Error}} = 24.60 \pm t_{6;.025}\sqrt{5.63} = (18.79, 30.40) \ .$$

The two degrees of freedom associated with the main-effect of A_5 can be partitioned into single degrees of freedom by using the components β_{00001} and β_{00002}. Other ways to partition these two degrees of freedom are possible, but the choice of the matrix Z_5 reflects, presumably, that these are two meaningful treatment contrasts. For the sum of squares corresponding to β_{00001} we obtain

$$\frac{N}{M}\widehat{\beta}^2_{00001} = 646.43 \ ;$$

while for β_{00002} this is only

$$\frac{N}{M}\widehat{\beta}^2_{00002} = 3.91 \ .$$

Together these sums of squares form the sum of squares for A_5. Only the sum of squares for β_{00001} is significant at $\alpha = 0.10$. A 95% confidence interval for β_{00001} is given by

$$\widehat{\beta}_{00001} \pm t_{6;.025}\sqrt{2 \text{ MS Error}} = (30.15, 41.76) \ .$$

The parameters β_{01000} and β_{00001} explain most of the variability in the data very adequately and can be used to provide a fairly simple explanation of how the response varies with changes in the levels of factors A_2 and A_5, especially since the estimators are uncorrelated.

It is a little more difficult to interpret the actual values of the estimators. What precisely does it mean to say that β_{01000} is estimated by 24.60? This number represents the estimate of the slope for the regression of the response variable on a scaled version of the levels of factor A_2. It may however be easier to interpret an estimate of the slope on the actual or coded levels of A_2. To illustrate this, if the levels of factor A_2 were temperature levels of 80 (coded as 0) and 90 (coded as 1) degrees Fahrenheit, then, using the notation for observed totals introduced prior to Example 11.4, the slope would be estimated by

$$\frac{5\ T(\{2\},(1)) - 5\ T(\{2\},(0))}{12\ (25+25)} = 0.71$$

for the actual temperature levels, and by

$$\frac{0.5\ T(\{2\},(1)) - 0.5\ T(\{2\},(0))}{12\ (0.25+0.25)} = 7.10$$

for the coded levels. The coefficients for the observed totals in these expressions are obtained by subtracting the means (85 and 0.5) from each of the levels: $90 - 85 = 5$, $80 - 85 = -5$; $1 - 0.5 = 0.5$, $0 - 0.5 = -0.5$. In the denominator we take the sum of squared coefficients multiplied by the replication for each of the levels. The value of 0.71 suggests that, on the average, for every increment of one unit on the temperature scale (i.e., one degree Fahrenheit) the response increases by 0.71. The second estimate leads (fortunately) to the same conclusion: for every increment of one unit on the coded scale (where one unit represents ten degrees Fahrenheit) the response increases by 7.10 units.

Estimates for the variances of these slope estimators are given by

$$\frac{\text{MS Error}}{12\ (25+25)} = 0.0047$$

and

$$\frac{\text{MS Error}}{12\ (0.25+0.25)} = 0.47\ .$$

Confidence intervals for these slopes can be obtained as for β_{01000}.

For formulating conclusions from an experiment we recommend that one of these latter slope estimators be used. For the general theory developed in this chapter the normalized parameters, like β_{01000}, are extremely convenient. ∎

The degrees of freedom for error in Example 11.6 would usually be considered too small. Two of the many reasons for this are the following. First, under the assumption that ϵ follows a multivariate normal distribution $N(0_N, \sigma^2 I_N)$, it is well known that SS Error$/\sigma^2$ follows a chi-square distribution with as many degrees of freedom as that of the error. Since the variance of a chi-square distribution is twice the number of its degrees of freedom, it follows that

$$\text{Var (MS Error)} = 2\sigma^4/df \text{ error}\ .$$

The variance of MS Error as an estimator of σ^2 is therefore reduced when the degrees of freedom of the error increases. Secondly, the vector e of residuals,

$$e = [I_N - P_{XU}]Y\ ,$$

can contain important information for checking model assumptions, such as normality and homogeneity of variances, as well as for detecting possible outliers in the data. A normal probability plot of the residuals and a plot of the residuals versus the fitted values $\widehat{Y}$, where

$$\widehat{Y} = XU\widehat{\gamma} = P_{XU}Y\ ,$$

can suggest possible violations of the model assumptions (see, for example, Montgomery, 1997). These residuals, however, are not statistically independent and satisfy multiple linear relationships. Indeed, it is obvious that

$$P_{XU}e = 0_{N\times 1}\ .$$

The plots are therefore rather meaningless if $N - \text{rank}(XU)$ is small. While we strongly recommend use of residual plots if $N - \text{rank}(XU)$ is larger, say at least 10, the user should not be too concerned with "minor departures" from a "perfect plot". Plots are a very useful, but also very subjective tool.

Other residual plots, such as plots of residuals versus the levels of a factor, can also shed light on questions of interest. These may for example reveal that level changes for a factor affect the variance of the response, and thus the assumption of homogeneous variances.

Two residual plots for Example 11.6 are presented in Figure 11.1, even though there are insufficient degrees of freedom for error for these plots to be truly meaningful.

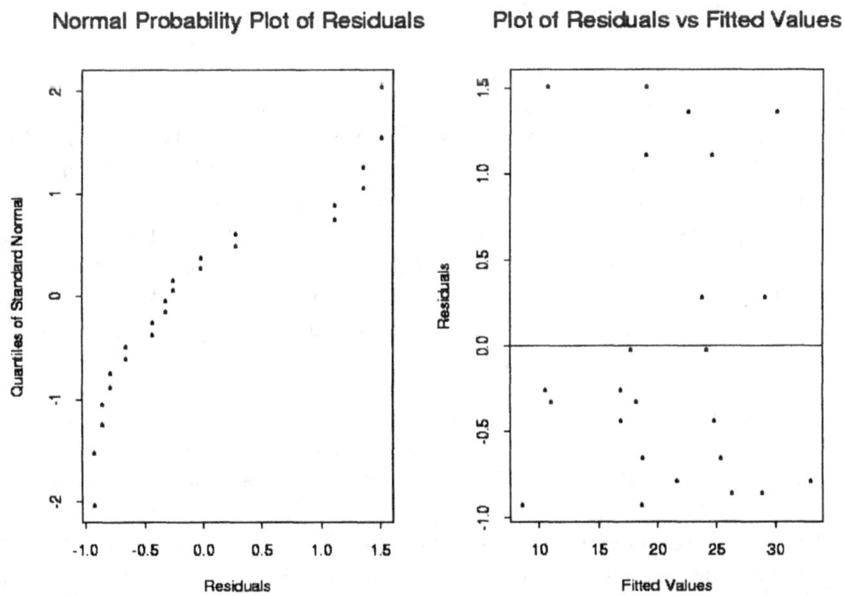

Figure 11.1: Residual plots for Example 11.6

It is worth observing that the results in this section provide a statistical interpretation for Rao's inequalities. If t is even, the length of the vector γ_1 in Theorem 11.3, say R_1, is precisely equal to the lower bound for the number of runs N in Rao's inequalities. Estimability of all elements of γ_1 requires R_1 degrees of freedom, while each run contributes one degree of freedom to the (uncorrected) total degrees of freedom. Thus the inequalities simply state that the number of observations must at least be equal to the number of unknown parameters that we require to be estimable. For odd t the interpretation is slightly more difficult (although these inequalities follow from those for even t, just as in Section 2.2). With $t = 2u + 1$, it follows readily from the proof of Theorem 11.3 that if γ_1 consist of the general mean, the components of all main-effects, the components of all interactions of at most u factors, and the components of all $u+1$ factor interactions that include one selected factor, and γ_2 is the empty vector, then all elements of γ_1 are estimable for an orthogonal array of strength $2u + 1$. The length of this vector γ_1 is precisely equal to Rao's bound for the number of runs when t is odd, providing an interpretation analogous to that for even t.

11.5 Two-Level Fractional Factorials with a Defining Relation

A subclass of orthogonal arrays, referred to as *fractions with a defining relation*, or *regular fractions*, has traditionally been popular among statisticians and other users of fractional factorials. Properties that make these fractions appealing will be discussed in this section, making use of terminology both from statistics and coding theory. We will restrict the discussion to fractions in which all factors are at 2 levels; extensions to s-level fractions, where s is a prime power, are straightforward. This is particularly true since fractions with a defining relation are nothing but linear orthogonal arrays or their translates. Regarding these fractions as codes then reveals many of their properties, besides making it easier to compute them.

In this section and in Section 11.6 we will use capital Latin letters A, B, $C, \ldots$ to denote the two-level factors in an experiment. The two levels will be called simply "low" and "high". These labels can be assigned arbitrarily for a qualitative factor, and are usually assigned according to the natural order for a quantitative factor. Level combinations will be denoted by lower case Latin letters, where the presence of a letter indicates that the factor is at its high level. Thus acf denotes the level combination in which factors A, C and F are at their high levels and all other factors in the experiment are at their low levels. The special symbol (1) is used for the level combination in which all factors are at their low levels.

The notation for treatment effects is exactly the same as that for level combinations. Thus under model (11.2) a random variable for an observation at level combination acf has expected value acf. The dual use of this notation can be confusing at first, but the advantage over having to introduce yet another set of symbols will rapidly become apparent.

With k two-level factors in the experiment, the $2^k - 1$ factorial effects are usually defined to be the treatment contrasts obtained as the last $2^k - 1$ elements of the vector $\underline{\beta}$, defined by

$$\underline{\beta} = \frac{1}{2^{k-1}}(H_2 \otimes H_2 \otimes \cdots \otimes H_2)\mu \ , \tag{11.18}$$

where

$$H_2 = \begin{bmatrix} 1 & 1 \\ -1 & 1 \end{bmatrix}$$

is used k times in (11.18), and where μ is the $2^k \times 1$ vector of treatment effects in the standard order:

$$(1), a, b, ab, c, ac, bc, abc, \ldots \ .$$

Equation (11.18) is similar but not identical to (11.5). The only difference, besides the fact that (11.5) is not restricted to two-level factors, is that the elements of $\underline{\beta}$ differ by a multiplicative factor from those of β. The coefficient for each treatment effect in an element of β is equal to $\pm 2^{-k/2}$; in $\underline{\beta}$ it is $\pm 2^{-(k-1)}$. While the former is mathematically convenient, the latter is more meaningful in the sense that the factorial effects are computed on a per unit basis. However, as already mentioned in Example 11.1, while the divisor of 2^{k-1} is appealing and commonly used in statistics, it is not universally adopted. In this section we will use $\underline{\beta}$ as in (11.18).

The first element of $\underline{\beta}$ is equal to $\frac{1}{2^{k-1}}1^T\mu$ and will be denoted by $2M$. Thus M is the average of all treatment effects. The next element of $\underline{\beta}$ is called the main-effect of A, and is equal to the difference between the average of all treatment effects with A at its high level and those with A at its low level. This main-effect is, just as the factor itself, denoted by A. The next element of $\underline{\beta}$ is B, the main-effect of factor B, which is followed by the interaction of factors A and B, denoted by AB. If $k \geq 3$, the next elements are C, $AC, \ldots$.

Example 11.9. Consider a factorial experiment with three two-level factors (also called a 2^3 factorial experiment). Then

$$\mu = ((1), a, b, ab, c, ac, bc, abc)^T$$

and

$$\underline{\beta} = (2M, A, B, AB, C, AC, BC, ABC)^T \ .$$

These two vectors are related by

$$\underline{\beta} \ = \ \frac{1}{4}(H_2 \otimes H_2 \otimes H_2)\mu$$

$$= \frac{1}{4}\begin{bmatrix} 1 & 1 & 1 & 1 & 1 & 1 & 1 & 1 \\ -1 & 1 & -1 & 1 & -1 & 1 & -1 & 1 \\ -1 & -1 & 1 & 1 & -1 & -1 & 1 & 1 \\ 1 & -1 & -1 & 1 & 1 & -1 & -1 & 1 \\ -1 & -1 & -1 & -1 & 1 & 1 & 1 & 1 \\ 1 & -1 & 1 & -1 & -1 & 1 & -1 & 1 \\ 1 & 1 & -1 & -1 & -1 & -1 & 1 & 1 \\ -1 & 1 & 1 & -1 & 1 & -1 & -1 & 1 \end{bmatrix} \mu \, .$$

This also implies

$$\mu = \frac{1}{2}(H_2 \otimes H_2 \otimes H_2)^T \underline{\beta} \, ,$$

which is again the crucial identity for arriving at a reparameterization of the original model (compare the steps that led to (11.6)).

A row in $H_2 \otimes H_2 \otimes H_2$ indicates how either $2M$ or a factorial effect can be expressed in terms of treatment effects. Note that the row corresponding to an interaction can easily be obtained by componentwise multiplication of the rows for the main-effects that appear in this interaction. For example, the row corresponding to ABC is

$$-1 \quad 1 \quad 1 \quad -1 \quad 1 \quad -1 \quad -1 \quad 1 \quad ;$$

the i-th element in this row is equal to the product of the i-th elements in the rows for the main-effects of A, B and C — the three factors occurring in the interaction. ■

The property observed in the previous paragraph holds for any $k \geq 2$, not just for $k = 3$. Knowing the rows corresponding to the k main-effects provides us trivially with the $2^k - k - 1$ rows for the interactions. The rows in the matrix

$$H_2 \otimes H_2 \otimes \cdots \otimes H_2$$

that correspond to the k main-effects can also be obtained easily. For the main-effect of A we write down a row that is alternately -1 and $+1$, starting with -1; for B alternate between groups of -1's and $+1$'s of size $2^1 = 2$, starting with two -1's; for C alternate between groups of -1's and $+1$'s of size $2^2 = 4$, again starting with a group of -1's; and so on. Table 11.10 shows the rows corresponding to the four main-effects A, B, C, D in a 2^4 factorial. The columns of this 4×16 submatrix of $H_2 \otimes H_2 \otimes H_2 \otimes H_2$ provide the level combinations (with levels coded as -1 and 1) in standard order.

Table 11.10. Standard order for a 2^4-factorial. The rows correspond to the factors A, B, C, D.

Level Combinations															
−1	1	−1	1	−1	1	−1	1	−1	1	−1	1	−1	1	−1	1
−1	−1	1	1	−1	−1	1	1	−1	−1	1	1	−1	−1	1	1
−1	−1	−1	−1	1	1	1	1	−1	−1	−1	−1	1	1	1	1
−1	−1	−1	−1	−1	−1	−1	−1	1	1	1	1	1	1	1	1

Let X be the $N \times 2^k$ matrix as in (11.2). Then we can express the model as

$$Y = X\mu + \epsilon = \frac{1}{2}X(H_2 \otimes H_2 \otimes \cdots \otimes H_2)^T \underline{\beta} + \epsilon \,. \tag{11.19}$$

Once again we will focus on using only a fraction of the complete 2^k factorial. An important method of forming a fraction is to require that the treatment combinations satisfy some *defining relation*.

Example 11.11. Continuing Example 11.9, we have

$$ABC = \frac{1}{4}(-(1) + a + b - ab + c - ac - bc + abc) \,.$$

Consider the half-replicate of the 2^3 factorial obtained by taking only those level combinations corresponding to treatment effects that appear with a minus sign in ABC, namely

$$(1), ab, ac, bc \,. \tag{11.20}$$

We denote this fraction by writing "$I_0 = -ABC$", where I_0 represents $+1$, the identity element. If we code the high levels of the factors as $+1$ and the low levels as -1, then the fraction $I_0 = -ABC$ consists precisely of those level combinations for which the product of the levels of A, B and C is equal to -1. For example ac is part of $I_0 = -ABC$, because

$$-1 = (1)(-1)(1) \,.$$

Similarly, the fraction "$I_0 = ABC$" consists of the level combinations for which the product of the levels for A, B and C is 1:

$$a, b, c, abc \,. \tag{11.21}$$

$I_0 = -ABC$ and $I_0 = ABC$ are called the *defining relations* for these two fractions. Observe that both fractions correspond to an $OA(4, 3, 2, 2)$, that $I_0 = -ABC$ is a linear orthogonal array, and that $I_0 = ABC$ is a translate of this linear array.

For both of the fractions, ABC is called the *defining contrast*. We will also say that ABC is a *word of length 3* in the defining relation for both of these fractions. (The defining relation may impose several simultaneous conditions on the treatment combinations, i.e. there may be several words in the defining relation.)

Of course, there are other half-replicates of the 2^3 factorial that possess a defining relation. For example we could take $I_0 = AC$, which consists of (1), b, ac, abc; or $I_0 = -A$, consisting of (1), b, c, bc. Both are linear orthogonal arrays, the first of strength 1 and the second of strength 0. ∎

Fractions with a defining relation have a simple description in terms of linear codes. More than that, they actually *are* linear codes, or translates of linear codes, and using this language to describe them makes it easier to construct them and to understand their properties. Only the most elementary concepts from coding theory are needed.

We remind the reader that a linear code C over a field S (of order s) is simply a linear subspace of S^k, and that the vectors in this subspace are called codewords. A code is specified by an $n \times k$ generator matrix whose rows form a basis, where n is the dimension of C. Alternatively, C may be specified by a $(k-n) \times k$ parity check matrix, whose rows span the orthogonal space. The minimal distance of the code is the smallest number of nonzero components in any nonzero codeword. Any linear code has a dual, obtained by using a parity check matrix for C as generator matrix. For further details see Section 4.2.

Returning to the previous example, where $s = 2$, we represent the low levels by 0 and the high levels by 1, regarded as elements of $S = GF(2)$. The fraction (11.20) defined by "$I_0 = -ABC$" consists precisely of the level combinations in the linear code with parity check matrix $[1\ 1\ 1]$. The fraction (11.21) defined by "$I_0 = ABC$" is a translate of this code by the vector 001.

While this alternative representation using error-correcting codes clearly demonstrates the correspondence with linear orthogonal arrays and their translates, and suggests how to generalize these definitions to fractions of an s^k factorial, for any prime power s, the representation of these fractions through their defining relations is far more common in the statistics literature.

A word in the defining relation of a fraction of a 2^k factorial may be represented by a binary vector of length k, with a 1 in the i-th coordinate if and only if the i-th factor appears in the word. Vector addition of the binary vectors then provides us with a way to multiply words. (Formally, the set of words based on k factors then becomes a multiplicative abelian group of order 2^k, say $\mathcal{G}_k$, with identity element I_0.) To illustrate the multiplication, ABD and ADE (with $k = 5$) correspond to the vectors 11010 and 10011, whose binary sum is

01001; equivalently, the product of ABD and ADE is

$$(ABD)(ADE) = A^2BD^2E = BE .$$

We call the product BE the *generalized interaction* of ABD and ADE.

The binary vectors corresponding to the defining relation for a fraction generate a linear code, the *relation code* $\mathcal{C}$. In Example 11.9, $\mathcal{C}$ has generator matrix [1 1 1]. If $\mathcal{C}$ has dimension n, we can select n generating vectors and obtain corresponding words $W_1, W_2, \ldots, W_n$. Then the defining relation

$$I_0 = \pm W_1 = \pm W_2 = \cdots = \pm W_n = \text{generalized interactions} \tag{11.22}$$

specifies a 2^{k-n} fractional factorial, for any choice of the n signs. In other words, for any choice of the n signs there are precisely 2^{k-n} level combinations among all 2^k possible combinations that satisfy all the equalities in the defining relation.

One of these fractions corresponds to the dual code $\mathcal{C}^\perp$, and the others are its $2^n - 1$ nonzero translates.[1] A fraction is *regular* if and only if its level combinations correspond to a translate of some linear code.

The minimal distance R (say) of the code $\mathcal{C}$ is the length of the shortest nonzero word defining the fraction. Statisticians refer to R as the *resolution* of the corresponding fraction. By Theorem 4.6, all the fractions obtained in this way are orthogonal arrays with parameters $OA(2^{k-n}, k, 2, R-1)$. We should therefore choose $\mathcal{C}$ to have as high a minimal distance as possible.

Example 11.12. Let $k = 6$. A 2^{6-2} fraction may be obtained from any code of length 6 and dimension 2, for instance those with generator matrices

$$\begin{bmatrix} 111000 \\ 000111 \end{bmatrix} \quad \text{or} \quad \begin{bmatrix} 111100 \\ 001111 \end{bmatrix} .$$

The first, $\mathcal{C}_1$ say, has minimal distance 3, while the second, $\mathcal{C}_2$ say, has minimal distance 4, giving fractions of resolutions 3 and 4 respectively. One of the 2^{6-2} fractions generated by $\mathcal{C}_2$ has as its defining relation

$$I_0 = ABCD = -ABEF = -CDEF . \tag{11.23}$$

The level combinations in this fraction, an $OA(16, 6, 2, 3)$, can be obtained by taking the appropriate translate of the dual code $\mathcal{C}_2^\perp$. Alternatively, they may be found as follows. We choose 4 $(= 6-2)$ factors that cannot be used to form a nonzero word that appears in $\mathcal{C}_2$. The factors A, B, C and E constitute such a

[1] If we had represented the high levels by -1 and the low levels by $+1$ (which would have seemed awkward), the linear fractional factorial would be obtained by taking all the signs in (11.22) to be positive.

choice in this example. (A, B, D and E could be used if the fraction was based on $\mathcal{C}_1$.) All possible level combinations will appear for these four factors, and we can list them in the standard order. The levels for the other factors, D and F, can now be computed by multiplication of the appropriate columns. Since $I_0 = ABCD$, we see that

$$D = ABC \ ,$$

and since $I_0 = -ABEF$ it follows that

$$F = -ABE \ .$$

The 2_{IV}^{6-2} fraction that is induced by (11.23) is shown in Table 11.13 (the Roman numeral subscript gives the resolution). ∎

Table 11.13. The 2_{IV}^{6-2} fractional factorial $I_0 = ABCD = -ABEF = -CDEF$.

A	B	C	E	$D = ABC$	$F = -ABE$	
−1	−1	−1	−1	−1	1	f
1	−1	−1	−1	1	−1	ad
−1	1	−1	−1	1	−1	bd
1	1	−1	−1	−1	1	abf
−1	−1	1	−1	1	1	cdf
1	−1	1	−1	−1	−1	ac
−1	1	1	−1	−1	−1	bc
1	1	1	−1	1	1	$abcdf$
−1	−1	−1	1	−1	−1	e
1	−1	−1	1	1	1	$adef$
−1	1	−1	1	1	1	$bdef$
1	1	−1	1	−1	−1	abe
−1	−1	1	1	1	−1	cde
1	−1	1	1	−1	1	$acef$
−1	1	1	1	−1	1	$bcef$
1	1	1	1	1	−1	$abcde$

Thus fractions with a defining relation simply form a subclass of the class of orthogonal arrays. As with linear orthogonal arrays, this subclass is somewhat restrictive because the number of runs is necessarily a power of 2, or more generally a power of s for fractions of an s^k factorial, if s is a prime power. Furthermore, even for a 2^{k-n} fractional factorial we could use an orthogonal array that does not correspond to a fraction with a defining relation (see also Chapter 7).

The popularity of fractions with a defining relation is in part due to the ease with which one can determine a maximal set of estimable functions of the factorial effects. We will first illustrate this by using the 2_{IV}^{6-2} fractional factorial described in Example 11.12.

Example 11.14. Multiplying each of the terms in (11.23) by A gives the relation

$$A = BCD = -BEF = -ACDEF \ . \tag{11.24}$$

This is one of the *alias relationships* for this fractional factorial. The message it conveys is that restricted to the 16 level combinations in this fraction, the four factorial effects in this relationship cannot be distinguished from each other. As a consequence, these effects are not estimable separately; what can be estimated unbiasedly, however, is

$$A + BCD - BEF - ACDEF \ .$$

The signs in the column for A in Table 11.13 suggest how to estimate this linear combination of factorial effects, namely as

$$\frac{1}{8}(Y_{abcde} - Y_{bcef} + Y_{acef} - Y_{cde} + \cdots) \ ,$$

that is, as the difference in average responses for level combinations with A at its high and low levels.

The complete set of all alias relationships can be obtained in a similar way and is given in Table 11.15.

If the fractional factorial is taken to be the linear array $C^{\perp}$, the alias relationships are in one-to-one correspondence with the nonzero translates of $C^{\perp}$.

Each of the equations in Table 11.15 induces an estimable linear combination of four factorial effects. These 15 estimable functions are pairwise orthogonal treatment contrasts, and, for an equal number of observations at each of the 16 level combinations, can be used to partition the treatment sum of squares with 15 degrees of freedom into single degrees of freedom.

Table 11.15. Alias relationships for $I_0 = ABCD = -ABEF = -CDEF$.

A	=	BCD	=	$-BEF$	=	$-ACDEF$
B	=	ACD	=	$-AEF$	=	$-BCDEF$
C	=	ABD	=	$-ABCEF$	=	$-DEF$
D	=	ABC	=	$-ABDEF$	=	$-CEF$
E	=	$ABCDE$	=	$-ABF$	=	$-CDF$
F	=	$ABCDF$	=	$-ABE$	=	$-CDE$
AB	=	CD	=	$-EF$	=	$-ABCDEF$
AC	=	BD	=	$-BCEF$	=	$-ADEF$
AD	=	BC	=	$-BDEF$	=	$-ACEF$
AE	=	$BCDE$	=	$-BF$	=	$-ACDF$
AF	=	$BCDF$	=	$-BE$	=	$-ACDE$
CE	=	$ABDE$	=	$-ABCF$	=	$-DF$
CF	=	$ABDF$	=	$-ABCE$	=	$-DE$
ACE	=	BDE	=	$-BCF$	=	$-ADF$
ACF	=	BDF	=	$-BCE$	=	$-ADE$

If interactions of three or more factors are assumed to be negligible, then all main-effects are estimable in this example. Two-factor interactions would still not be estimable, but certain sums of them, such as $AB + CD - EF$ and $AC + BD$, would be. This information is readily obtained from Table 11.15.

Also, again under the assumption that interactions of three or more factors are negligible, the last two alias relationships in Table 11.15 induce two (non-trivial) unbiased estimators of 0. This implies that there would be two degrees of freedom for error with the 2_{IV}^{6-2} fraction under the model obtained from (11.19) by deleting all interactions of three or more factors. Alternatively, these two contrasts could be used to judge (at least to some extent) the reasonableness of the assumption that interactions of three or more factors are negligible. A normal probability plot of estimates of the 15 estimable functions suggested by Table 11.15 would be useful in this example. ■

In general, in a 2^{k-n} fractional factorial each of the $2^{k-n} - 1$ alias relationships involves 2^n factorial effects, possibly with a minus sign. Sums of these signed effects provide $2^{k-n} - 1$ estimable functions. The full model in (11.19) can be reduced by assuming that certain factorial effects are negligible. A full set of linearly independent estimable functions is then obtained by taking the sums of the signed effects in the alias relationships after deleting all effects that are assumed to be negligible. This will also identify the degrees of freedom that are available for error under this fraction when using the reduced model. If this is small, possibly zero, a normal probability plot of the estimates of signed sums can be used, as discussed earlier, to identify the most important sets of effects.

The resolution of a fractional factorial with a defining relation is an important characteristic of the fraction. It is easily seen that the following conclusions hold for a fraction of resolution $R \geq 3$.

(i) If R is even, the mean M, main-effects and interactions of at most $(R-2)/2$ factors are estimable if interactions of $(R+2)/2$ or more factors are negligible.

(ii) If R is odd, the mean M, main-effects and interactions of at most $(R-1)/2$ factors are estimable if interactions of $(R+1)/2$ or more factors are negligible.

Since a fraction of resolution R is an orthogonal array of strength $R - 1$, these statements follow also from Theorem 11.3. In view of this, we extend the concept of resolution to arbitrary (possibly mixed) fractions as follows. A fractional factorial is said to be of *resolution* R if

(1) condition (i) or (ii) in Theorem 11.3 holds for $t = R - 1$;

(2) for the model referred to in (1), the components required to be estimable in (1) are all estimated with a relative efficiency of 1 (see Section 11.4), and the corresponding ordinary least squares estimators are uncorrelated.

The results in Section 11.4 show that any orthogonal array of strength t is of resolution $t+1$. It will follow from Theorem 11.30 that orthogonal arrays of strength t are the only fractions of resolution $t+1$. (Some authors only require (1) to hold for a general definition of resolution; see, for example, Raktoe, Hedayat and Federer, 1981, p. 88. In that case, however, orthogonal arrays of strength t would not be the only fractions of resolution $t+1$.)

Our arguments in the previous paragraphs have not made any direct reference to the model (11.19). It is however easy to see what happens to the matrix

$$X(H_2 \otimes H_2 \otimes \cdots \otimes H_2)^T \tag{11.25}$$

in (11.19) when X is a $2^{k-n} \times 2^k$ matrix corresponding to a 2_R^{k-n} fractional factorial with a defining relation. First, since the matrix in (11.25) consists of 2^{k-n} distinct rows of the Sylvester-type Hadamard matrix of order 2^k, it has rank 2^{k-n} (see Chapter 7). Second, if the defining relation looks like

$$I_0 = -ABD = \cdots ,$$

say, this means that all level combinations in this fraction have the property that the product of the levels for factors A, B and D, when coded as -1 and 1, is equal to -1. Therefore the column in (11.25) corresponding to $2M$ differs by only a multiplicative constant (namely -1) from the column for the ABD interaction.

Since there are $2^n - 1$ nontrivial words in the defining relation, the defining relation identifies 2^n columns in (11.25) that are either identical or differ by a multiplicative factor -1. Similarly, each of the other alias relationships identifies a group of 2^n such columns.

Since each factorial effect appears either in the defining relation or in exactly one of the alias relationships, it follows that the 2^k columns in (11.25) can be partitioned into 2^{k-n} groups of 2^n columns each, such that any two columns within a group are identical or differ by a multiplicative factor -1. Consequently, for any model that contains $2M$ and at most one factorial effect from each of the alias relationships, all parameters in the model are estimable under this fraction. Alternatively, if we do not assume any effects are negligible, under such a fraction the full model can be reparameterized by replacing $2M$ and the factorial effects by sums of signed effects. For example, if

$$I_0 = -ABD = \cdots ,$$

then $2M$, $ABD, \ldots$ are replaced by

$$2M - ABD + \cdots ;$$

similarly, using the alias relationship $A = -BD = \cdots$, another parameter in this reparameterization is

$$A - BD + \cdots .$$

It is obvious that, under such a fraction, each of the 2^{k-n} signed sums of effects obtained in this way is estimable. This also shows that, without any further model assumptions, all of the $2^{k-n}-1$ degrees of freedom for treatment contrasts are exhausted by the $2^{k-n} - 1$ alias relationships, not leaving any degrees of freedom for error.

11.6 Blocking for a 2^{k-n} Fractional Factorial

One of the most basic principles in the design of experiments is *blocking.* Experimental units, the possibly hypothetical units to which treatments are assigned, are never the same, and can sometimes be very different from each other. In an agricultural experiment, soil conditions and fertility for two different plots will never be identical, and may be quite different for plots that are widely separated. Furthermore, although each plot should be handled in exactly the same way, for example with respect to seeding, growing or harvesting of a variety, none of these routines can ever be carried out in a completely uniform manner.

For almost any proposed experiment there are hundreds of potential reasons for variability in a response variable, in addition to possible differences between the treatments. Intuitively it is obvious that it will be difficult or impossible to recognize differences between treatments if these are masked by large differences between the experimental units.

The use of a *Completely Randomized Design* (or CRD), in which the treatments are assigned to the experimental units at random, is therefore appropriate if there are no known or suspected reasons for differences between the units, that is, if the units are thought to be fairly homogeneous. If, however, there is a variable that might help to explain major differences between the units, then it is advisable to group the experimental units into smaller sets, called *blocks*, in such a way that units within a block are fairly homogeneous. For example, plots at the same gradient level in a sloping field may be more alike than plots at different gradient levels; or in studying the effect of different music programs on the daily production in a factory, production on Mondays may be quite different from that on Fridays.

Block effects will now appear in the statistical model, and the blocking variable will appear in the ANOVA table and will account for part of the variability caused by differences between the experimental units. The source error in the ANOVA table will now only need to account for differences between units within blocks, and if the blocking was effective, the mean squared error will be smaller than it would have been under the CRD design. As a result the experiment will be more capable of detecting possible differences between the treatments.

If there are b blocks, and the j-th block contains n_j units, for $j = 1, \ldots, b$, we must select n_j (not necessarily distinct) treatments to be used in the j-th block. Within each block, the n_j units are randomly assigned to the selected treatments.

We will only consider the case when all blocks have the same number of units, i.e. $n_j = n_0$ for $j = 1, \ldots, b$. If $n_0 = m2^k$, for a positive integer m, and if the treatments are the level combinations in a 2^k factorial, we can use each level combination m times in each block. If $m = 1$ this is known as a *Randomized Complete Block Design* (or RCBD); for other values of m this design has also been called a *Generalized Randomized Block Design* (or GRBD)—see for example Hinkelmann and Kempthorne (1994). In an RCBD or GRBD inference for factorial effects is exactly the same as it would have been under a CRD; the presence of block effects in the model does not affect this inference. Block effects do, of course, affect inference for experimental error variance, and thus also estimates of standard errors.

Since our emphasis in this chapter is on fractional factorials, we will not give any further details about the use of an RCBD or GRBD.

The situation that we will consider here is that of using a 2^{k-n} factorial in 2^p blocks of $n_0 = 2^{k-n-p}$ units each, for $1 \le p < k - n \le k$. The case $n = 0$ corresponds to the use of a complete factorial in 2^p incomplete blocks of size 2^{k-p}. A small example will illustrate the basic idea.

Example 11.16. For one replicate of a complete 2^3 factorial, suppose that the experiment will be conducted over two days, using four level combinations on each day. To account for possible day-to-day variability, we could run this experiment in two blocks of four units each. Ignoring order, there are $\binom{8}{4}/2 = 35$ ways to divide the eight level combinations into two groups of four level combinations. Are any of these decompositions better than others?

Without any assumptions about negligible factorial effects, the statistical model may be written as

$$Y = \frac{1}{2}X(H_2 \otimes H_2 \otimes H_2)^T \underline{\beta} + X_B\delta + \epsilon \,, \tag{11.26}$$

where X_B is an 8×2 $(0,1)$-matrix with a 1 in position (i,j) if and only if the i-th element in Y corresponds to a response in block j, and where $\delta = (\delta_1, \delta_2)^T$ is the parameter vector of block effects. All other notation is as in (11.19).

Consider first the following design (as we shall see, this can be improved). For one of the blocks, say the first, take the four level combinations (1), ab, bc and abc; this leaves a, b, c and ac for the second block.

By $\underline{\widehat{\beta}} = (2\widehat{M}, \widehat{A}, \ldots, \widehat{ABC})^T$ we denote the vector of least squares estimators that would have been appropriate if the data had been obtained by using a CRD. Thus $\underline{\widehat{\beta}}$ is the vector of least squares estimators under model (11.19).

By evaluating the expected values of these estimators under the selected design and model (11.26), the appropriate model here, we obtain

$$\begin{aligned}
E(2\widehat{M}) &= 2M + \delta_1 + \delta_2 \ , \\
E(\widehat{A}) &= A \ , \\
E(\widehat{B}) &= B + \frac{1}{2}(\delta_1 - \delta_2) \ , \\
E(\widehat{AB}) &= AB + \frac{1}{2}(\delta_1 - \delta_2) \ , \\
E(\widehat{C}) &= C \ , \\
E(\widehat{AC}) &= AC \ , \\
E(\widehat{BC}) &= BC + \frac{1}{2}(\delta_1 - \delta_2) \ , \\
E(\widehat{ABC}) &= ABC - \frac{1}{2}(\delta_1 - \delta_2) \ .
\end{aligned}$$

If we were to make the additional model assumption that the three-factor interaction is negligible, so that $ABC = 0$, we see that

$$\widehat{A}, \widehat{B} + \widehat{ABC}, \ \widehat{AB} + \widehat{ABC}, \ \widehat{C}, \ \widehat{AC}, \text{ and } \widehat{BC} + \widehat{ABC}$$

are unbiased estimators, and indeed least squares estimators, of

$$A, B, AB, C, AC, \text{ and } BC \ ,$$

respectively. Note, however, that

(i) our estimators are biased if the assumption that $ABC = 0$ is false, and

(ii) some parameters are estimated more precisely than others; we see that

$$\text{Var }(\widehat{A}) = \text{Var }(\widehat{C}) = \text{Var }(\widehat{AC}) = \sigma^2/2, \quad \text{and}$$

$$\text{Var }(\widehat{B} + \widehat{ABC}) = \text{Var }(\widehat{AB} + \widehat{ABC}) = \text{Var }(\widehat{BC} + \widehat{ABC}) = \sigma^2 \ .$$

Can we select a better partition of the eight level combinations into two groups of four? We need to realize that blocking has a price attached to it. With two blocks, one degree of freedom is reserved for blocks, so that only six degrees of freedom remain for treatments. Since main-effects and two-factor interactions represent six pairwise orthogonal treatment contrasts, we might ask if we can partition the eight treatments so that

(iii) unbiased estimators for main-effects and two-factor interactions are available, whether or not $ABC = 0$, and

(iv) the least squares estimators for all main-effects and two-factor interactions are obtained as under a CRD, all with variance $\sigma^2/2$.

A block design possessing properties (iii) and (iv) would overcome the deficiencies noted in (i) and (ii). The two major differences between such a block design and a CRD would be that inferences for ABC are not possible under the block design and, as a result of reducing the experimental error variance, if the blocking was effective then inferences for main-effects and two-factor interactions could be conducted with increased precision under the block design.

A block design for which (iii) and (iv) hold does indeed exist. In one block we use the level combinations a, b, c and abc, and in the other (1), ab, ac and bc. Note that, if the levels are coded as -1 and 1, the first block consists precisely of those level combinations for which the product of the levels for factors A, B and C is 1; the second block therefore consists of the combinations for which this product is -1. We express this by saying that the three-factor interaction ABC is *confounded with blocks*; we will also say that ABC is the *generating contrast* for the two blocks. ■

In general, if we use one replicate of a complete 2^k factorial in 2^p blocks of size 2^{k-p} each, there are $2^p - 1$ degrees of freedom for blocks. We would like to identify $2^p - 1$ factorial effects to be confounded with blocks, so that the other $2^k - 2^p$ factorial effects can be estimated as under a CRD. Since effects that are confounded with blocks are indistinguishable from block contrasts, we would like the confounded effects to consist only of the interactions of many factors. However, we cannot choose these $2^p - 1$ effects arbitrarily; the unit element I_0 and the $2^p - 1$ effects must form a subgroup of order 2^p of $\mathcal{G}_k$. Equivalently, the binary vectors corresponding to these effects, taken together with the zero vector, must form a binary linear code $\mathcal{B}$ (say) of dimension p. As in Section 11.5 we let $W_1, W_2, \ldots, W_p$ denote words corresponding to generators for this code.

Thus the problem of choosing a desirable confounding pattern for a 2^k factorial in 2^p blocks of size 2^{k-p} is analogous to choosing a desirable defining relation for a 2^{k-p} fractional factorial. In both cases we need to choose a binary linear code of dimension p with as large a minimal distance as possible.

The 2^p blocks of size 2^{k-p} that confound $W_1, W_2, \ldots, W_p$ and all their generalized interactions are obtained by taking all the cosets of the dual code $\mathcal{B}^\perp$. Thus each block consists of the level combinations corresponding to a 2^{k-p} fractional factorial of the form

$$I_0 = \pm W_1 = \pm W_2 = \cdots = \pm W_p = \text{generalized interactions },$$

for all 2^p possible sign combinations.

Finally, let us consider the use of a 2^{k-n} fractional factorial in 2^p blocks of size 2^{k-n-p} each, where now $1 \leq p < k - n \leq k - 1$. In this case two codes of length k (i.e. two subgroups of $\mathcal{G}_k$) need to be chosen: a code $\mathcal{C}$ of dimension n to define a fractional factorial, and a code $\mathcal{B}$ of dimension p to select generating contrasts for the 2^p blocks. In selecting $\mathcal{B}$ it should however be observed that, since we are using a fractional factorial, factorial effects have aliases, and what is now confounded with blocks are $2^p - 1$ signed sums of factorial effects that are aliases of each other. All effects in any such sum should ideally be interactions of a large number of factors.

$\mathcal{C}$ and $\mathcal{B}$ need to have independent generators, so that the smallest code of length k that contains them both has dimension $n + p$.

The above ideas are illustrated in Example 11.17. A more elaborate introduction to two-level fractional factorials with a defining relation, including the idea of confounding some contrasts with blocks when needed (but without mentioning linear codes), can be found in Box and Hunter (1961a,b). Tables of fractional factorials with suggestions for contrasts to be confounded with blocks, if needed, are presented in Montgomery (1997), Appendix XII. Table 8.7 in Montgomery (1997) provides some confounding patterns for complete 2^k factorials.

The reader who has grasped the relationship between fractional factorials with a defining relation and linear codes will have no difficulty in generalizing the ideas in this section to s^{k-n} fractional factorials in s^p blocks of size s^{k-n-p}, for any prime power s.

Example 11.17. Consider using one replicate of a 2^6 factorial in 2^3 blocks of size 2^3 each. Since there are 2^3 blocks, we need to choose a three-dimensional code $\mathcal{B}$ of length 6. An obvious choice is the code with generator matrix

$$\begin{bmatrix} 111100 \\ 110011 \\ 101010 \end{bmatrix} .$$

This corresponds to using the three effects $ABCD$, $ABEF$ and ACE to define the blocks. If these effects are used to generate the blocks, then the effects confounded with blocks are

$$ABCD, \quad ABEF, \quad ACE$$

and the generalized interactions

$$\begin{aligned} (ABCD)(ABEF) &= CDEF , \\ (ABCD)(ACE) &= BDE , \end{aligned}$$

$$\begin{aligned} (ABEF)(ACE) &= BCF \text{ , and} \\ (ABCD)(ABEF)(ACE) &= ADF \text{ .} \end{aligned}$$

All other effects, including all main-effects and two-factor interactions, can be estimated as under a CRD with this particular choice. To obtain the eight blocks under this choice we will first determine the block that includes (1). This is the dual code $\mathcal{B}^{\perp}$, which has generator matrix

$$\begin{bmatrix} 111100 \\ 110011 \\ 010101 \end{bmatrix}$$

and so the eight level combinations in this block are

$$(1),\ ade,\ bdf,\ abef,\ cdef,\ acf,\ bce,\ abcd\ .$$

Other blocks can now simply be found as cosets of this block. Each time we find a treatment that has not yet appeared in any of the blocks and multiply this treatment by all the treatments in the block that contains (1). For example, the block that contains a is

$$a,\ de,\ abdf,\ bef,\ acdef,\ cf,\ abce,\ bcd\ .$$

As a second scenario, consider the use of the 2^{6-2} fractional factorial of Table 11.13 in two blocks of size 2^3. To partition the 16 level combinations in this fraction into two blocks of size 8 we could use the contrast $ACE + BDE - BCF - ADF$. One block would now consist of those eight level combinations in the selected fraction that appear with a + sign in ACE, the other would consist of those with a $-$ sign. The resulting two sets of eight treatments would be

$$ad,\ abf,\ cdf,\ bc,\ e,\ bdef,\ acef,\ abcde$$

and

$$f,\ bd,\ ac,\ abcdf,\ adef,\ abe,\ cde,\ bcef\ .$$

An easy way to obtain this partition is to start with Table 11.13 (to identify the level combinations that belong to the selected fraction) and to add a column for ACE to the table, to identify those level combinations that appear with a + sign in ACE and those with a $-$ sign.

In this design, main-effects and two-factor interactions are not part of the contrast that is confounded with blocks. If interactions of three or more factors are negligible, all main-effects are estimable while two-factor interactions are aliased with each other. The contrast confounded with blocks then consists only of negligible effects. ∎

With or without blocking, normal probability plots of factorial effect estimates are a useful tool for identifying the most important factorial effects (or components of factorial effects if $s > 2$), provided that the effects sparsity assumption is reasonable. Standard statistical software packages, such as *SAS* and *S-Plus*, facilitate producing these plots. The estimates of the effects that are not important will roughly fall along a straight line near the center of the plot. Estimates of effects that are important will fall away from this line at the extremes: they will be more extreme in magnitude than one would expect in a random sample from a normal distribution. The estimate of the intercept parameter should not be included when making this plot; neither should estimates of effects that are confounded with blocks (if blocking was used). If a regular fractional factorial is used, estimates of linear combinations of the factorial effects (as suggested by the defining relation) can be plotted. Based on the results of the plot a simpler model can then be fitted, one that only includes the few effects that appear to be important in the plot (and the block effects if blocking was used). This simpler model will have more degrees of freedom for error, and a residual analysis can now be used to study its adequacy.

11.7 Orthogonal Main-Effects Plans and Orthogonal Arrays

Orthogonal arrays of strength 2 are part of the larger class of orthogonal main-effects plans. Two examples of the latter that are only orthogonal arrays of strength 0 were presented in Tables 1.13 and 1.15. A formal definition of an orthogonal main-effects plan may be given as follows.

Definition 11.18. Consider the model (11.8) with γ_1 consisting of all components of the k main-effects and γ_2 of the intercept parameter alone. Let an $N \times k$ array with symbols $0, 1, \ldots, s_l - 1$ in column l, $s_l \geq 2$, $l = 1, 2, \ldots, k$, be used to select N level combinations. If, under the stated model and with this choice of level combinations,

(i) all components of the k main-effects are estimable, and

(ii) the ordinary least squares estimators of any two components from two different main-effects are uncorrelated,

then we say that the $N \times k$ array is an *orthogonal main-effects plan.*

That an orthogonal array of strength 2 is indeed an orthogonal main-effects plan follows immediately from Theorem 11.3 and Equation (11.12). The model in Definition 11.18 is also known as a *main-effects model.*

Example 11.19. Let $k = 4$, $s_1 = s_2 = s_3 = s_4 = 2$, and $N = 9$. Let the level combinations be those in Table 1.13, rearranged according to the standard order as shown in Table 11.20.

Table 11.20. An orthogonal main-effects plan.

0	0	0	0
0	0	0	0
0	1	0	0
1	1	0	0
0	0	1	0
1	0	1	0
0	0	0	1
1	0	0	1
0	1	1	1

With $\gamma_1 = (\beta_{1000}, \beta_{0100}, \beta_{0010}, \beta_{0001})^T$ and $\gamma_2 = \beta_{0000}$ we find that

$$XU_1 = \frac{1}{4}\begin{bmatrix} -1 & -1 & -1 & -1 \\ -1 & -1 & -1 & -1 \\ -1 & 1 & -1 & -1 \\ 1 & 1 & -1 & -1 \\ -1 & -1 & 1 & -1 \\ 1 & -1 & 1 & -1 \\ -1 & -1 & -1 & 1 \\ 1 & -1 & -1 & 1 \\ -1 & 1 & 1 & 1 \end{bmatrix}, \quad XU_2 = \frac{1}{4}1_9 .$$

(Since the discussion in this section is not restricted to two-level factors, we are using the vector β of Section 11.3 and not $\underline{\beta}$ of Section 11.5. This explains why there is a divisor of 4 rather than 2 in the above expressions.) For the information matrix C of γ_1 we obtain

$$C = U_1^T X^T (I_9 - \frac{1}{9} 1_9 1_9^T) X U_1 = \frac{1}{2} I_4 .$$

This implies that γ_1 is estimable and Var $(\widehat{\gamma}_1) = 2\sigma^2 I_4$. Therefore the ordinary least squares estimators $\widehat{\beta}_{1000}, \widehat{\beta}_{0100}, \widehat{\beta}_{0010}$, and $\widehat{\beta}_{0001}$ are uncorrelated, and the array is indeed an orthogonal main-effects plan.

Of course, whether the nine runs are ordered as in Table 1.13 or 11.20, or in any other way, is irrelevant in deciding whether the array is an orthogonal main-effects plan. The order in Table 11.20 is merely chosen to be consistent with the general discussion in Section 11.2 on forming the matrix X. ∎

A fairly simple characterization of orthogonal main-effects plans can be given through the so-called condition of proportional frequencies. According to Addelman and Kempthorne (1961b) this characterization was first stated by Plackett (1946).

In the following definition we will use the notation n_l^j to denote the frequency with which a particular symbol j, $0 \le j \le s_l - 1$, appears in the l-th column of a given array, a frequency assumed to be at least equal to 1.

Definition 11.21. An $N \times k$ array with symbols $0, 1, \ldots, s_l - 1$ in the l-th column, $s_l \geq 2$, $l = 1, 2, \ldots, k$, is said to satisfy the *condition of proportional frequencies* if for any two columns l and l', $l \neq l'$, and for any two symbols $j \in \{0, 1, \ldots, s_l - 1\}$ and $j' \in \{0, 1, \ldots, s_{l'} - 1\}$ the frequency with which the ordered pair (j, j') appears as a row of the $N \times 2$ subarray formed by columns l and l' is equal to

$$n_l^j n_{l'}^{j'} / N \ .$$

Theorem 11.22. *An $N \times k$ array is an orthogonal main-effects plan if and only if it satisfies the condition of proportional frequencies.*

Proof: From (11.9), the information matrix for γ_1 of Definition 11.18 is, under the main-effects model, given by

$$C = U_1^T X^T (I_N - \frac{1}{N} J_N) X U_1 \ .$$

With

$$U_1 = [U_{11} \ \ldots \ U_{1k}] \ ,$$

where U_{1l} is the $M \times (s_l - 1)$ matrix corresponding to the components of the main-effect of factor A_l, the information matrix C consists of blocks of the form

$$C_{ll'} = U_{1l}^T X^T (I_N - \frac{1}{N} J_N) X U_{1l'} \ . \tag{11.27}$$

It follows that an $N \times k$ array as in Definition 11.18 is an orthogonal main-effects plan if and only if

$$\text{rank } (C_{ll}) = s_l - 1$$

and $C_{ll'} = 0$ for all $l \neq l'$, $l, l' = 1, 2, \ldots, k$.

For the remainder of this proof we will use the notation $b^{(l)}(i, j)$ and Z_l as defined in (11.4). We will also use the symbols n_l^j as in Definition 11.21, and we will use $n_{ll'}^{jj'}$ to denote the number of level combinations in a given array for which the l-th factor is at level j and the l'-th factor at level j'.

For $j \in \{0, 1, \ldots, s_l - 1\}$, the vector

$$\frac{1}{\prod_{m \neq l} \sqrt{s_m}} (b^{(l)}(1, j), \ldots, b^{(l)}(s_l - 1, j))$$

appears n_l^j times as a row of the matrix XU_{1l}. Since $n_l^j > 0$ it follows that

$$\text{rank } ([1_N \ \ XU_{1l}]) = \text{rank } (Z_l) = s_l,$$

and thus also that rank $(C_{ll}) = s_l - 1$.

Thus the array is an orthogonal main-effects plan if and only if $C_{ll'} = 0$ for all $l \neq l'$, $l, l' = 1, 2, \ldots, k$.

For fixed $l \neq l'$, an element of this matrix is, for some $i \in \{1, 2, \ldots, s_l - 1\}$ and $i' \in \{1, 2, \ldots, s_{l'} - 1\}$, equal to

$$\sum_{j=0}^{s_l-1} \sum_{j'=0}^{s_{l'}-1} n_{ll'}^{jj'} \left(b^{(l)}(i,j) - \frac{1}{N} \sum_{h=0}^{s_l-1} n_l^h b^{(l)}(i,h) \right) \times$$

$$\times \left(b^{(l')}(i',j') - \frac{1}{N} \sum_{h'=0}^{s_{l'}-1} n_{l'}^{h'} b^{(l')}(i',h') \right) \Big/ \left(\prod_{m \neq l} \sqrt{s_m} \right) \left(\prod_{m \neq l'} \sqrt{s_m} \right) .$$

Observe that the numerator of this expression is equal to

$$\sum_{j=0}^{s_l-1} \left(b^{(l)}(i,j) - \frac{1}{N} \sum_{h=0}^{s_l-1} n_l^h b^{(l)}(i,h) \right) \sum_{j'=0}^{s_{l'}-1} \left(n_{ll'}^{jj'} \quad n_l^j n_{l'}^{j'}/N \right) b^{(l')}(i',j')$$

$$= \sum_{j=0}^{s_l-1} b^{(l)}(i,j) \sum_{j'=0}^{s_{l'}-1} (n_{ll'}^{jj'} - n_l^j n_{l'}^{j'}/N) b^{(l')}(i',j') .$$

Since 1_{s_l} spans the null space of the matrix B_l in (11.4) it follows that the array is an orthogonal main-effects plan if and only if

$$\sum_{j'=0}^{s_{l'}-1} (n_{ll'}^{jj'} - n_l^j n_{l'}^{j'}/N) b^{(l')}(i',j')$$

is independent of $j \in \{0, 1, \ldots, s_l - 1\}$. Since summing over j gives zero, this holds if and only if

$$\sum_{j'=0}^{s_{l'}-1} (n_{ll'}^{jj'} - n_l^j n_{l'}^{j'}/N) b^{(l')}(i',j') = 0 .$$

Because $1_{s_{l'}}$ spans the null space of $B_{l'}$ in (11.4), this is true if and only if

$$n_{ll'}^{jj'} - n_l^j n_{l'}^{j'}/N$$

is independent of $j' \in \{0, 1, \ldots, s_{l'} - 1\}$, for all $j \in \{0, 1, \ldots, s_l - 1\}$. Since summing over j' gives zero, this holds if and only if

$$n_{ll'}^{jj'} - n_l^j n_{l'}^{j'}/N = 0$$

for all $j \in \{0, 1, \ldots, s_l - 1\}$ and $j' \in \{0, 1, \ldots, s_{l'} - 1\}$. This argument holds for all $l, l' \in \{1, 2, \ldots, k\}$, $l \neq l'$, and the desired result follows. ∎

Observe that a diagonal element of the matrix C in this proof, say c_{li}, can be written as

$$c_{li} = \frac{1}{\prod_{m \neq l} s_m} \sum_{j=0}^{s_l-1} n_l^j \left[b^{(l)}(i,j) - \frac{1}{N} \sum_{h=0}^{s_l-1} n_l^h b^{(l)}(i,h) \right]^2 . \tag{11.28}$$

Note also that for an $OA(N, s_1 s_2 \ldots s_k, 2)$

$$n_{ll'}^{jj'} = N/(s_l s_{l'}) = (N/s_l)(N/s_{l'})/N = n_l^j n_{l'}^{j'}/N ,$$

again confirming that orthogonal arrays of strength 2 are orthogonal main-effects plans.

Example 11.23. Continuing Example 11.19, for the array in Table 11.20 we see that

$$n_1^0 = 6 \;,\; n_1^1 = 3 \;,\; n_2^0 = 6 \;,\; n_2^1 = 3 \;,\; N = 9 \,.$$

Indeed, we also see that

$$\begin{aligned} n_{12}^{00} &= n_1^0 n_2^0/N = 4 \;, & n_{12}^{10} &= n_1^1 n_2^0/N = 2 \;, \\ n_{12}^{01} &= n_1^0 n_2^1/N = 2 \;, & n_{12}^{11} &= n_1^1 n_2^1/N = 1 \;. \end{aligned}$$

Since similar results hold for any other pair of factors, the array in Table 11.20 is an orthogonal main-effects plan. ■

Thus, for a main-effects model and any orthogonal main-effects plan, the ordinary least squares estimators of two components of different main-effects are uncorrelated. One difference between orthogonal arrays of strength 2 and other orthogonal main-effects plans becomes apparent when considering the precision with which components of main-effects are estimated.

Consider the component $\beta_{0\ldots 0i0\ldots 0}$, with i appearing as the l-th subscript, $l \in \{1, 2, \ldots, k\}$, $i \in \{1, 2, \ldots, s_l - 1\}$. The variance of $\widehat{\beta}_{0\ldots 0i0\ldots 0}$ under an orthogonal main-effects plan is then equal to $\sigma^2 c^{li}$, where c^{li} denotes the diagonal element of C^{-1}, the inverse of the information matrix, corresponding to the component $\beta_{0\ldots 0i0\ldots 0}$. With N as the number of level combinations in the plan, and $M = \prod_{l=1}^k s_l$, the relative efficiency (see Section 11.4) of the plan for $\beta_{0\ldots 0i0\ldots 0}$ is therefore equal to

$$RE(\beta_{0\ldots 0i0\ldots 0}) = \frac{M/N}{c^{li}} \,. \tag{11.29}$$

If the matrix C_{ll} defined in (11.27) is a diagonal matrix, say $C_{ll} = \text{diag}\,\{c_{l1}, \ldots, c_{l,s_l-1}\}$, then $c^{li} = 1/c_{li}$, and it follows from (11.28) that

$$RE(\beta_{0\ldots 0i0\ldots 0}) = \frac{s_l}{N} \sum_{j=0}^{s_l-1} n_l^j \left[b^{(l)}(i,j) - \frac{1}{N} \sum_{h=0}^{s_l-1} n_l^h b^{(l)}(i,h) \right]^2 . \tag{11.30}$$

Some relative efficiencies of two- and three-level orthogonal main-effects plans are presented in Table 11.24. For a two-level factor the matrix B_l was chosen to be

$$B_l = \frac{1}{\sqrt{2}}[-1 \ \ 1] ;$$

for a three-level factor we used

$$B_l = \frac{1}{\sqrt{6}}\begin{bmatrix} -\sqrt{3} & 0 & \sqrt{3} \\ 1 & -2 & 1 \end{bmatrix} ,$$

meaning that the two components of the main-effect correspond to the linear and quadratic components for a three-level factor with equally spaced levels. For all cases considered in Table 11.24, the matrix C_{ll} is a diagonal matrix, so that all relative efficiencies are immediately obtained from (11.30).

Table 11.24. Relative efficiencies of two- and three-level orthogonal main-effects plans.

s_l	n_l^0/N	n_l^1/N	n_l^2/N	$RE(Linear)$	$RE(Quadratic)$
2	1/3	2/3	–	8/9	–
2	1/4	3/4	–	3/4	–
2	1/5	4/5	–	16/25	–
2	2/5	3/5	–	24/25	–
3	1/4	2/4	1/4	3/4	9/8
3	1/5	3/5	1/5	3/5	27/25
3	2/5	1/5	2/5	6/5	18/25

It follows from (11.30) that if $n_l^0 = n_l^1 = \ldots = n_l^{s_l-1}$, a case not considered in Table 11.24, then all relative efficiencies for components of the main-effect of factor l are equal to 1. We will soon see that the converse of this statement is also true, and that orthogonal arrays of strength 2 are the only plans for which the relative efficiency for every component of every main-effect is equal to 1. We will also see that for any plan the average of the relative efficiencies for the components of any main-effect is at most 1, with equality holding for every main-effect if and only if the plan is an orthogonal array of strength 2.

Orthogonal main-effects plans that are not orthogonal arrays of strength 2 can still be important if no orthogonal array of strength 2 exists that meets the requirements for a proposed experiment, for example in terms of the number of runs. They can also be important if we wish to increase the precision of estimating a linear or quadratic component at the expense of the precision of estimating higher order components.

Some basic methods for the construction of orthogonal main-effects plans and tables of such plans can be found in Addelman and Kempthorne (1961a) and Dey (1985).

The following result provides a characterization of orthogonal arrays of strength 2 based on their information matrix, under the main-effects model, for the vector consisting of all the components of all the main-effects.

Theorem 11.25. *Consider an $N \times k$ array with symbols $0, 1, \ldots, s_l - 1$ in its l-th column, $s_l \geq 2$, $l = 1, 2, \ldots, k$. The array is an orthogonal array of strength 2 (or higher) if and only if the information matrix C for γ_1, the vector of all $R_1 = \sum_{l=1}^{k}(s_l - 1)$ components of the main-effects, is equal to*

$$C = (N/M)I_{R_1}$$

under the main-effects model.

Proof: If the array is an orthogonal array of strength 2, then it follows from Theorem 11.3 that all elements of γ_1 are estimable. Moreover, it follows, as for (11.12), that

$$\mathrm{Var}\ (\widehat{\gamma}_1) = \sigma^2(M/N)I_{R_1}\ ,$$

and thus that

$$C = (N/M)I_{R_1}\ .$$

(Note that γ_1 and R_1 are here not quite the same as in (11.12) since the intercept parameter is now not included in γ_1.)

For the converse, from (11.9) we see that

$$C = U_1^T X^T \left(I_N - \frac{1}{N} J_N \right) X U_1\ .$$

Hence, using (11.28),

$$\begin{aligned} Tr(C) &= \sum_{l=1}^{k}\sum_{i=1}^{s_l-1} c_{li} \leq \sum_{l=1}^{k}\sum_{i=1}^{s_l-1}\sum_{j=0}^{s_l-1} n_l^j [b^{(l)}(i,j)]^2/(\prod_{m\neq l} s_m) \\ &= \sum_{l=1}^{k}\sum_{j=0}^{s_l-1} n_l^j (1 - 1/s_l)/(\prod_{m\neq l} s_m) = \frac{N}{M}\sum_{l=1}^{k}(s_l - 1) = NR_1/M \qquad (11.31) \end{aligned}$$

with equality if and only if

$$\sum_{j=0}^{s_l-1} n_l^j b^{(l)}(i,j) = 0$$

for all $i \in \{1, 2, \ldots, s_l - 1\}$ and $l \in \{1, 2, \ldots, k\}$. By arguments as in the proof of Theorem 11.22, this implies that $n_l^0 = n_l^1 = \ldots = n_l^{s_l-1} (= N/s_l)$ for all $l = 1, 2, \ldots, k$.

Since C is a nonsingular diagonal matrix, so that the array is an orthogonal main-effects plan, it follows now from Theorem 11.22 that for all $l \neq l'$

$$n_{ll'}^{jj'} = n_l^j n_{l'}^{j'}/N = N/(s_l s_{l'}) ,$$

for all $j \in \{0, 1, \ldots, s_l - 1\}$ and $j' \in \{0, 1, \ldots, s_{l'} - 1\}$. The array is thus an orthogonal array of strength 2. ∎

Observe that the proof of this theorem implies that

$$U_1^T X^T \left(I_N - \frac{1}{N} J_N\right) X U_1 = (N/M) I_{R_1}$$

if and only if

$$\begin{bmatrix} \frac{1}{\sqrt{M}} 1_N^T \\ U_1^T X^T \end{bmatrix} \begin{bmatrix} \frac{1}{\sqrt{M}} 1_N & X U_1 \end{bmatrix} = (N/M) I_{R_1+1} .$$

We could therefore have also stated the result in Theorem 11.25 as follows.

Corollary 11.26. *Consider an $N \times k$ array with symbols $0, 1, \ldots, s_l - 1$ in its l-th column, $s_l \geq 2$, $l = 1, 2, \ldots, k$. Let γ_1 consist of all the $R_1 = 1 + \sum_{l=1}^{k} (s_l - 1)$ components in the main-effects model, while γ_2 is empty. The array is an orthogonal array of strength 2 (or higher) if and only if the information matrix C for γ_1 is*

$$C = (N/M) I_{R_1}$$

under the main-effects model.

The following corollaries substantiate the claims in the text immediately after Table 11.24.

Corollary 11.27. *For any $N \times k$ plan, the average of the relative efficiencies for the components of the main-effect of factor l is at most 1, for $l \in \{1, 2, \ldots, k\}$.*

Proof: Using the notation of (11.29), the average in the statement of the corollary is equal to

$$\frac{1}{s_l - 1} \sum_{i=1}^{s_l - 1} \frac{M/N}{c^{li}} .$$

Since $c^{li} c_{li} \geq 1$ (cf. Graybill, 1983, Theorem 12.2.10), we see from (11.28) that

$$\frac{1}{c^{li}} \leq c_{li} \leq \frac{s_l}{M} \sum_{j=0}^{s_l - 1} n_l^j [b^{(l)}(i, j)]^2 ,$$

and that

$$\sum_{i=1}^{s_l-1} \frac{1}{c^{li}} \leq \frac{s_l}{M} \sum_{j=0}^{s_l-1} n_l^j \sum_{i=1}^{s_l-1} [b^{(l)}(i,j)]^2 = \frac{s_l}{M} \sum_{j=0}^{s_l-1} n_l^j (1 - 1/s_l) = (s_l - 1)N/M \ .$$

The result follows now immediately. ∎

Corollary 11.28. *For any $N \times k$ plan, if the average of the relative efficiencies for the components of the main-effects of factor l, $l \in \{1, 2, \ldots, k\}$, is equal to 1, then $n_l^0 = n_l^1 = \ldots = n_l^{s_l-1}$ and all of these relative efficiencies are equal to 1.*

Proof: In order for this average to be 1, it follows from the proof of Corollary 11.27 and (11.28) that we must have

$$\sum_{j=0}^{s_l-1} n_l^j b^{(l)}(i,j) = 0 \ ,$$

for all $i \in \{1, 2, \ldots, s_l - 1\}$. As in the proof of Theorem 11.25, this leads to the conclusion that

$$n_l^0 = n_l^1 = \ldots = n_l^{s_l-1} \ .$$

The latter implies that the relative efficiency for each of the components of the main-effect of factor l must be equal to 1, as was already observed in the text after Table 11.24. ∎

Corollary 11.29. *For any $N \times k$ plan, if, for every factor $l \in \{1, 2, \ldots, k\}$, the average of the relative efficiencies for the components of the main-effect of factor l is 1, then the array must be an orthogonal array of strength 2 (or higher).*

Proof: It follows from Corollary 11.28 that all the relative efficiencies are equal to 1. From the proof of Corollary 11.27 this implies that $1/c^{li} = c_{li} = N/M$ for all $i = 1, 2, \ldots, s_l - 1$, $l = 1, 2, \ldots, k$, which means that

$$C = (N/M)I_{R_1} \ .$$

The desired conclusion follows now from Theorem 11.25. ∎

In the special case $s_1 = s_2 = \ldots = s_k = 2$, the result of Theorem 11.25 was already stated in Problem 1.8. Moreover, Problem 1.9 suggests that a similar characterization of orthogonal arrays may be possible for arrays of higher strength. Such a characterization is formalized in the following result.

Theorem 11.30. *Consider an $N \times k$ array with symbols $0, 1, \ldots, s_l - 1$ in its l-th column, $s_l \geq 2$, $l = 1, 2, \ldots, k$. For $u \geq 1$, define the following two models, using the notation of (11.8):*

(i) γ_1 consists of the intercept parameter, all the components of all the main-effects and all the interactions of at most u factors, while γ_2 is empty;

(ii) γ_1 is the same as in (i), while γ_2 consists of all the components of all the interactions of precisely $u + 1$ factors.

Let R_1 denote the length of the vector γ_1, and let C_1 and C_2 denote the information matrices of γ_1 under models (i) and (ii), respectively. Then

(1) $C_1 = (N/M)I_{R_1}$ if and only if the array is an orthogonal array of strength $t = 2u$ (or higher);

(2) $C_2 = (N/M)I_{R_1}$ if and only if the array is an orthogonal array of strength $t = 2u + 1$ (or higher).

Proof: If the array is an orthogonal array of strength t, as in (1) or (2), then the desired conclusion follows easily, as indicated in the paragraphs after the proof of Theorem 11.3. For the converse, rows and columns of C_1 can be labeled by $\beta_{i_1 i_2 \ldots i_k}$, where $i_l \in \{0, 1, \ldots, s_l - 1\}$, $l = 1, 2, \ldots, k$, and at most u of the i_l's are nonzero. Let

$$n_{l_1 l_2 \ldots l_v}^{j_1 j_2 \ldots j_v},$$

$1 \leq v \leq k$, $1 \leq l_1 < \ldots < l_v \leq k$, $j_w \in \{0, 1, \ldots, s_{l_w} - 1\}$, denote the numbers of runs in the array in which factor l_1 is at level j_1, factor l_2 is at level j_2, and so on. Then for any factors $1 \leq l_1 < l_2 < \ldots < l_v \leq k$, with $1 \leq v \leq t, t = 2u$, and for every nonzero vector $(i_1, i_2, \ldots, i_k)$, with $i_l \in \{0, 1, \ldots, s_l - 1\}$ and $i_l = 0$ if $l \notin \{l_1, l_2, \ldots, l_v\}$, the matrix C_1 has an element in an off-diagonal position that is equal to

$$\frac{\prod_{w=1}^{v} s_{l_w}}{M} \sum_{j_1=0}^{s_{l_1}-1} \cdots \sum_{j_v=0}^{s_{l_v}-1} n_{l_1 l_2 \ldots l_v}^{j_1 j_2 \ldots j_v} b^{(l_1)}(i_{l_1}, j_1) \ldots b^{(l_v)}(i_{l_v}, j_v) . \tag{11.32}$$

Since $C_1 = (N/M)I_{R_1}$, all such elements must be equal to 0. By considering the matrix

$$Z_{l_v} \otimes Z_{l_{v-1}} \otimes \ldots \otimes Z_{l_1} , \tag{11.33}$$

where Z_l is as defined in (11.4), we see that all elements in (11.32) are 0 precisely if the vector of $n_{l_1 l_2 \ldots l_v}^{j_1 j_2 \ldots j_v}$'s, where the superscripts take all possible values in standard order, is orthogonal to all rows of the matrix in (11.33) except the first. Since this matrix is an orthogonal matrix, this means that the vector of $n_{l_1 l_2 \ldots l_v}^{j_1 j_2 \ldots j_v}$'s must be a multiple of the first row of the matrix in (11.33), implying that the value of $n_{l_1 l_2 \ldots l_v}^{j_1 j_2 \ldots j_v}$ does not depend on the choice of $j_w \in \{0, 1, \ldots, s_{l_w} - 1\}$, $w = 1, \ldots, v$. Since the argument is valid for any $v \leq t$ and any factors $l_1, l_2, \ldots, l_v$, the array is an orthogonal array of strength t.

The argument for C_2 is similar, with a minor additional difficulty at the start. First, using

$$C_2 = U_1^T X^T (I_{R_1} - P_{XU_2}) X U_1 ,$$

we observe that

$$Tr(C_2) \leq Tr(U_1^T X^T X U_1) = N R_1 / M ,$$

with equality if and only if $U_1^T X^T X U_2 = 0$. Since $C_2 = (N/M) I_{R_1}$, this means that

$$U_1^T X^T X U_2 = 0 \tag{11.34}$$

and

$$U_1^T X^T X U_1 = (N/M) I_{R_1} . \tag{11.35}$$

By an argument similar to that for C_1, (11.34) and (11.35) imply that terms of the form (11.32) must all be zero if $1 \leq v \leq 2u + 1$. The conclusion that the array is an orthogonal array of strength $t = 2u + 1$ follows as before. ■

The value of characterizing orthogonal arrays as in Theorem 11.30 is that it enables us to test whether a given array has a particular strength simply by performing matrix multiplications. The reader may want to consult Section 3.5 for characterizations of orthogonal arrays that are not based on statistical models.

11.8 Robust Design

Designed experiments have always been an integral part of science, starting with the need for experiments in agriculture. The use of designs in industry, while somewhat more recent, also has a long tradition, as exemplified by articles in the journal *Technometrics*. The explosive increase in the use of design for product quality improvement and monitoring in industrial processes during the last 15 years (certainly in the United States) is however staggering. The trigger for this explosion was the tremendous success of Japanese industry, with W. E. Deming's total quality management ideas as a key ingredient. It was not until the early 1980's, after a visit by G. Taguchi to AT&T, that Taguchi's contributions to the Japanese successes drew attention in the United States, both among engineers and statisticians. In this section we will give a brief overview of Taguchi's contributions and subsequent developments.

Main-effects and interactions that explain how the average response is affected by level changes of one or more factors are also known as *location effects.* The discussion in the previous sections centered around the identification and interpretation of location effects under the assumption of equality of variances as stated after model (11.1). However, if multiple measurements for a variable of interest were made at each level combination, we might see not only differences in average response but also in the sample variances. Parameters to explain possible differences in variances are often referred to as *dispersion effects*. Stressing the importance of studying both the process mean and process variance is perhaps the single most important contribution of Taguchi to quality improvement techniques. However, the methods that Taguchi proposes for this (cf. Taguchi, 1987) have received much criticism, with regard to both design and analysis.

The recent literature contains multiple proposals for studying a process mean and variance simultaneously, but it is clear that this problem will require more attention in years to come. Two excellent sources for further reading are Nair (1992) and Steinberg (1996).

Taguchi distinguishes between various types of factors, the main types being *control factors* and *noise factors*. Control factors are factors that can be set at specified levels during the production process, while noise factors can be fixed at selected levels during an experiment but not during the production or later use of the product. For example, the amount of sugar, sweetener or any other ingredient in a cake mix corresponds to a control factor. On the other hand, when set at a recommended temperature level, actual oven temperatures can vary greatly from one oven to another, and even within an oven, so that a good cake mix needs to possess a certain robustness with respect to oven temperature. Oven temperature is a noise factor. Of course, in an experiment to determine an ideal cake mix, oven temperature can be controlled very precisely.

Thus settings for the control factors typically determine process conditions. Noise factors are often environmental factors, such as humidity or temperature (cf. Taguchi, 1986), or factors associated with use of the product. They could also correspond to deviations from nominal settings for control factors. There are many processing variables (control factors) in the production process of automobiles that need to be set at optimal levels. In addition though, one wants the car that one buys to perform well whether one lives in Florida or Alaska. The performance should also not depend on how the car is driven.

In an experiment we would like to learn how changes in settings of the control factors affect the average quality of a product, and how they affect the product's robustness to variations in the noise factors. Taguchi uses the term *parameter design* for experiments aimed at selecting optimal levels for the control factors, although the term *robust design* is now more commonly used.

Having identified the control factors and noise factors in an experiment, Taguchi suggests using two orthogonal arrays, possibly with mixed levels, one for the control factors and one for the noise factors, with say N_1 and N_2 level combinations respectively. He then forms the *product array*, consisting of the

$N_1 N_2$ level combinations obtained by combining each level combination for the control factors in the first array with each level combination for the noise factors in the second array. For ease of notation we will use Y_{ij}, $i = 1, \ldots, N_1$, $j = 1, \ldots, N_2$ to denote the response variable corresponding to the level combination consisting of the ith level combination in the control factors and the jth one in the noise factors. By $\bar{Y}_i$ and S_i^2 we will mean

$$\bar{Y}_i = \frac{1}{N_2} \sum_{j=1}^{N_2} Y_{ij}, \quad S_i^2 = \frac{1}{N_2 - 1} \sum_{j=1}^{N_2} (Y_{ij} - \bar{Y}_i)^2 , \tag{11.36}$$

the sample mean and variance of the responses for the ith level combination in the control factors. Depending on the objective (e.g. finding settings for the control factors that maximize the response, or finding settings to attain a certain target value for the response), Taguchi advocates the use of certain functions of the Y_{ij}'s, usually referred to as *signal-to-noise ratios*, for analyzing the data and obtaining optimal settings for the control factors. Examples of such signal-to-noise ratios are

$$SN_1 = -10 \log(\sum_{j=1}^{m} Y_{ij}^2 / m) ,$$

$$SN_2 = -10 \log(\sum_{j=1}^{m} (1/Y_{ij})^2 / m) ,$$

$$SN_3 = 10 \log(\bar{Y}_i^2 / S_i^2) .$$

For each run i in the orthogonal array for the control factors, the value of a selected signal-to-noise ratio is computed. This signal-to-noise ratio is used as the response variable in an analysis of variance (typically using the main-effects model) to identify the important control factors. These factors are set at levels to maximize the signal-to-noise ratio. As already mentioned, the choice of a signal-to-noise ratio depends on the objective: if the response Y is to be minimized, SN_1 is used; for maximization, SN_2; and for the target-is-best objective, SN_3. In the latter case an analysis of variance is also performed with $\bar{Y}_i$ as the response variable. We wish to find settings for the control factors that yield the desired target with a small value for SN_3.

There are strong arguments against Taguchi's approach (cf. Steinberg, 1996). These include:

1. Use of the signal-to-noise ratios does not reveal how control factors affect the response. If a control factor is identified as an important factor, we would still not know whether it primarily affects the process mean, the process variance, or both.

2. There is no consideration for the possibility of interactions among the factors. In fact, since Taguchi almost exclusively uses orthogonal arrays of strength 2, such interactions would be difficult or impossible to detect.

3. Despite the N_1N_2 observations in a product array, there are only N_1 different level combinations for the control factors. The alias structure for the control factors is identical to what it would have been without the replication generated by using the different level combinations for the noise factors.

4. We do not learn how the noise factors affect the response variable. Are only some of the noise factors creating noise? Do they interact with some of the control factors? An analysis based on a summary statistic (such as signal-to-noise ratio) cannot answer such questions.

A simple alternative to Taguchi's approach that addresses the first criticism, and that is now often viewed as preferable, is to perform separate factorial analyses on the sample means and variances, or on a transformation thereof, along the lines discussed in the previous sections. In particular, it is not uncommon to see one analysis with the $\bar{Y}_i$'s and one with the $\log(S_i^2)$'s as the responses, where the former identifies the factorial effects based on the control factors that explain variability in the sample means (location effects) and the latter those that explain variability in the logarithms of the sample variances (dispersion effects).

Rosenbaum (1994, 1996) proposes the use of compound orthogonal arrays (see Chapter 10) instead of product arrays for this alternative. Using the notation from Section 10.7, the array B provides the level combinations for the control factors, while the C_i's do the same for the noise factors. If $C_1 = \cdots = C_{N_1}$ we are back to the use of product arrays. Starting from a model that includes effects associated with control factors and noise factors, Rosenbaum provides a justification for the analyses of the $\bar{Y}_i$'s and the S_i^2's; whether such analyses are meaningful depends on the model terms and the orthogonal arrays used to form the compound orthogonal array. In particular, for the product arrays that Taguchi's approach normally uses, even for the simplest model the analyses may give misleading results. For details see Rosenbaum (1996). Rosenbaum (1997) considers the use of compound orthogonal arrays when there is a need for blocking.

Another alternative to Taguchi's approach is the analysis of all the Y_{ij}'s, using a model based on both control and noise factors. Under this alternative there is no immediate need to compute S_i^2's, and consequently no immediate need to use compound orthogonal arrays. This is very appealing and can address all the criticisms of Taguchi's methods. Instead of using separate orthogonal arrays for the control and noise factors, one *combined array* is selected for both types of factors. If we use N_1N_2 runs, we no longer need to replicate each of

N_1 level combinations for the control factors N_2 times, but can in principle choose an array that includes $N_1 N_2$ different level combinations for the control factors. A possible model can consist of an intercept parameter, main-effects of all factors, and some two-factor interactions. All two-factor interactions between a control factor and noise factor are normally included, since these can provide information on how to set the control factors so that the product becomes more robust to changes in the noise factors. For more information on these ideas see Shoemaker, Tsui and Wu (1991) and Welch, Yu, Kang and Sacks (1990). Vining and Myers (1990), using a similar starting point, propose a dual response approach to the problem, developing a response surface for both the process mean and process variance. See also Myers, Khuri and Vining (1993) and Myers and Montgomery (1995), Chapter 10.

A compound orthogonal array is sometimes preferred over a combined array. This is the case if there are really only N_1, rather than $N_1 N_2$, experimental units for the control factors, with each observed under N_2 different conditions for the noise factors. The experiment has then a so-called split-plot nature, which needs to be considered in the analysis. See Steinberg (1996), Section 8.7, for a brief discussion and some references.

A very interesting article is that of Box and Meyer (1986). In the absence of noise factors, they suggest a method for studying location and dispersion parameters simultaneously, based on data from a regular two-level fractional factorial, using each of the selected level combinations for only one run. Their method is appealing in that it requires only a small number of level combinations but relies heavily on the assumption of effects sparsity, and there is the danger of confounding between location and dispersion effects. In the context of replicated (fractional) factorials this is further discussed in Nair and Pregibon (1988), who conclude that an extension of Box and Meyer's method to replicated designs has undesirable properties. But, of course, Box and Meyer's method was intended for unreplicated fractional factorials, and in the absence of clearly superior alternatives, using their ideas as part of an exploratory analysis remains worthwhile.

Other relevant references for this section include Abt and Pukelsheim (1995), Box, Bisgaard and Fung (1988), Engel (1992), Engel and Huele (1996), Steinberg and Bursztyn (1994). While various interesting and creative contributions have been made to date, the conclusion must be that these represent only the first steps towards achieving the full potential of robust design.

11.9 Other Types of Designs

While the class of orthogonal arrays is an extremely rich class of designs, with a great number of important applications, there are many other important classes of designs These include balanced arrays, central composite designs, supersaturated designs, covering designs and computer-generated designs. We will now say a few words about these.

Balanced arrays. A *balanced array of strength t* relaxes the requirement on an orthogonal array of strength t that all level combinations must appear equally often for any t factors. For a precise definition and additional references see Raktoe, Hedayat and Federer (1981), Section 13.7, or Srivastava (1990), p. 339. Orthogonal arrays are balanced arrays, but the latter is a much larger class of designs.

Central composite designs. *Central composite designs* (cf. Box and Draper, 1987, Section 15.3; Myers and Montgomery, 1995, Section 7.4) are extremely popular and useful for certain response surface problems. The levels of a quantitative factor can be coded through a linear transformation of the actual levels by mapping two selected levels to -1 and 1. Any other coded level, such as 0, can now be associated with an actual level of the factor through this linear transformation: 0 would simply correspond to the midpoint of the levels corresponding to -1 and 1. If all factors are quantitative, then the *center point* is the point with all factors (in terms of coded levels) at level 0. An *axial point* is a point where one factor is at level a or $-a$, usually for some $a > 1$, and all other factors are at level 0. A central composite design consists normally of a two-level orthogonal array (almost always a regular fraction) with levels -1 and 1, several copies of the center point, and all axial points for some value of a, $a > 1$. These designs are useful for fitting a so-called second-order model. They are especially appealing in the context of a sequential approach (using, for example, steepest ascent methods) for finding desirable level combinations for the factors. Initially the design would consist only of the two-level orthogonal array and some center points; if the initial experiment provides strong evidence for curvature in the response surface, axial points and additional center points are added to form a central composite design.

Supersaturated designs. Some authors use the term *saturated orthogonal array* for what we have called a tight orthogonal array. More generally, for a specified vector γ in (11.7), a *saturated design* is a design for which γ is estimable and the number of observations, N, is equal to the rank of the model matrix XU in (11.7). Thus there are zero degrees of freedom for error for this combination of model and design. The term saturated orthogonal array is therefore somewhat ambiguous: whether the array is saturated does not just depend on the array but also on the model for which it is used. It is for this reason that we prefer the term *tight orthogonal array* for an array that achieves equality in Rao's inequalities (see Chapter 2). The discussion in the last paragraph of Section 11.4 suggests a model for which a tight orthogonal array is also a saturated design.

If, for a model such as (11.7), the number of observations in a design is less than the length of the vector γ, then the design is known as a *supersaturated design.* These designs have received attention for use in screening experiments. The model is taken to be the main-effects model, and all factors are used at two levels. The objective of the experiment is to identify the most important main-effects. A supersaturated design with k factors is any design with $N < 1 + k$. This means however that we cannot estimate all main-effects unbiasedly. The appeal of these designs (a small number of runs despite a large number of factors) comes therefore with a hefty price (the increased peril of not being able to identify the important factors). These designs should only be used if practical considerations (such as cost, time, and available resources) necessitate their use. Some recent results on supersaturated designs appear in Lin (1993), Chen and Lin (1998) and Cheng (1997).

Covering designs. Covering designs are somewhat similar to supersaturated designs. A *covering design with N runs, k factors, s levels and strength t* is an $N \times k$ matrix with symbols from an alphabet of size s (which need not be a prime power) such that every $N \times t$ subarray contains each possible t-tuple *at least once.* Other names that have been used for covering designs are *t-independent arrays, t-surjective arrays*, and *qualitatively independent* families.

The advantage of these designs is that N can be very much smaller than the number of factors, in fact need only be proportional to a small multiple of $\log k$. The problem of finding good covering designs is an attractive one, with connections to many branches of combinatorics, and many results are known. For more information on the mathematical aspects of this question see for example Section 20.3 of Cohen et al. (1997), and Sloane (1973).

Several authors have developed algorithms and software for constructing reasonably efficient covering designs that apply even when different factors may have different numbers of levels: see Dalal and Mallows (1998), Dunietz et al. (1997), Mallows (1997) and Sherwood (1994).

If we replaced "at least once" in the definition by "equally often", we would of course have an orthogonal array. For some applications, however—such as testing software, compilers, etc.—it is enough that each possible combination of the symbols in any t variables gets exercised at least once. As an example, several statistical software packages allow the user to call multiple procedures when handling a data set, such as reading, printing, plotting or analyzing the data. These procedures sometimes have subcommands for which one of several options must be chosen. It is not uncommon for a software bug to show up only if a certain option or combination of options for one or more subcommands is selected. Since exhaustive testing of all combinations of options for all subcommands is often not practicable, a covering design of strength t, with subcommands of various procedures as factors and their options as levels, may be used to assure that all combinations of t options are tested. The hope is that using such a design will reveal a pattern of errors, if a bug is present in the software. Orthogonal arrays are also used for this purpose, though these generally require more runs than covering designs (cf. Mandl, 1985).

A class of matrices that somewhat resemble covering designs are matrices with "Property P_t". An $N \times k$ matrix is said to have *Property* P_t if any t columns are linearly independent. Here the linear independence could be over a finite field or over the real numbers. If the independence is over $GF(s)$ then such matrices are precisely the class of parity check matrices for linear codes with minimal distance $t+1$. Classical results from coding theory show that for fixed s and t, such designs exist with N proportional to $\log k$ (MacWilliams and Sloane, 1977, Chapter 17). Less is known about the real case. Matrices with Property P_t have been used by Srivastava (1976) and Srivastava and Ghosh (1977) in connection with search designs (see the end of Section 11.10 for references).

Computer-generated designs. Many computer programs now exist that will produce designs tailored to particular applications (cf. Piepel, 1997). The *Gosset* program of Hardin and Sloane (1992, 1993) is especially flexible and powerful. With this program the user can specify the individual factors, which may be discrete and/or continuous, the factors may be required to satisfy linear equalities or inequalities (so mixture designs and constrained measurement regions can be handled) and the model to be fitted can be any low degree polynomial or other simple function. The number of runs is also specified by the user, who can request an I-, A-, D- or E-optimal design. The region where the model is to be fitted need not be the same as the measurement region. If no model is available the program will look for a packing, i.e. will attempt to place the points uniformly over the measurement region. Other features include the possibility of constructing sequential designs, or of finding designs which are optimal given that a run may be lost, etc.

With the widespread use of computers for analyzing experimental results, some of the advantages of classical designs such as orthogonal arrays are less important, and the greater flexibility of computer-generated designs together with their smaller size is making them increasingly important for industry.

11.10 Notes on Chapter 11

There are other applications of orthogonal arrays, some of which we now briefly discuss.

As a generalization of Latin hypercube sampling in d-dimensional space, the use of orthogonal arrays has been investigated for numerical integration and computer experiments. Orthogonal arrays share certain desirable properties with Latin hypercubes while allowing for a multivariable stratification. For details we refer to Owen (1994); see also Koehler and Owen (1996). Further references on designs for computer experiments are Currin et al. (1991), Johnson, Moore and Ylvisaker (1990), McKay, Beckman and Conover (1979), Morris (1991), Sacks, Schiller and Welch (1989), Sacks et al. (1989), Welch et al. (1990), Welch et al. (1992).

Owen (1992) also suggests the use of orthogonal arrays for fitting models such as Multivariate Adaptive Regression Splines, or MARS (Friedman, 1991). Additional results on this topic are reported in Chen (1999), who is also interested in space-filling properties of orthogonal arrays.

Another application of orthogonal arrays, and more generally orthogonal multi-arrays (see Chapter 10), is in survey sampling. Orthogonal (multi-)arrays facilitate the efficient estimation of the variance of nonlinear statistics in stratified sampling with possibly unequal numbers of primary sampling units per stratum. The crux of the idea for this application is due to McCarthy (1969), who considered the case of two primary sampling units per stratum. Extensions in order to remove the latter restriction were made by Wu (1991) and Sitter (1993).

Orthogonal arrays have additional desirable statistical properties. Most noteworthy is that under certain models they are universally optimal for the estimation of some or all factorial effects that appear in the model. In particular, for the models and notation of Theorem 11.3, an orthogonal array of strength t is universally optimal for the effects in γ_1. For more details see Cheng (1980) and Mukerjee (1982).

Fractional factorials with the same resolution may, under certain models, still have very different properties when it comes to estimability of factorial effects. For that reason further criteria have been introduced to distinguish between such fractions, such as minimal aberration (cf. Chen and Hedayat, 1996, and Suen, Chen and Wu, 1997) and estimation capacity (cf. Cheng, Steinberg and Sun, 1999). The problem of finding a minimal aberration design is equivalent to that of finding a maximin distance design (see Section 4.7) that contains the smallest number of pairs of points that are the minimal distance apart. See Duckworth (1999) for more details.

If an orthogonal array is of strength t, but not of strength $t+1$, then it may occasionally still possess the property that for *some* sets of $t+1$ factors all possible level combinations appear equally often. Hedayat (1990) gives examples of this phenomenon, and orthogonal arrays of strength t with this property may be preferable to arrays of strength t that do not possess this property. This preference can in part be motivated through projection properties of orthogonal arrays (cf. Cheng, 1995). The MacWilliams transform of the weight distribution of the orthogonal array provides a convenient way to study this question.

Another topic that has received considerable attention is that of choosing run orders with desirable properties. In an experiment where the level combinations are used sequentially, as is the case in many experiments in industry, the order in which the runs are used is normally determined by randomization. But occasionally a judiciously selected run order is preferable. The design is then called a *systematic design*. There are various reasons why one might wish to select a systematic run order. If it is difficult or expensive to change the levels of one or more of the factors, a run order that minimizes the number of level changes with respect to these factors can be preferable. Another reason might be the presence of a time trend during the experiment as the result of tiring operators, deteriorating equipment, and so on. It may then be possible to select the run order in such a way that the usual ordinary least squares estimators for the components of effects of most interest (for example, main-effects and two-factor interactions) remain unbiased estimators of these components when a polynomial of specified degree is used to model the time trend. The run order is also of great importance with a temporal or spatial dependency of the error terms in model (11.7).

For a systematic design, data analysis may require a leap of faith. If, for example, a run order is used that changes the levels of a hard-to-change factor as little as possible, could not a difference in response at the different levels of this factor also be due to some unknown extraneous source of variability that changes gradually during the experiment?

More on the selection of run orders in factorial experiments can be found in Cheng (1990), Cheng and Steinberg (1991), Coster (1993), Jacroux (1990), and the references mentioned there. For some of the philosophical issues concerning systematic designs and for additional references, see also Lin and Stufken (1999).

We have briefly discussed aliasing of factorial effects for a regular fraction in Section 11.5. Problem 11.7 will address how the concept of aliasing can be defined for arbitrary orthogonal arrays. The alias relations can become rather complex in general, and this continues to be thought of as a reason to avoid such fractions (cf. Montgomery, 1997, p. 419). Hamada and Wu (1992) demonstrate however that the more complicated alias structure can be a blessing. We return to this in Problem 11.8.

An issue that can occasionally be important is the analysis of categorical data from a designed experiment. We refer the reader to Chipman and Hamada (1996) for an approach to this problem.

For an entirely different perspective on the use of orthogonal arrays in factorial experiments see Srivastava (1987, 1990, 1996). This work strongly advocates the use of so-called *search designs*.

In connection with Section 11.5, it should be pointed out that since the alias relationships are essentially the same as the nonzero cosets of $\mathcal{C}^{\perp}$, the "*standard array*" of coding theory provides a convenient way to describe these relationships: see MacWilliams and Sloane (1977), page 16. Furthermore, the weight distribution of the coset leaders is helpful in determining which sums of interactions are estimable, especially if the fraction is taken to be the linear array $\mathcal{C}^{\perp}$.

11.11 Problems

11.1. The data in Table 11.31 are taken from an injection molding experiment discussed in Box, Hunter and Hunter (1978). All eight factors in the experiment have two levels, and the 16 level combinations in Table 11.31 were used in a completely randomized design. You can assume throughout this question that interactions of three or more factors are negligible; that is, the vector γ in (11.7) consists only of the intercept parameter, the main-effects, and the two-factor interactions.

(i). Verify that the fraction in Table 11.31 is an orthogonal array with a defining relation, and determine this defining relation and the resolution.

(ii). Provide the alias relationships for this fraction. (Since interactions of three or more factors are assumed to be negligible, it suffices to specify the relations among the main-effects and two-factor interactions.)

(iii). Write the model as in (11.8), where γ_1 contains all main-effects and γ_2 the intercept parameter and all two-factor interactions. Compute the information matrix for γ_1.

(iv). Obtain the degrees of freedom and sum of squares for the following sources: (a) each of the main-effects; (b) a remainder term for the two-factor interactions; (c) error; and (d) the corrected total. Summarize your computations in an ANOVA table.

(v). Certain linear combinations of the two-factor interactions are estimable for this fraction and model. One such linear combination is $\beta_{11000000} + \beta_{00100010} + \beta_{00010001} + \beta_{00001100}$, which in the notation of Section 11.5 is equivalent to $AB + CG + DH + EF$ (the two expressions being identical except for a multiplicative constant). Using either of the two forms, show that this linear combination is indeed estimable, provide its ordinary least squares estimate, and obtain the corresponding sum of squares. (When using the first form, you can compute the sum of squares as in Example 11.6, that is, as $\frac{N}{M}$ times the square of the ordinary least squares estimate. Alternatively, the sum of squares can be computed as $\frac{N}{4}$ times the square of the ordinary least squares estimate for the second expression.)

Table 11.31. Injection molding data. Factor levels are denoted by − (for low) and + (for high).

Run	Mold temperature A	Moisture content B	Holding pressure C	Cavity thickness D	Booster pressure E	Cycle time F	Gate size G	Screw speed H	Shrinkage Y
1	−	−	−	+	+	+	−	+	14.0
2	+	−	−	−	−	+	+	+	16.8
3	−	+	−	−	+	−	+	+	15.0
4	+	+	−	+	−	−	−	+	15.4
5	−	−	+	+	−	−	+	+	27.6
6	+	−	+	−	+	−	−	+	24.0
7	−	+	+	−	−	+	−	+	27.4
8	+	+	+	+	+	+	+	+	22.6
9	+	+	+	−	−	−	+	−	22.3
10	−	+	+	+	+	−	−	−	17.1
11	+	−	+	+	−	+	−	−	21.5
12	−	−	+	−	+	+	+	−	17.5
13	+	+	−	−	+	+	−	−	15.9
14	−	+	−	+	−	+	+	−	21.9
15	+	−	−	+	+	−	+	−	16.7
16	−	−	−	−	−	−	−	−	20.3

(vi). Provide a set of seven linear combinations of the two-factor interactions (using the form in (v) you prefer) such that: (a) one of the linear combinations is that of part (v); (b) each of the seven linear combinations is estimable; and (c) any two of the linear combinations form orthogonal contrasts. (Hint: The alias relationships should suggest the seven combinations needed.)

(vii). Compute the sum of squares for each of the seven contrasts in part (vi). If your computations are correct, you should see that these seven sums of squares add to the sum of squares for the remainder in the ANOVA table in part (iv). Can you provide an algebraic justification for this?

(viii). For the analysis of variance table in (iv), there are zero degrees of freedom for error. Make a normal probability plot of the ordinary least squares estimates of the eight main-effects (A, B, ...) and the seven linear combinations of the two-factor interactions ($AB+CG+DH+EF$, ...). Based on this plot, do you agree that C, E and $AE+BF+CH+DG$ (or their equivalents in terms of the alternative notation) appear to be the most important factorial effects?

(ix). Now consider the model that contains only the factorial effects C, E and AE. For this model there are 12 degrees of freedom for error. Prepare an ANOVA table for this model, identify which of the three factorial effects are significantly different from 0, and study residual plots (as in Example 11.6) to assess model adequacy. (While we are using AE as a model term, the estimate for this effect is really an estimate for $AE+BF+CH+DG$, something you may want to keep in mind when answering part (x).)

(x). Based on the analysis in the previous parts, formulate how level changes for the various factors affect the response.

(xi). If you were given the resources to make eight additional observations, which level combinations would you use, and why? What if you could only make four additional observations?

11.2. Consider the fraction in Table 11.31. For a different experiment that uses the same fraction, if blocking the experimental units into two blocks of eight units had been desirable, how would you have divided the sixteen runs over the two blocks? Explain your answer.

11.3. The principal reason for using one replicate of a regular 2^{k-n} fractional factorial in 2^p blocks, each with 2^{k-n-p} units, was discussed in Section 11.6. Can this be done as easily for an arbitrary orthogonal array? We will not attempt to answer this question in general, but will consider one special case.

Consider a mixed $OA(18, 3^6 6^1, 2)$ (see Example 9.17), and suppose that five three-level factors are to be used in an experiment. The level combinations for the experiment are selected by assigning each of the five factors to a different three-level column of the orthogonal array, say A_1 through A_5. We wish to partition the 18 level combinations into six blocks of three units each.

Option 1: The 6-level column in the orthogonal array, say A_7, is used to partition the 18 level combinations. (Thus a level of 0 for A_7 puts the corresponding level combination for A_1 through A_5 in the first block, a 1 moves it to the second block, and so on.)

Option 2: A_7 is converted to a two-level column (by replacing 0, 1 and 2 by 0, and 3, 4 and 5 by 1), again called A_7. The level combinations of the sixth three-level column, A_6, and A_7 are now used to partition the 18 level combinations for the five three-level factors. (A combination of 00 for A_6 and A_7 puts the corresponding level combination for A_1 through A_5 in block 1, 01 moves it to block 2, and so on.)

If the model uses only the intercept parameter, all the components of the main-effects, and the six block effects, which of these two options do you prefer? Are there any differences in terms of estimability, precision of estimation, ... ? (You may see an important difference simply by writing out the six blocks for each option.)

11.4. Do the following fractions exist? Explain your answers.

(i). A regular 2^{8-3} fraction of resolution 5.

(ii). A regular 2^{9-5} fraction of resolution 4.

(iii). A regular 2^{7-3} fraction of resolution 4 in two equal blocks such that no main-effects or two-factor interactions appear in the contrast that is confounded with blocks.

11.5. This problem suggests a simple construction for an orthogonal main-effects plan from an orthogonal array of strength 2. For an s-level factor in the orthogonal array, replace its levels by those for an s'-level factor (with $s' < s$) using a mapping f from $\{0, 1, \ldots, s-1\}$ *onto* $\{0, 1, \ldots, s'-1\}$.

(i). Show that the resulting array is an orthogonal main-effects plan.

(ii). Under what conditions is it an orthogonal array of strength 2?

(iii). Search for orthogonal main-effects plans with only two- and three-level factors and between 10 and 20 runs. Present the parameters of the arrays that you have found in a table with the columns labeled by N (for the number of runs), k_1 (for the number of two-level factors), and k_2 (for the number of three-level factors). Mark the parameters that correspond to an orthogonal array of strength 2 by an asterisk.

(iv). For the main-effects plans in your table, what is the smallest relative efficiency for a two-level factor main-effect?

11.6. The data in this problem are presented in Engel (1992). They are also considered by various other authors, including Engel and Huele (1996), Steinberg (1996), and Steinberg and Bursztyn (1994). Engel and Huele (1996) identify a 1987 course manual for an *Introduction to Quality Engineering* as the source for the data. As presented here, the data contain a small "correction" suggested by Steinberg and Bursztyn (1994).

The data are from another injection molding experiment, with (as far as we know) no relationship to that in Problem 11.1. The experiment used a product array with seven control factors and three noise factors. The array for the control factors is an $OA(8, 7, 2, 2)$, while that for the noise factors is an $OA(4, 3, 2, 2)$.

The response variable was a measurement of percent shrinkage. The ten factors in the experiment are shown in Table 11.32 and the design and data in Table 11.33.

Table 11.32. Control factors and noise factors.

Control factors	Noise factors
A Cycle time	M Percentage regrind
B Mold temperature	N Moisture content
C Cavity thickness	O Ambient temperature
D Holding pressure	
E Injection speed	
F Holding time	
G Gate size	

Table 11.33. Product array and data for injection molding experiment.

							Noise factors M, N, O					
							−1	−1	1	1		
		Control factors					−1	1	−1	1		
A	B	C	D	E	F	G	−1	1	1	−1	$\bar{Y}_i$	S_i
−1	−1	−1	−1	−1	−1	−1	2.2	2.1	2.3	2.3	2.225	0.10
−1	−1	−1	1	1	1	1	2.5	0.3	2.7	0.3	1.450	1.33
−1	1	1	−1	−1	1	1	0.5	3.1	0.4	2.8	1.700	1.45
−1	1	1	1	1	−1	−1	2.0	1.9	1.8	2.0	1.925	0.10
1	−1	1	−1	1	−1	1	3.0	3.1	3.0	3.1	3.025	0.05
1	−1	1	1	−1	1	−1	2.1	4.2	1.0	3.1	2.600	1.37
1	1	−1	−1	1	1	−1	4.0	1.9	4.6	2.2	3.175	1.33
1	1	−1	1	−1	−1	1	2.0	1.9	1.9	1.8	1.900	0.08

(i). Determine the defining relation and alias relationships for the $OA(8, 7, 2, 2)$ used for the control factors. (For the alias relationships you can assume that interactions of three or more factors are negligible.)

(ii). Identify the most important control factors by analyzing the signal-to-noise ratio SN_3 of Section 11.8 and the observed means $\bar{Y}_i$, in both cases using a main-effects model based on the control factors. What are your conclusions?

(iii). Analyze the observed means $\bar{Y}_i$ and $\log(S_i)$, again using main-effects models in the control factors, and formulate your conclusions based on these analyses.

(iv). Analyze the individual Y_{ij}'s, as if the data resulted from a combined array, using a model that contains the intercept parameter, all ten main-effects, and all two-factor interactions between a control and noise factor. What are your conclusions?

(v). The larger values in the column for S_i in Table 11.33 correspond precisely to the runs with factor F at its high level. It would thus appear "obvious" that the main-effect of F is an important dispersion effect. However, although the analyses in (ii) and (iii) support this logic, that in (iv) does not. It has been argued that the analysis in (iv) is probably correct in not identifying the main-effect of F as an important dispersion effect. Provide an argument to support this claim, and explain the apparent contradiction concerning the importance of F for explaining variance in the response.

11.7. We introduced alias relationships in Section 11.5, but did not go beyond fractions that have a defining relation. In this problem we extend the notion of alias relationships to arbitrary fixed- or mixed-level fractions. Suppose we want to find the alias relationships that involve the components of the main-effects for a given fraction, such as an orthogonal array. Write the model in the form of (11.8) with γ_1 consisting of all the components of the main-effects and γ_2 of all other components in the model (including the intercept parameter). If we ignore the effects in γ_2 and estimate the elements of γ_1 based on the model

$$Y = XU_1\gamma_1 + \epsilon ,$$

then the resulting ordinary least squares estimator for γ_1 is

$$\widehat{\gamma}_1 = (U_1^T X^T X U_1)^{-1} U_1^T X^T Y .$$

For the model that only contains γ_1, this estimator is an unbiased estimator of γ_1. But for the model

$$Y = XU_1\gamma_1 + XU_2\gamma_2 + \epsilon ,$$

the expected value of this estimator is

$$\delta = \gamma_1 + (U_1^T X^T X U_1)^{-1} U_1^T X^T X U_2 \gamma_2 ,$$

which is an unbiased estimator of γ_1 only if $U_1^T X^T X U_2 = 0$. The matrix $(U_1^T X^T X U_1)^{-1} U_1^T X^T X U_2$ is known as the *alias* or *bias matrix*. The vector δ provides us with the alias relationships between the components of the main-effects (the elements of γ_1) and the elements of γ_2.

(i). Let γ_1 and γ_2 be as above, and consider the fraction $I_0 = ABC$ used in Example 11.11. By computing the vector δ above, show that $A + BC$, $B + AC$ and $C + AB$ are estimable. (This is precisely the conclusion we would have obtained from the alias relationships if we had computed them from the defining relation.)

(ii). Consider the $OA(12, 7, 2, 2)$ formed by the first seven columns (labeled A through G) of the $OA(12, 11, 2, 2)$ in Table 11.34. Suppose that γ_1 is as above, and that γ_2 contains the intercept parameter and the three two-factor interactions between factors A, B and C. Study the vector δ above, and comment on the alias relationship for the main-effects under this model and with this design. (Of course, since 12 is not a power of 2, this design does not have a defining relation.)

Table 11.34. Data from Hunter, Hodi and Eager (1982).

A	B	C	D	E	F	G	8	9	10	11	*Meas.*
+	+	−	+	+	+	−	−	−	+	−	6.058
−	+	+	−	+	+	+	−	−	−	+	5.863
+	−	+	+	−	+	+	+	−	−	−	5.917
−	+	−	+	+	−	+	+	+	−	−	5.818
−	−	+	−	+	+	−	+	+	+	−	6.607
−	−	−	+	−	+	+	−	+	+	+	5.682
+	−	−	−	+	−	+	+	−	+	+	5.752
+	+	−	−	−	+	−	+	+	−	+	7.000
+	+	+	−	−	−	+	−	+	+	−	5.899
−	+	+	+	−	−	−	+	−	+	+	4.652
+	−	+	+	+	−	−	−	+	−	+	4.733
−	−	−	−	−	−	−	−	−	−	−	4.809

11.8. The more complicated alias relationships that you should have discovered in Problem 11.7 (ii) are sometimes seen as undesirable. On the other hand, Hamada and Wu (1992) argue that they can actually be advantageous. For a regular fraction of resolution 3, if $A = BC$, say, there is no hope of disentangling the possible effects of this main-effect and the two-factor interaction. For a design with a more complicated alias structure, it may be possible to recognize important two-factor interactions even if the fraction is only of resolution 3.

Following Hamada and Wu (1992), we consider the data from an experiment described in Hunter, Hodi and Eager (1982). The experiment addresses the effect of seven factors on fatigue life of weld repair castings (the measurements are logarithms of fatigue life). The factors, all at two levels, were assigned to the first seven columns of the $OA(12, 11, 2, 2)$ in Table 11.34, and are denoted by A through G.

(i). Using the main-effects model for factors A through G, which factors appear to be important for modeling the expected response?

(ii). Hunter, Hodi and Eager (1982) were not satisfied with the model that contains only the intercept parameter and the two most important main-effects from (i). This displeasure was partly based on considerations of the subject matter, and partly on the fit of the model for these data. From a residual analysis, and possibly other considerations, do you agree that this model is inadequate? Explain.

(iii). Now consider the model that contains the intercept parameter, the main-effect of F, and the two-factor interaction FG. Do you agree that this model provides a better fit? Explain.

(iv). The important question is of course whether, using the data and our knowledge of the problem, we can formulate a procedure that leads us towards a reasonable model, such as that in (iii). Just looking at a main-effects model will not be enough if there are important interactions; including all two-factor interactions in the model is not tenable, since there are simply too many of them. How would you try to resolve this? Before looking at the solution by Hamada and Wu, which is partly sketched below, formulate your own procedure and apply it to the data in Table 11.34. (Hamada and Wu propose an iterative procedure. Using the data and possibly knowledge of the subject matter, they start with a model that includes some of the main-effects. Using a forward selection procedure, they then try to improve this model, allowing the inclusion of one or more two-factor interactions. After some further iterations, they hope to discover a model with a reasonable fit that is consistent with what is known about the problem.)

Chapter 12

Tables of Orthogonal Arrays

This chapter contains several tables: (a) Tables showing the smallest possible index (and hence the smallest number of runs) in 2-, 3- and 4-level orthogonal arrays with at most 32 factors and strengths between 2 and 10. (b) Tables summarizing most of the arrays constructed in this book, including a table of both mixed- and fixed-level orthogonal arrays of strength 2 with up to 100 runs. (c) A table summarizing the connections between orthogonal arrays and other combinatorial structures. We also discuss what can be done if the orthogonal array you want does not exist, or is not presently known to exist, or exists but is too large for your application.

12.1 Tables of Orthogonal Arrays of Minimal Index

Tables 12.1–12.3 give the smallest index λ that is possible in an orthogonal array $OA(\lambda s^t, k, s, t)$ with s (the number of levels) equal to 2, 3 or 4, respectively, k (the number of factors) at most 32, and t (the strength) between 2 and 10. The typical entry gives either the exact value of λ if that is known, or otherwise a range λ_0–λ_1, indicating that any such orthogonal array must have index at least λ_0, and that an orthogonal array with index λ_1 is known.

For example, the entry for $k = 10$, $t = 6$ in Table 12.1 reads 6–8, which indicates that an $OA(N, 10, 2, 6)$ is known with $N = 8 \cdot 2^6 = 512$ runs, that any such array must contain at least $6 \cdot 2^6 = 384$ runs, but that the existence of an array with fewer than $8 \cdot 2^6$ runs is an open question.

In view of Theorem 11.3, if the (k, t) entry in the table of orthogonal arrays with s levels is λ_0–λ_1, then there is an s-level fractional factorial of resolution $t+1$ with k factors and $N = \lambda_1 s^t$ runs.

The entries in Tables 12.1–12.3 are discussed in more detail in Section 12.2. The additional information provided in that section will direct the reader to the chapter of the book (or a place in the literature) where the arrays are described.

12.2 Description of Tables 12.1–12.3

Some entries in Tables 12.1–12.3 carry a label, taken from the list of symbols given in the Key at the end of this section.

The first two columns of Table 12.1, describing binary orthogonal arrays of strengths 2 and 3, are special in that the minimal index is known exactly for any k (assuming, that is, that Hadamard matrices exist of all possible orders — see Chapter 7).

Apart from these two columns, all *unlabeled* lower bounds are obtained either from the linear programming bound — see Theorem 4.15, or from the trivial observation that the nonexistence of an $OA(N,k,s,t)$ of index λ implies the nonexistence of an $OA(N,k+1,s,t)$ of index λ.

All unlabeled upper bounds are consequences of the trivial observations that (a) an $OA(s^t,t+1,s,t)$ with index 1 always exists (see Section 2.1 and Problem 4.5), and (b) an $OA(N,k,s,t)$ of index λ implies the existence of an $OA(N,k-1,s,t)$ of index λ, an $OA(N/s,k-1,s,t-1)$ of index λ, and an $OA(sN,k+1,s,t)$ of index $s\lambda$ (by Properties 6, 7 and 3 of Chapter 1).

In practice this means that if an upper bound λ_1 is unlabeled, then the justification for it can be obtained by following the table downwards and possibly diagonally downwards to the right, until an entry λ_1 is reached that does carry a label.

Labels on entries in the last ($k = 32$) row of these tables may refer to constructions with larger values of k. For example the entry 27^{Hl} in the last row of Table 12.2 indicates an $OA(3^6,32,3,3)$ that is a shortened version of the $OA(3^6,56,3,3)$ given in Section 5.12.

Most of the labels for upper bounds refer to constructions using linear codes. We remind the reader that a linear code of length k, dimension n and minimal distance d over the field $GF(s)$ produces a linear $OA(s^{k-n},k,s,d-1)$ (by taking the codewords of the dual code to be the runs, see Theorem 4.6). These arrays are therefore regular fractional factorials (see Section 11.5).

Table 12.1. Minimal possible index of orthogonal arrays with 2 levels, k factors and strength t.

k, t	2[Hd]	3[Hd]	4	5	6	7	8	9	10
4	2	1	1						
5	2	2	1	1					
6	2	2	2	1	1				
7	2	2	[SZ]4	2	1	1			
8	3	2	4[c]	[SZ]4	2	1	1		
9	3	3	6-8	4[c]	4	2	1	1	
10	3	3	6–8	6–8	[SZ]6–8	4	2	1	1
11	3	3	6–8	6–8	8[c]	[SZ]6–8	4	2	1
12	4	3	7–8	6–8	12–16	8[c]	6–8	4	2
13	4	4	8	7–8	16	12–16	10–16	6–8	4
14	4	4	8	8	16	16	16[c]	10–16	6–8
15	4	4	8[NR]	8	16[RH]	16	26–32	16[c]	11–16
16	5	4	10–16	8[NR]	21–32	16[RH]	39–64	26–32	19–32
17	5	5	12–16	10–16	26–32	21–32	52–64[jx]	39–64	32[c]
18	5	5	13–16	12–16	29–32	26–32	52–128	52–64[jx]	54–64
19	5	5	14–16[X4]	13–16	29–32	29–32	52–128	52–128	86–128
20	6	5	15–32	14–16[X4]	29–32	29–32	64–128[c]	52–128	128[c]
21	6	6	17–32	15–32	32	29–32	86–256	64–128[c]	171–256
22	6	6	20–32	17–32	32	32	108–256	86–256	171–256
23	6	6	22–32[W]	20–32	32[Go]	32	118–256[HP]	108–256	171–256[uv]
24	7	6	22–64	22–32[W]	41–64	32[Go]	119–512	118–256[HP]	219–512
25	7	7	23–64	22–64	51–128	41–64	127–512	119–512	290–512
26	7	7	26–64	23–64	58–128	51–128	149–512	127–512	384–512[re]
27	7	7	29–64	26–64	66–128[Ka]	58–128	164–512[Pi]	149–512	456–1024[N1]
28	8	7	29–64	29–64	73–256	66–128[Ka]	165–1024	164–512[Pi]	458–1024
29	8	8	29–64	29–64	74–256	73–256	168–1024	165–1024	464–1024
30	8	8	33–64	29–64	87–256	74–256	189–1024	168–1024	570–1024
31	8	8	37–64	33–64	96–256[CS]	87–256	199–1024[Sh]	189–1024	681–1024[BC]
32	9	8	37–64[BC]	37–64[BC]	108–512	96–256[CS]	209–2048	199–1024[Sh]	721–2048[N2]

Table 12.2. Minimal possible index of orthogonal arrays with 3 levels, k factors and strength t.

k, t	2	3	4	5	6	7	8	9	10
4	1[BI]	1	1						
5	2	2[FN]	1	1					
6	2	[HS]3	[HU]3	1	1				
7	2[AK]	3	3	[HU]3	1	1			
8	3	3	3	3	3	1	1		
9	3	3	3	3	4–9	3	1	1	
10	3	3	3	3	4–9	5–9	3	1	1
11	3	[Se]4–9	3	3	4–9	7–9	5–9	3	1
12	3	4–9	5–9	3[Go]	6–27	7–9	9	6–9	3
13	3[RH]	4–9	5–9	5–9	7–27	8–27	9[RH]	11–27	6–9
14	4–6	4–9	6–9[KP]	6–9[qr]	8–27	10–27	9–27	17–27	13–27
15	4–6	4–9	6–27	6–27	9–27[LZ]	10–27	14–27	17–81	23–27[HN]
16	4–6	4–9	7–27	6–27	9–81	10–81	15–27[HN]	18–81	23–81
17	5–6	4–9	9–27	7–27	10–81	11–81	15–81	25–81	23–81
18	5–6	5–9	9–27	9–27	12–81	11–81	20–81	25–81	34–81
19	5–6	5–9	10–27	9–27	14–81	12–81	26–81	25–243	40–81[GB]
20	[Z2]6	5–9[KP]	11–27	10–27	16–81	15–81	27–81	31–243	40–243[N3]
21	[Z2]6	5–27	12–27	11–27	18–81	16–81	29–81	32–243[GB]	46–243[KP]
22	[Z1]6	[B0]6–27	13–27	12–27	21–81	19–81	35–81	33–729	55–729
23	6	[B0]6–27	15–27	13–27	24–243	22–81	39–81	37–729	66–729
24	6	6–27	16–27	15–27	25–243	24–243	41–81[qr]	40–729	71–729[KP]
25	6[AK]	6–27	17–27	16–27	31–243	25–729	45–243	43–729[BE]	83–2187
26	7–9	6–27	18–27	17–27	34–243	31–729	49–729	49–2187	95–2187
27	7–9	7–27	20–27	18–27	37–243[EB]	34–729	55–729	51–2187	106–2187
28	7–9	7–27	21–81	20–27[KP]	42–729	37–729	63–729[qr]	57–2187	126–2187
29	7–9	7–27	21–81	21–81	47–729	42–729[EB]	72–2187	66–2187	152–2187
30	7–9	7–27	24–81	21–243	52–729	47–2187	80–2187	74–2187	158–2187
31	7–9	[B0]8–27	26–81	24–243	55–729	52–2187	96–2187	85–2187	182–2187
32	8–9[RH]	[B0]8–27[Hl]	27–81[BC]	26–243[BZ]	63–729[EB]	55–2187	104–2187	96–2187	213–2187[Pl]

Table 12.3. Minimal possible index of orthogonal arrays with 4 levels, k factors and strength t.

k, t	2	3	4	5	6	7	8	9	10
4	1	1	1						
5	1	1	1	1					
6	2	1[B1]	2–4	1	1				
7	2	2–4	2–4	2–4	1	1			
8	2	2–4	2–4	2–4	3–4	1	1		
9	2[AK]	2–4	2–4	2–4	3–4	3–4	1	1	
10	3	2–4	2–4	2–4	3–16	4[KP]	3–4	1	1
11	3	3–4	3–4	2–4	4–16	4–16	6–16	3–4	1
12	3	3–4	4–16	3–4[qr]	4–16	5–16	6–16	7–16	3–4
13	3[ds]	3–4	4–16	4–16	4–16	5–16	6–16	7–16	8–16
14	3–4	3–4	4–16	4–16	4–16	5–16	7–16	7–16	12–16
15	4	3–4	5–16	4–16	4–16	5–16	7–64	11–16	12–16
16	4	4	6–16	5–16	6–16	5–16	8–64	11–64	14–16[BC]
17	4	4[KP]	6–16	6–16	7–16	6–16	8–64	11–64	16–64
18	4	4–16	7–16	6–16	7–64	7–16[qr]	8–256	11–64[Gu]	16–256
19	4	4–16	7–16	7–16	8–64	7–64	8–256	11–256	19–256
20	4	4–16	8–16	7–16[MP]	9–64	8–64[qr]	9–256	12–256	19–256
21	4[RH]	4–16	9–16[KP]	8–64	11–64[KP]	9–256	11–256	13–256	19–256
22	5–8	[No]5–16	10–64	9–64	13–256	11–256	13–256	14–256	19–256
23	5–8	5–16	10–64	10–64	15–256	13–256	15–256	15–256	20–256
24	5–8	5–16	11–64	10–64	16–256	15–256	17–256	17–256	23–256
25	5–8	5–16	12–64	11–64	19–256	16–256	18–256	19–256	25–256
26	6–8	5–16	13–64	12–64	20–256[EB]	19–256	24–256	22–256	27–256
27	6–8	6–16	14–64	13–64[EB]	22–1024	20–1024	28–256	27–256	30–256
28	6–8	6–16	15–64	14–256	26–1024	22–1024	31–1024	29–256	34–256
29	6–8	6–16	16–64	15–256	29–1024	26–1024	37–4096	31–1024	36–256[qr]
30	6–8	6–16	17–64	16–256	31–1024	29–1024	41–16384	37–4096	41–1024
31	7–8	6–16	18–64	17–256	33–1024	31–1024	44–16384	41–16384	51–4096
32	7–8[AK]	7–16[IN]	20–64[GB]	18–256[EB]	37–1024	33–1024[KP]	53–16384	44–16384	58–16384[qr]

So if the (k,t) entry in the table for orthogonal arrays with s levels reads λ_0–λ_1^{ABC}, the upper bound λ_1^{ABC} often indicates that an $OA(\lambda_1 s^t, k, s, t)$ can be constructed from a linear code of type ABC with length k, dimension $k - t - \log_s \lambda_1$ and minimal distance $t+1$ over $GF(s)$. If the number of runs is less than 1000 then the array will usually be found in Chapter 5. Otherwise the code must be found from the references given below.

For example, the entry for $k = 21$, $t = 5$ in Table 12.1 reads 15–32. The upper bound 32 is unlabeled, but following the column downwards we find an entry 22–32^{W} at $k = 24$, $t = 5$. The corresponding code is a binary code of length $k = 24$, dimension $24 - 5 - 5 = 14$ and minimal distance $t + 1 = 6$. By referring to the Key below we see that this is a Wagner code, and the reference there sends us to Table 5.11 where a generator matrix for the orthogonal array (i.e. the dual code) may be found.

There are many embarrassingly large gaps in these tables, such as that between the lower bound $\lambda \geq 58$ and the upper bound $\lambda \leq 16384$ for an $OA(\lambda 4^{10}, 32, 4, 10)$, the final entry in Table 12.3.

Research Problem 12.4. Reduce the gaps in Tables 12.1–12.3. Find better lower bounds. Apart from the Addelman-Kempthorne construction, almost all the upper bounds in Tables 12.2 and 12.3 are achieved by linear arrays (usually obtained as the duals of linear codes). This is surely a reflection of our ignorance. Find more constructions for nonlinear orthogonal arrays with small numbers of levels that are better than the best linear arrays presently known.

Research Problem 12.5. Even using linear codes there are many places where it may be possible to improve our tables by finding better codes. For example, as far as we know today, linear codes could exist with any of the following parameters: $(24, 4^{12}, 10)_4$, $(24, 4^{13}, 9)_4$, $(24, 4^{14}, 8)_4$, $(24, 4^{15}, 7)_4$, $(24, 4^{17}, 6)_4$, $(24, 4^{18}, 5)_4$. If any one of these codes could be found, it would improve an entry in Table 12.3 (and if any of these could be shown not to exist it would improve an entry in the tables of codes given by Brouwer, 1998). Dozens of similar examples can be found by comparing our tables with those of Brouwer (1998).

Key to Tables 12.1–12.3

AK Addelman and Kempthorne (1961a) construction (Section 3.3).

B0 Bose and Bush (1952) bound (Theorem 2.8).

B1 Bush (1952b) or Reed and Solomon (1960) construction (Sections 3.2, 5.5).

BC BCH code (Section 5.4).

BE Bierbrauer and Edel (1997) code.

BZ Blokh and Zyablov (1974) or concatenated code construction — see MacWilliams and Sloane (1977), Chap. 10, §11; Brouwer (1998), Section 4.1.9.

c Cyclic code (Section 5.2).

CS Cheng and Sloane (1989) code.

ds From a difference scheme via Corollary 6.20 and Table 6.67.

EB Edel-Bierbrauer (1999) code.

FN Fujii, Namikawa and Yamamoto (1987) construction (Example 2.16).

GB Gulliver and Bhargava (1992) code.

Go Golay code (Sections 5.7, 5.11, 5.12).

Gu Gulliver and Östergård (1998) quasicyclic code (Section 5.13).

Hd Exact answer given by Hadamard array (Chapter 7).

Hl Hill (1973) code (Section 5.12).

HN Hill and Newton (1988) code.

HP Hashim and Pozdniakov (1976) code.

HS Hedayat, Seiden and Stufken (1997) bound (Example 2.16).

HU Hedayat, Stufken and Su (1997) bound (Example 2.16).

IN Itoh and Nakamichi (1983) code. The reference is to a $(41, 4^{36}, 4)_4$ code.

jx Juxtaposition construction (Section 10.2).

Ka Karlin (1969) code (MacWilliams and Sloane, 1977, page 509).

KP Kschischang and Pasupathy (1992) code (Sections 5.12, 5.13).

MP Monroe and Pless (1993) lexicographic code.

N1 There exists an $OA(2^{20}, 27, 2, 11)$, obtained by shortening a $(50, 2^8, 23)_2$ cyclic code and taking the dual code.

N2 There exists an $OA(2^{21}, 32, 2, 11)$, from the dual to a $(32, 2^{11}, 12)_2$ extended BCH code.

N3 There exists an $OA(3^{15}, 20, 3, 11)$, obtained by shortening a $(54, 4^6, 34)_4$ code (Gulliver and Östergård, 1998) and taking the dual code.

No Both the linear programming bound and the Rao bound give $\lambda \geq 4$ here. But $\lambda = 4$ is impossible by Noda (1986).

NR Nordstrom-Robinson (1967) code. (Note that the version in Section 5.10 due to Forney, Sloane and Trott, 1993, and Hammons, Kumar, Calderbank, Sloane and Solé, 1994, is much simpler than the original version.)

Pi Piret (1980) code.

Pl Pless (1972) double circulant code. The reference is to a $(36, 3^{18}, 12)_3$ code.

qr Quadratic residue code (Section 5.7).

re Residual code construction (Helgert and Stinaff, 1973; MacWilliams and Sloane, 1977, p. 593, §9.2).

RH Rao-Hamming construction (Sections 3.4, 5.3).

Se Seiden (1955a,b) bound (Section 2.3).

Sh Shearer (1988) code. The reference is to a $(32, 2^{13}, 10)_2$ code.

SZ Seiden and Zemach (1966) bound (Section 2.5).

uv $(u, u+v)$ construction (Section 10.3).

W Wagner (1965) code (Table 5.11).

X4 Construction X4 (Section 10.4).

Z1 Corollary 4.27.

Z2 Corollary 4.27 and Shrikhande and Singhi (1979a).

12.3 Index Tables

Tables 12.6 and 12.7 provide an index to many of the orthogonal arrays constructed in this book. Table 12.6 deals with fixed-level orthogonal arrays and Table 12.7, which includes mixed-level arrays, with arrays containing up to 100 runs. In Table 12.6 s is a prime power unless noted otherwise. The column headed "Reference" gives the section where the array can be found, or if the array is given implicitly rather than explicitly (as in the case of arrays based on Hadamard matrices), a specific reference to an appropriate table.

Table 12.6. Index table to (fixed-level) orthogonal arrays constructed in this book.

(a) **General Constructions** (s must be a prime power except for the cases marked †).

N	k	s	t	Reference	Remarks
4λ	$4\lambda-1$	2	2	7.3	Hadamard
2^{m+1}	2^m	2	3	5.3	Rao-Hamming
8λ	4λ	2	3	7.3	Hadamard
2^{2m}	2^m-1	2	4	5.4	BCH ($m \ge 4$)
2^{2m+1}	2^m+1	2	5	5.10	BCH ($m \ge 5$)
2^{2m}	2^m	2	5	5.10	Kerdock (even $m \ge 4$)
2^{2m+1}	2^m+m	2	5	10.4	X4 (even $m \ge 4$)
2^{3m-1}	2^m	2	7	5.10	Delsarte-Goethals (even $m \ge 6$)
2^{k-2m}	2^m	2	$2^{m-1}-2^{(m-2)/2}-1$	5.10	Preparata (even $m \ge 4$)
2^{k-2m}	2^m-1	2	$2^{m-1}-2^{\lfloor m/2 \rfloor}-1$	5.4	BCH ($m \ge 4$)
2^{k-1-m}	2^m	2	$2^{m-1}-1$	5.3	Rao-Hamming
$2^{1+\binom{m}{1}+\cdots+\binom{m}{r}}$	2^m	2	$2^{r+1}-1$	5.8	Reed-Muller ($0 \le r \le m$)
2^{3m}	2^m+2	2^m	3	3.2, 5.5	Bush/Reed-Solomon code
2^{k-3}	2^m+2	2^m	2^m-1	3.2, 5.5	Bush/Reed-Solomon code
N	k	s	0	1	trivial †
s^m	$\frac{s^m-1}{s-1}$	s	2	3.4, 5.3	Rao-Hamming
$2s^2$	s	s	2	10.9	Grieg ($s \pm 1$ prime powers) †
λs^2	$\lambda\frac{s^{d+1}-1}{s^d-s^{d-1}}+1$	s	2	6.3	Bose-Bush ($s=p^v, \lambda=p^u, d=\lfloor u/v \rfloor$)
$4s^m$	$\frac{4(s^m-1)}{s-1}-3$	s	2	6.6	difference schemes
$2s^m$	$\frac{2(s^m-1)}{s-1}-1$	s	2	3.3, 6.4	Addelman-Kempthorne
s^4	s^2+1	s	3	5.9	ovoid
s^{k-m}	$\frac{s^m-1}{s-1}$	s	$s^{m-1}-1$	5.3	Rao-Hamming
λs^t	$t+1$	s	t	2.1	zero-sum †
s^t	$s+1$	$s \ge t$	t	3.2, 5.5	Bush/Reed-Solomon code

Table 12.6 (continued)

(b) Binary arrays ($s = 2$) with strength 2

N	k	s	t	Reference	Remarks
4	3	2	2	1, 2.3	zero-sum
8	7	2	2	3.4, 4.2, Table 7.9	Rao-Hamming
12	11	2	2	4.4, Table 7.15	Hadamard
16	15	2	2	5.3	Rao-Hamming
16	15	2	2	7.5, Table 7.26	Hadamard
20	19	2	2	7.5	Hadamard
28	27	2	2	Table 7.19	Hadamard
32	31	2	2	5.3, 5.11	Rao-Hamming
36	35	2	2	Table 7.33	Hadamard
44	43	2	2	Table 7.33	Hadamard
52	51	2	2	Table 7.33	Hadamard
60	59	2	2	Table 7.33	Hadamard
68	67	2	2	Table 7.33	Hadamard
76	75	2	2	Table 7.33	Hadamard
84	83	2	2	Table 7.32	Hadamard
92	91	2	2	Table 7.32	Hadamard
100	99	2	2	Table 7.32	Hadamard
108	107	2	2	Table 7.32	Hadamard
116	115	2	2	Table 7.34	Hadamard
124	123	2	2	Table 7.34	Hadamard
132	131	2	2	Table 7.34	Hadamard
148	147	2	2	Table 7.34	Hadamard
156	155	2	2	Table 7.34	Hadamard
172	171	2	2	7.5	Hadamard
188	187	2	2	Table 7.35	Hadamard

(c) Binary arrays ($s = 2$) with strength 3

N	k	s	t	Reference	Remarks
8	4	2	3	1, 5.3	zero-sum
16	8	2	3	5.3, 5.11, 6.5, Table 7.9	Rao-Hamming
24	12	2	3	4.4, Table 7.15	Hadamard
32	16	2	3	5.3, 5.11	Rao-Hamming
32	16	2	3	7.5, Table 7.26	Hadamard
40	20	2	3	7.5, Table 7.23	Hadamard
56	28	2	3	Table 7.19	Hadamard
64	32	2	3	5.3, 5.11	Rao-Hamming
72	36	2	3	Table 7.33	Hadamard
88	44	2	3	Table 7.33	Hadamard
104	52	2	3	Table 7.33	Hadamard
120	60	2	3	Table 7.33	Hadamard
136	68	2	3	Table 7.33	Hadamard
152	76	2	3	Table 7.33	Hadamard
168	84	2	3	Table 7.32	Hadamard
184	92	2	3	Table 7.32	Hadamard
200	100	2	3	Table 7.32	Hadamard
216	108	2	3	Table 7.32	Hadamard
232	116	2	3	Table 7.34	Hadamard
248	124	2	3	Table 7.34	Hadamard
264	132	2	3	Table 7.34	Hadamard
296	148	2	3	Table 7.34	Hadamard
312	156	2	3	Table 7.34	Hadamard
344	172	2	3	7.5	Hadamard
376	188	2	3	Table 7.35	Hadamard

Table 12.6 (continued)

(d) Binary arrays ($s = 2$) with strength ≥ 4

N	k	s	t	Reference	Remarks
16	5	2	4	2.2	zero-sum
80	6	2	4	Problem 2.17	Seiden-Zemach
128	9	2	5	5.11	cyclic
256	16	2	5	5.10	Nordstrom-Robinson
512	20	2	5	10.4	X4
2^{10}	24	2	5	5.11	Wagner
2^{11}	32	2	5	5.11	dual BCH
2^{10}	12	2	7	5.11	cyclic
2^{11}	16	2	7	5.3	Rao-Hamming
2^{12}	24	2	7	5.11	Golay
2^{15}	18	2	9	10.2	juxtaposition
2^{18}	23	2	10	10.3	$u, u+v$
2^{19}	24	2	11	10.3	$u, u+v$
2^{26}	32	2	15	5.3	Rao-Hamming

(e) Ternary arrays ($s = 3$)

N	k	s	t	Reference	Remarks
9	4	3	2	2.2, 5.12	Bush (tetracode)
18	7	3	2	2.3, 3.3	Addelman-Kempthorne
27	13	3	2	3.4, 5.12	Rao-Hamming
45	10	3	2	Table 6.67	difference scheme
45	22	3	2	4.7	Does not exist
54	25	3	2	3.3	Addelman-Kempthorne
81	40	3	2	5.12	Rao-Hamming
27	4	3	3	3.4	Rao-Hamming
54	5	3	3	2.3	4 arrays exist
81	10	3	3	5.9, 5.12	ovoid
243	20	3	3	5.12	Kschischang-Pasupathy
729	56	3	3	5.12	Hill
243	11	3	4	2.2, 5.12	Golay
729	14	3	4	5.12	Kschischang-Pasupathy
729	12	3	5	5.12	Golay

Table 12.6 (concluded)

(f) Quaternary arrays ($s = 4$)

N	k	s	t	Reference	Remarks
64	21	4	2	5.13	Rao-Hamming
160	17	4	2	Table 6.67	difference scheme
64	6	4	3	3.2, 5.2	Bush (hexacode)
256	17	4	3	5.9, 5.13	ovoid
4^5	11	4	4	5.13	quadratic residue
4^6	21	4	4	5.13	Kschischang-Pasupathy
4^6	12	4	5	5.13	quadratic residue
4^6	12	4	5	5.13	dodecacode

(g) Arrays with $s \geq 5$

N	k	s	t	Reference	Remarks
75	8	5	2	Table 6.67	difference scheme
150	16	5	2	Table 6.67	difference scheme
175	18	5	2	Table 6.67	difference scheme
36	3	6	2	Section 2.1	zero-sum
72	7	6	2	Table 6.67	difference scheme
288	11	6	2	Table 6.67	difference scheme
360	12	6	2	Table 6.67	difference scheme
245	12	7	2	Table 6.67	difference scheme
294	19	7	2	Table 6.67	difference scheme
441	29	7	2	Table 6.67	difference scheme
576	22	8	2	Table 6.67	difference scheme
100	4	10	2	Theorem 8.28	$POL(10, 2)$
200	10	10	2	Section 10.9	Grieg
800	11	10	2	Table 6.67	difference scheme
144	7	12	2	Table 6.11	difference scheme
144	7	12	2	Theorem 8.28	$POL(12, 5)$
288	12	12	2	Section 10.9	Grieg
225	6	15	2	Section 6.5	difference scheme
225	6	15	2	Theorem 8.28	$POL(15, 4)$
648	18	18	2	Section 10.9	Grieg

Table 12.7 attempts to give a fairly complete list of the most important mixed- or fixed-level orthogonal arrays with at most 100 runs. We use the notation $OA(N, s_1^{k_1} s_2^{k_2} \cdots, 2)$ as in Chapter 9.

Table 12.7. Index table to (mixed- or fixed-level) orthogonal arrays of strength 2 with at most 100 runs

Runs	Levels^{Factors}	Reference
4	2^3	Chapter 1; Sections 2.3, 3.4, 5.3, 5.5, 6.3,
6	$2^1 3^1$	Complete factorial
8	2^7	Chapter 1; Sections 3.4, 4.2, 5.3, 6.3; §
	$2^4 4^1$	Examples 9.11, 9.16
9	3^4	Sections 2.2, 3.2, 3.4, 5.3, 5.5, 5.12, 6.3
10	$2^1 5^1$	Complete factorial
12	2^{11}	Section 4.4; Table 7.15
	$2^4 3^1$	Example 9.2
	$2^2 6^1$	Problem 9.12; Example 9.19
	$3^1 4^1$	Complete factorial
14	$2^1 7^1$	Complete factorial
15	$3^1 5^1$	Complete factorial
16	2^{15}	Sections 3.4, 5.3, 6.3, 7.5, 8.3; §
	$2^{12} 4^1$	Example 9.9; Example 9.12; §
	$2^9 4^2$	Example 9.9; Example 9.12; §
	$2^8 8^1$	Example 9.16
	$2^6 4^3$	Example 9.9; Example 9.12; §
	$2^3 4^4$	Example 9.9; Example 9.12; §
	4^5	Sections 3.2, 3.4, 5.3, 5.5, 6.2, 6.3; Taguchi, 1987
18	$2^1 3^7$	Example 9.19; Taguchi, 1987; §
	3^7	Sections 2.3, 3.3, 6.4; §
	$3^6 6^1$	Example 9.19
20	2^{19}	Section 7.5
	$2^8 5^1$	Wang and Wu, 1992; Dey and Midha, 1996
	$2^2 10^1$	Problem 9.12; Example 9.19
24	2^{23}	§
	$2^{20} 4^1$	Example 9.17; Wang and Wu, 1991
	$2^{13} 3^1 4^1$	Wang and Wu, 1991

Table 12.7 (continued)

Runs	Levels^{Factors}	Reference
24	$2^{12}12^1$	Example 9.16
(cont.)	$2^{11}4^1 6^1$	Wang and Wu, 1991
25	5^6	Sections 3.2, 3.4, 5.3, 5.5, 6.3; Taguchi, 1987
27	3^{13}	Sections 3.4, 5.3, 5.12, 6.3; Taguchi, 1987; §
	$3^9 9^1$	Example 9.19
28	2^{27}	Table 7.19
	$2^{12}7^1$	Dey and Midha, 1996
	$2^2 14^1$	Problem 9.12; Example 9.19
32	2^{31}	Sections 3.4, 5.3, 6.3, 7.5; §
	$2^{28}4^1$	Example 9.17; Wu, Zhang and Wang, 1992; §
	$2^{24}8^1$	Example 9.17; §
	$2^{16}16^1$	Example 9.16
	$2^4 4^9$	Example 9.19; Hedayat, Pu and Stufken, 1992; Wang, 1996b; §
	$4^8 8^1$	Example 9.19
36	2^{35}	Table 7.33
	$2^{27}3^1$	Example 9.30
	$2^{20}3^2$	Example 9.30
	$2^{18}3^1 6^1$	Example 9.30
	$2^{13}9^1$	Dey and Midha, 1996
	$2^{11}3^{12}$	Example 9.19; Taguchi, 1987; Wang and Wu, 1991; Sloane and Stufken, 1996; §
	$2^{11}3^2 6^1$	Example 9.30
	$2^9 3^1 6^2$	Example 9.30
	$2^4 3^{13}$	Example 9.19; Taguchi, 1987; Wang and Wu, 1991; §
	$2^4 3^3 6^1$	Example 9.30
	$2^2 3^{12}6^1$	Example 9.19; Wang and Wu, 1991; §
	$2^2 3^2 6^2$	Example 9.30
	$2^2 18^1$	Problem 9.12; Example 9.19
	$2^1 3^8 6^2$	Zhang, Lu and Pang, 1999; §
	$3^{13}4^1$	Dey, 1985, p. 62; Wang and Wu, 1991; §
	3^{13}	Seiden, 1954; §
	$3^{12}12^1$	Example 9.19; Wang and Wu, 1991
	$3^7 6^3$	Finney, 1982

Table 12.7 (continued)

Runs	Levels^{Factors}	Reference
40	$2^{36}4^1$	Ex. 9.17; Dey–Ramakrishna, 1977; Wang–Wu, 1991
	$2^{28}5^1$	§
	$2^{25}4^15^1$	Wang and Wu, 1991
	$2^{20}20^1$	Example 9.16; Dey, 1985, p. 67; Wang and Wu, 1991
	$2^{19}4^110^1$	Agrawal and Dey, 1982; Wang and Wu, 1991
44	2^{43}	Table 7.33
	$2^{12}11^1$	Juxtaposition
	2^222^1	Problem 9.12; Example 9.19
45	3^915^1	Example 9.19
48	$2^{40}8^1$	Dey, 1985, p. 72; Wang and Wu, 1991
	$2^{38}4^3$	Ex. 9.17; Chacko et al., 1979; Wang and Wu, 1991; §
	$2^{33}3^18^1$	Wang and Wu, 1991
	$2^{31}3^14^3$	Wang and Wu, 1991; §
	$2^{31}6^18^1$	Wang and Wu, 1991
	$2^{29}4^36^1$	Wang and Wu, 1991; §
	$2^{28}3^14^4$	§
	$2^{27}4^312^1$	Agrawal and Dey, 1982; Wang and Wu, 1991; §
	$2^{24}24^1$	Example 9.16
	$4^{12}12^1$	Example 9.19; Suen, 1989b; Wang and Wu, 1991
49	7^8	Sections 3.2, 3.4, 5.3, 5.5
50	2^15^{11}	Taguchi, 1987; §
	$5^{10}10^1$	Example 9.19
52	2^{51}	Table 7.33
	$2^{12}13^1$	Juxtaposition
	2^226^1	Problem 9.12; Example 9.19
54	2^13^{25}	Example 9.19; Taguchi, 1987; §
	3^{25}	Section 3.3, 6.4; §
	$3^{20}6^19^1$	Wang and Wu, 1991
	$3^{18}18^1$	Example 9.19
56	$2^{52}4^1$	Example 9.17 ($\lambda = 7, \mu = 1/2, f = 1$)
	$2^{28}28^1$	Example 9.16

Table 12.7 (continued)

Runs	Levels^{Factors}	Reference
60	2^{59}	Table 7.33
	$2^{2}30^{1}$	Problem 9.12; Example 9.19
63	$3^{11}21^{1}$	Example 9.19
64	2^{63}	Sections 3.4, 5.3, 6.3, 7.5; §
	$2^{49}8^{2}$	Table 8.21; Wu, Zhang and Wang, 1992; §
	$2^{32}32^{1}$	Example 9.16; Hedayat, Pu and Stufken, 1992
	4^{21}	Sections 3.4, 5.3, 5.13; Taguchi, 1987; §
	$4^{16}16^{1}$	Example 9.19; Hedayat, Pu and Stufken, 1992
	$4^{15}8^{2}$	Table 8.21
	8^{9}	Sections 3.2, 3.4, 5.3, 5.5
68	2^{67}	Table 7.33
	$2^{13}17^{1}$	Juxtaposition
	$2^{2}34^{1}$	Problem 9.12; Example 9.19
72	$2^{68}4^{1}$	Example 9.17 ($\lambda = 9, \mu = 1/2, f = 1$)
	$2^{47}3^{12}$	Wang, 1996b; Sloane and Stufken, 1996; §
	$2^{44}3^{12}4^{1}$	Wang and Wu, 1991
	$2^{38}3^{12}6^{1}$	Wang and Wu, 1991; §
	$2^{38}18^{1}$	§
	$2^{37}3^{13}4^{1}$	Wang and Wu, 1991
	$2^{36}36^{1}$	Example 9.16
	$2^{35}3^{12}4^{1}6^{1}$	Wang and Wu, 1991
	$2^{23}3^{24}$	Dey, 1985; Sloane and Stufken, 1996; §
	$2^{20}3^{24}4^{1}$	Wang, 1996b; §
	$2^{18}3^{7}6^{5}$	Wang, 1996a; §
	$2^{17}3^{3}6^{6}$	Wang, 1996a; §
	$2^{13}3^{25}4^{1}$	Wang, 1996b; §
	$2^{11}3^{24}4^{1}6^{1}$	Wang, 1996b; §
	$2^{11}3^{8}6^{5}$	Wang, 1996a; §
	$2^{10}3^{4}6^{6}$	Wang, 1996a; §
	$2^{8}3^{3}6^{7}$	Wang, 1996a; §
	$2^{7}3^{7}6^{5}12^{1}$	Wang, 1996a
	$2^{6}3^{4}4^{1}6^{6}$	Wang, 1996a; §
	$2^{6}3^{3}6^{6}12^{1}$	Wang, 1996a
	$3^{24}24^{1}$	Example 9.19

Table 12.7 (continued)

Runs	Levels^{Factors}	Reference
75	5^8	Example 9.19; §
	$5^7 15^1$	Example 9.19
76	2^{75}	Table 7.33
	$2^2 38^1$	Problem 9.12; Example 9.19
80	$2^{70} 4^3$	Example 9.19; Chacko et al., 1979; Wang, 1996b; §
	$2^{61} 5^1 8^1$	Wang, 1996b
	$2^{59} 4^3 5^1$	Wang, 1996b; §
	$2^{55} 8^1 10^1$	Wang and Wu, 1989
	$2^{53} 4^3 10^1$	Wang and Wu, 1989; §
	$2^{51} 4^3 20^1$	Dey, 1985; Wang and Wu, 1989
	$2^{40} 40^1$	Example 9.16
	$4^8 20^1$	Example 9.19
81	3^{40}	Sections 3.4, 5.3, 5.12, 6.3; §
	$3^{32} 9^2$	Table 8.21; §
	$3^{27} 27^1$	Example 9.19
	9^{10}	Sections 3.2, 3.4, 5.3, 5.5
84	2^{83}	Table 7.32
	$2^2 42^1$	Problem 9.12; Example 9.19
88	$2^{84} 4^1$	Example 9.17
	$2^{44} 44^1$	Example 9.16
90	$3^{30} 30^1$	Example 9.19
92	2^{91}	Table 7.32
	$2^2 46^1$	Problem 9.12; Example 9.19
96	$2^{92} 4^1$	Example 9.17; §
	$2^{88} 8^1$	Example 9.17; §
	$2^{80} 16^1$	Dey and Midha, 1996; Wang and Wu, 1989
	$2^{77} 8^1 12^1$	Wang and Wu, 1989
	$2^{73} 3^1 16^1$	Wang and Wu, 1989
	$2^{71} 6^1 16^1$	Wang and Wu, 1989

Table 12.7 (continued)

Runs	Levels^{Factors}	Reference
96	$2^{48}4^{12}12^1$	Example 9.19; Wang and Wu, 1989; §
(cont.)	$2^{48}48^1$	Example 9.16
	$2^{47}4^{16}$	Wang and Wu, 1991; Wang, 1996b; §
	$2^{46}3^14^{11}6^18^1$	Wang, 1996b
	$2^{44}4^{14}6^1$	Wang and Wu, 1991; Wang, 1996b
	$2^{43}4^{15}8^1$	Wang, 1996b
	$2^{43}4^{12}6^18^1$	Wang, 1996b
	$2^{40}3^14^{16}$	Wang and Wu, 1991; Wang, 1996b
	$2^{39}3^14^{14}8^1$	Wang, 1996b
	$2^{36}4^{12}24^1$	Dey and Midha, 1996
	$4^{16}24^1$	Example 9.19
98	$7^{14}14^1$	Example 9.19
100	2^{99}	Table 7.32
	2^250^1	Problem 9.12; Example 9.19
	$5^{20}20^1$	Example 9.19
	5^910^2	Wang, 1996a; §
	5^810^3	Mandeli, 1995; Wang, 1996a
	10^4	Theorem 8.28; Mandeli, 1995

Comments on Table 12.7

There are two trivial ways by which one can modify an orthogonal array to obtain further arrays with the same number of runs.

(a) Deleting factors. For example, a 40-run 2^{36} array exists, and can be obtained from the 40-run $2^{36}4^1$ array in Table 12.7. Arrays that can be obtained in this way have generally not been included in the table.

(b) Expansive replacement. A factor with s levels can be replaced by any of the arrays found in the table under the entry "Runs = s", as explained in Section 9.3. It can also be replaced by an s'-level factor if s' divides s (using an $OA(s, (s')^1, 1)$). For example, a 12-level factor can be replaced by any of 2^{11}, 2^43^1, 2^26^1, 3^14^1, or any 12-run array obtained from one of these by deleting factors (such as a 2^{10}, 2^13^1, 4^1, and so on). For the larger numbers of runs such "child" arrays have been included in the table only if they are especially interesting: they are indicated by the symbol §.

The arrays in Table 12.7 are arranged in a self-evident "lexicographic" order. With this ordering "child" arrays always appear before their "parents". Even so, child-parent relationships are not always obvious. For example, the parent of the 48-run $2^{28}3^1 4^4$ array is the $4^{12}12^1$ array three lines below it; the former can be obtained from the latter by replacing 12^1 by $2^4 3^1$ and each of eight 4-level factors by 2^3.

Complete factorials have been included only at the beginning of the table, to help when using the expansive replacement method. Copies of arrays have never been included.

For the fixed-level arrays in the table we mostly refer to earlier chapters. We have not attempted to trace the history of all the mixed arrays in the table, but have just given some recent references. We apologize if we have given insufficient credit, and would appreciate being informed of this so we can make corrections in later editions. We would of course also like to hear of any improvements to this or any other of the tables in the book.

A much more extensive table than Table 12.7 will be found on the web site for the book (see the Preface), and includes many more children. Such a table is convenient for applications, but occupies too much space to be included here.

The following definition extends the notions of "parent" and "child" arrays, but applies to the parameters of arrays rather than to the arrays themselves. A *parameter set* for an orthogonal array of strength 2 is of the form $(N, s_1^{k_1} s_2^{k_2} \cdots s_v^{k_v})$, where $v \ge 1$, $2 \le s_1 < s_2 < \cdots < s_v$, $k_i > 0$ for all i, and signifies that an $OA(N, s_1^{k_1} s_2^{k_2} \cdots s_v^{k_v}, 2)$ exists. We also allow the parameter set $(N, 1^1)$ for all N, corresponding to the trivial array consisting of a single column of N 1's.

Definition 12.8. Let A and B be parameter sets for orthogonal arrays of strength 2 with N runs. We say that A is *dominated* by B, written $A \prec B$, if an orthogonal array with parameter set A can be obtained from an orthogonal array with parameter set B by a combination of deleting factors and using the expansive replacement method (this includes replacing an s-level factor by an s'-level factor if s' divides s). In this way the parameter sets for orthogonal arrays with N runs become a partially ordered set (or lattice).

Example 12.9. The parameter set $(12, 2^2 6^1)$ for an $OA(12, 2^2 6^1, 2)$ dominates the parameter sets $(12, 2^1 3^1)$ and $(12, 2^3 3^1)$. But it neither dominates nor is dominated by the parameter set $(12, 3^1 4^1)$. ▪

The parameter set (N, N^1) (corresponding to the trivial array with one factor taking levels $0, 1, \ldots, N-1$) dominates every N-run parameter set, and $(N, 1^1)$ is dominated by everything. The parameter sets (N, p^1) for primes p dividing N are the "atoms" in the partially ordered set—they dominate only $(N, 1^1)$. The interesting question is to find all "dual atoms", the parameter sets that lie just below (N, N^1).

Research Problem 12.10. For each $N \leq 100$, find all parameter sets that lie just below (N, N^1). Once we know these, we can find the parameters of all N-run orthogonal arrays by the operations mentioned in Definition 12.3.

For $N = 8$ runs, for example, there is a single dual atom, $(8, 2^4 4^1)$, as in Table 12.7.

12.4 If No Suitable Orthogonal Array Is Available

There are several possible reasons why you may not have found the orthogonal array you wanted in this book.

The array exists, but is outside the range where we give explicit constructions. In this case Tables 12.1–12.3, 12.6 and 12.7 should point you to the appropriate section or reference where a recipe for constructing the array can be found.

It is not known if the array exists. It may happen that the orthogonal array you were hoping to find is not presently known to exist, but at the same time does not violate any known bound. The lower bounds in Tables 12.1 to 12.3 specify putative arrays of that type. For example, the first case where there is a question in Table 12.1 concerns a possible $OA(96, 9, 2, 4)$ or $OA(112, 9, 2, 4)$: we do not know if either exists.

This brings us to the topic of computer searches for orthogonal arrays or approximations to orthogonal arrays. At present there seems to be no effective algorithm that will decide the existence of orthogonal arrays with even a couple of hundred runs. For strength 2 the approach via difference schemes (see Chapter 6) seems the most promising for the construction of arrays, though failing to find an array by this method will still not answer the existence question.

Research Problem 12.11. Develop an effective algorithm for finding an orthogonal array if it exists, or for establishing its nonexistence if it does not exist.

Little seems to be know about the complexity of these questions.

Research Problem 12.12. Investigate the complexity of the problems mentioned in Research Problem 12.11. (The survey article by Barg, 1998, on complexity issues in coding theory may be helpful.)

In the absence of a specific algorithm for finding orthogonal arrays, one can try applying more general tools. One approach is to apply algorithms intended for finding response surface designs, such as the *Gosset* program of Hardin and Sloane (1992, 1993).

Except for a few small cases, this approach will most likely not give an orthogonal array. It can however result in very efficient designs that are (in terms of their information matrices) very close to orthogonal arrays. In what follows we briefly indicate a possible approach for fixed-level orthogonal arrays; the extension to mixed-level arrays is straightforward.

Let $S = \{0, 1, \ldots, s-1\}$ and $-1 \leq x_l \leq 1$ for $l = 1, \ldots, k$. When trying to find an $OA(N, k, s, t)$, use the following model:

(i). If $t = 2u$, model the expected response at $(x_1, x_2, \ldots, x_k)$ as

$$\sum_{e=(e_1,\ldots,e_k)} \beta_e \, x_1^{e_1} \, x_2^{e_2} \cdots x_k^{e_k} \,, \tag{12.1}$$

where the sum extends over all $e = (e_1, \ldots, e_k) \in S^k$ with at most u nonzero elements (i.e., with $0 \leq w(e) \leq u$), and the β_e are unknown parameters.

(ii). If $t = 2u + 1$, model the expected response at $(x_1, x_2, \ldots, x_k)$ as

$$\sum_{e=(e_1,\ldots,e_k)} \beta_e \, x_1^{e_1} \, x_2^{e_2} \cdots x_k^{e_k} \,, \tag{12.2}$$

where the sum extends now over all $e = (e_1, \ldots, e_k) \in S^k$ with $0 \leq w(e) \leq u+1$ and $e_1 \geq 1$ if $w(e) = u + 1$. The β_e are again the unknown parameters.

We could then ask *Gosset* (or another response surface design program) to search for a D-optimal design for the model (12.1) or (12.2), with values for the x_l's taken from the set $\{-1, -1 + 2/(s-1), \ldots, 1 - 2/(s-1), 1\}$.

In view of the universal optimality properties of orthogonal arrays mentioned in Section 11.10, such a search should in principle converge to the desired array.

Other search methods may also be tried. Wang and Safadi (1990) discuss the use of simulated annealing to construct mixed orthogonal arrays (see also Safadi and Wang, 1991, Fuji-Hara, 1987). However, in our experience simulated annealing is not very successful at handling this type of problem.

The group-theoretic approach of Kreher and Radziszowski (1986, 1987, 1990) and Kreher (1990) (see also Kreher and Stinson 1998), already mentioned in Section 6.6, could also be tried.

No array exists, or the appropriate array is too large. It may happen that the only orthogonal array that is appropriate for your problem has 96 runs, but your budget will only support a 32-run experiment.

In this case one may consider using a main-effects plan (see Section 11.7), one of the alternative families of designs discussed in Section 11.9, or a *nearly orthogonal* array (see for example Nguyen, 1996, Wang and Wu, 1992).

An alternative solution (as discussed in the final paragraphs of Section 11.9) is to use an experimental design program such as Gosset (mentioned above) to construct a design with exactly the number of runs required. We prefer the designs produced by Gosset, since they can be specifically tailored to the problem at hand.

12.5 Connections with Other Structures

The reader may find it helpful to have a summary of the principal relationships between orthogonal arrays and other combinatorial structures:

- A linear code over the field $GF(s)$, of length k, dimension n and minimal distance d, is (through its dual) equivalent to a linear $OA(s^{k-n}, k, s, d-1)$ (Theorem 4.6).
- A (linear or nonlinear) code over the field $GF(s)$ with N codewords of length n and dual distance $d^{\perp}$ is equivalent to an $OA(N, k, s, d^{\perp} - 1)$ (Theorem 4.9).
- A (linear or nonlinear) MDS code over an alphabet of size s with s^{k-d+1} codewords of length k and minimal distance d is equivalent to an $OA(s^t, k, s, t)$ of index unity, where $t = k - d + 1$.
- A difference scheme $D(r, c, s)$ implies the existence of an $OA(rs, c+1, s, 2)$ (Corollary 6.20), as well as a mixed-level $OA(rs, s^c r^1, s)$ (Example 9.19). In some cases, the expansive replacement method (Section 9.3) can transform the latter to an $OA(rs, k, s, 2)$ with $k > c + 1$.
- A Hadamard matrix of order 4λ is equivalent to an $OA(4\lambda, 4\lambda - 1, 2, 2)$ and to an $OA(8\lambda, 4\lambda, 2, 3)$ (Theorem 7.5).
- A set of $k - 2$ pairwise orthogonal Latin squares of order s is equivalent to an $OA(s^2, k, s, 2)$ (Theorem 8.28).
- A set of k pairwise orthogonal $n \times n$ F-squares based on s symbols is equivalent to a mixed-level $OA(n^2, s^k n^2, 2)$ (Example 9.6).
- A projective plane of order s is equivalent to an $OA(s^2, s + 1, s, 2)$ (Theorem 8.43).

- A transversal design $TD_\lambda(k, s)$ or an $(s, k; \lambda)$-net is equivalent to an $OA(\lambda s^2, k, s, 2)$ (Section 10.9).
- An $OA(N, k, s, t)$ is a fractional factorial design of resolution $t+1$ (Section 11.5).

12.6 Other Tables

We mention here some related tables in the literature.

Taguchi (1987) contains a very small table of orthogonal arrays, including some with mixed levels. Dey (1985) has an extensive table of fractional factorial designs. The *Gosset* computer program (Hardin and Sloane, 1992, 1993) contains a built-in library of over 2300 optimal or close to optimal response surface designs for a variety of models.

Colbourn and Dinitz (1996a) contains tables of many different kinds of combinatorial designs. Beth, Jungnickel and Lenz (1986), Hall (1985) and Mathon and Rosa (1985) give tables of block designs, while Doyen and Rosa (1980, 1989) give tables of Steiner systems. Jungnickel (1992) includes a table of abelian difference sets.

Raghavarao (1971), Table 3.7.1, Brouwer (1979) and, most recently, Abel, Brouwer, Colbourn and Dinitz (1996), give tables of pairwise orthogonal Latin squares. Seberry and Yamada (1992) give extensive tables of Hadamard matrices.

Several tables of linear codes are available: for example, Verhoeff (1987), Brouwer and Verhoeff (1993) and, most recently, Brouwer (1998, 1999), which supersedes the earlier tables. See also Jaffe (1999). Tables of nonlinear binary codes can be found in MacWilliams and Sloane (1977, Appendix A) and, most recently, Litsyn (1998) and Litsyn, Rains and Sloane (1999). Tables of nonlinear ternary codes and mixed binary/ternary codes are given in Brouwer, Hämäläinen, Östergård and Sloane (1998).

Appendix A

Galois Fields

This Appendix provides a concise introduction to the theory of Galois fields, emphasizing the concepts that are needed in this book. The theory of Galois fields is a basic and indispensable tool in mathematics and is used in a large number of constructions of discrete structures, including of course orthogonal arrays. Readers who would like to learn more about Galois fields and their applications may consult the excellent books by Jacobson (1964), Lidl and Niederreiter (1983, 1986), McCoy (1948) or van der Waerden (1950, 1993).

A.1. Definition of a Field

The key point is that Galois fields are *finite* sets where we can add, subtract, multiply and divide. To put this another way: everyone knows how to add and subtract vectors with a specified number of components; the virtue of Galois fields is that they make it possible to also multiply and divide vectors. We now make this more precise.

We begin with the notion of a binary operation. If F is a nonempty set, then a binary operation on F is any map from $F \times F$ to F. For example, if F is the set of all integers, addition, subtraction and multiplication are binary operations, but division is not (because $1/0$ isn't an integer). Division is however a binary operation on the set of all nonzero rational numbers.

A *field* is a triplet $(F, +, *)$, often designated by F when it will not cause any confusion, where F is a nonempty set, possibly finite, and $+$ and $*$ are two binary operations on F such that the following nine properties hold:

1. $a + b = b + a$ for all $a, b \in F$;

2. $(a+b)+c = a+(b+c)$ for all $a, b, c \in F$;
3. There is a unique element $0 \in F$, the zero element, such that $a + 0 = a$ for all $a \in F$;
4. For every $a \in F$ there is a unique element $-a \in F$ such that $a+(-a) = 0$; we will call $-a$ the additive inverse of a and write $b-a$ instead of $b+(-a)$;
5. $a * b = b * a$ for all $a, b \in F$;
6. $(a * b) * c = a * (b * c)$ for all $a, b, c \in F$;
7. There is a unique element $1 \in F$, the unity element, such that $a * 1 = a$ for all $a \in F$;
8. For every $a \in F$, $a \neq 0$, there is a unique element $a^{-1} \in F$ such that $a*a^{-1} = 1$; we will call a^{-1} the multiplicative inverse of a and occasionally write b/a instead of $b * a^{-1}$;
9. $a * (b + c) = a * b + a * c$ for all $a, b, c \in F$.

The operations $+$ and $*$ will be called addition and multiplication, respectively. (These may however be completely different from the operations with these names that are usually associated with the real numbers.) Instead of $a * b$ we will often simply write ab. Further, observe that $0 \neq 1$ unless $F = \{0\}$. To see this, assume that $a \in F$, $a \neq 0$. If $0 = 1$, then $a = a * 0 = a * (0 + 0) = a * 0 + a * 0 = a + a$, so $a = 0$, which is a contradiction. Likewise, if a and b are nonzero elements of F then $ab \neq 0$ (or else $0 = aba^{-1}b^{-1} = a^{-1}ab^{-1}b = 1$). It is also immediately obvious that $-(-a) = a$ and $(a^{-1})^{-1} = a$.

Note that properties 1 through 4 and 5 through 8 are equivalent to requiring that $(F, +)$ and $(F \setminus \{0\}, *)$ are both abelian groups.

Simple examples of fields are the rational numbers, the real numbers or the complex numbers with the usual operations of addition and multiplication. These are all examples of infinite fields. An example of a finite field is $F = \{0, 1\}$ with addition and multiplication modulo 2. More generally, the set of residues modulo a prime number p, $\{0, 1, \ldots, p-1\}$, forms a field of p elements under addition and multiplication modulo p. This is false if p is not a prime, since in that case not every nonzero residue will have a multiplicative inverse. (Working modulo $p = 6$, for example, there is no multiplicative inverse for 2.)

We are primarily interested in finite fields, or *Galois fields*, as they are usually called, after the French mathematician Evariste Galois (1812–1832). The number of elements in the set F is then called the *order* of the field.

The elements

$$1, 1+1, 1+1+1, \ldots$$

are said to be the *integral elements* of the field, and will be denoted by $1, 2, 3, \ldots$. In the same way we will abbreviate $a + a + \cdots + a$ (k times) by ka, for any positive integer k. The number of integral elements in the field is called the *characteristic* of the field, and is obviously finite for any Galois field. If F is the field of residues modulo a prime number p with the usual addition and multiplication modulo p, then both the order and the characteristic of F are equal to p. In particular, all elements of this field are integral elements. We denote this field by $GF(p)$.

Among the questions that may now come to mind are such questions as whether the order of a Galois field can only be a prime number, whether the characteristic of a Galois field can only be a prime number, or whether the order and the characteristic of a Galois field are always equal. We will first present an example that provides a negative answer to two of these questions. We will then continue with a result that gives an affirmative answer to the remaining question.

Example A.1. Consider the set $F = \{0, 1, 2, 3\}$ with addition and multiplication defined as follows:

+	0	1	2	3
0	0	1	2	3
1	1	0	3	2
2	2	3	0	1
3	3	2	1	0

*	0	1	2	3
0	0	0	0	0
1	0	1	2	3
2	0	2	3	1
3	0	3	1	2

It can easily be verified that the triplet $(F, +, *)$ is a Galois field with order 4 and characteristic 2. ∎

Thus the order of a Galois field is not necessarily a prime and the order and characteristic of a Galois field are not necessarily equal. However, the characteristic of this Galois field is still a prime number, a result which holds for any finite field.

Theorem A.2. *The characteristic of a Galois field F is a prime number, p say, and p is the smallest positive number k such that $ka = 0$ for all $a \in F$. The integral elements in F can be identified with the field of residues modulo p.*

Proof: There are infinitely many sums 1, $1+1$, $1+1+1, \ldots$, so two of them must be equal, i.e. $k1 = 0$ for some positive integer k. Let p be the smallest such k. Then p must be a prime number, for if $p = p'p''$ with p' and p'' less than p, we would have $0 = p'p''1 = p'1p''1$ and so either $p'1 = 0$ or $p''1 = 0$, a contradiction. Thus the integral element $p = p1 = 1 + \cdots + 1$ (p times) is equal to $0 \in F$. The integral elements are then 0 $(= p)$, 1, 2, $\ldots$, $p - 1$, and behave exactly like the field of residues modulo p. ∎

Although the order of a Galois field is not necessarily a prime number, it cannot just be any positive integer.

Theorem A.3. *If a Galois field has order s and characteristic p, then $s = p^n$ for some positive integer n.*

Proof: As in Theorem A.2, let $0\ (= p), 1, \ldots, p-1$ denote the integral elements of the field. Let ω_1 be any nonzero element of the field, and consider

$$F_1 = \{\lambda_1\omega_1 : \lambda_1 = 0, 1, \ldots, p-1\} .$$

This is a set of p distinct elements of the field. If F_1 is the entire field, then $s = p$ and the claimed result holds with $n = 1$. If F_1 is not the entire field, then let ω_2 be an element of the field that is not in F_1. Define

$$F_2 = \{\lambda_1\omega_1 + \lambda_2\omega_2 : \lambda_1, \lambda_2 = 0, 1, \ldots, p-1\} .$$

F_2 contains p^2 distinct elements and if F_2 is the entire field, then the claim of the theorem holds with $n = 2$. If $s > p^2$, then we can choose an element ω_3 in the field that is not in F_2. Since there are only finitely many elements in the field, we eventually wind up representing all elements of the field by

$$\lambda_1\omega_1 + \lambda_2\omega_2 + \cdots + \lambda_n\omega_n, \quad \lambda_i = 0, 1, \ldots, p-1, \quad i = 1, 2, \ldots, n ,$$

for some positive integer n. It follows that $s = p^n$ for that integer n. ∎

Definition A.4. Rings. A *ring* is a set where we can add, subtract and multiply, but cannot necessarily divide. The formal definition is similar to that of a field, but omitting properties 5, 7 and 8 (so multiplication need not be commutative). Of course any field is a ring. Familiar examples of rings that are not fields are: the integers modulo m, where m is not prime; polynomials; matrices. If properties 1 through 7 and 9 hold, we speak of a *commutative ring with unity*.

A.2. The Construction of Galois Fields

Knowing that the order of a Galois field must be a power of a prime, we will now develop the tools to obtain a Galois field of order $s = p^n$ for any prime number p and any positive integer n. We will stop short of proving that the method is applicable for every such s, although that is indeed the case. We will then know all possible combinations of characteristic and order for which a Galois field exists, namely the characteristic must be a prime p and the order is some power p^n of that prime.

To begin, we will use $F[x]$ to denote all polynomials in the variable x with coefficients from a field F. Thus

$$F[x] = \{a_0 + a_1x + \cdots + a_kx^k : k \text{ a nonnegative integer, } a_0, a_1, \ldots, a_k \in F\} .$$

If some $a_i = 0$, we will usually omit the term a_ix^i. Two polynomials in $F[x]$ are said to be equal if and only if all the corresponding coefficients are equal. The zero polynomial in $F[x]$ is the polynomial with all coefficients equal to 0, and will also be denoted by 0. The degree of a nonzero polynomial $f(x) \in F[x]$, denoted by $\deg(f)$, is defined to be k if $f(x) = a_0 + a_1x + \cdots + a_kx^k$ and $a_k \neq 0$. In that case, a_k is called the leading coefficient of $f(x)$. If the leading coefficient is 1 the polynomial is called *monic.* The degree of the zero polynomial is defined to be $-\infty$.

The binary operations $+$ and $*$ on F induce binary operations on the elements of $F[x]$. These binary operations on $F[x]$ are extensions of those on F, and we will again denote them by $+$ and $*$. These operations are defined by analogy with the usual rules for addition and multiplication of polynomials. Thus

$$\begin{aligned}(a_0 + a_1x + \cdots + a_kx^k) &+ (b_0 + b_1x + \cdots + b_kx^k) \\ &= (a_0 + b_0) + (a_1 + b_1)x + \cdots + (a_k + b_k)x^k ,\end{aligned}$$

and

$$\begin{aligned}(a_0 + a_1x + \cdots + a_kx^k) &* (b_0 + b_1x + \cdots + b_kx^k) \\ &= (a_0 * b_0) + (a_0 * b_1 + a_1 * b_0)x + \cdots + (a_k * b_k)x^{2k} ,\end{aligned}$$

where a_k or b_k may be taken to be zero to allow the polynomials to be of different degrees. The triplet $(F[x], +, *)$ satisfies properties 1 through 7 and 9 of the requirements for a field, but does not satisfy 8, and so is a commutative ring with unity. In the case of $(F[x], +, *)$, or just $F[x]$, we will simply refer to it as the *polynomial ring* induced by the field F. For $f, g \in F[x]$, we write fg rather than $f * g$.

The following result gives some elementary properties of elements of a polynomial ring $F[x]$. The proof is left to the reader.

Lemma A.5. *If $f, g \in F[x]$, where F is a field, then*

(i) $\deg(f + g) \leq \max(\deg(f), \deg(g))$*;*

(ii) $\deg(fg) = \deg(f) + \deg(g)$*, provided that* $f \neq 0$, $g \neq 0$*;*

(iii) if $g \neq 0$, then there are unique polynomials $q, r \in F[x]$ such that

$$f(x) \equiv q(x)g(x) + r(x) \quad \text{and} \quad \deg(r) < \deg(g) .$$

The polynomial $r(x)$ in property (iii) is also called the residue of $f(x)$ with respect to $g(x)$, and we will write

$$f(x) \equiv r(x) \pmod{g(x)}$$

to express this.

For any fixed nonzero polynomial $g \in F[x]$ we can now define an equivalence relation $\sim$ on $F[x]$ by saying that if $f_1, f_2 \in F[x]$ then

$$f_1 \sim f_2 \quad \text{if} \quad f_1(x) - f_2(x) \equiv 0 \pmod{g(x)} .$$

Thus f_1 and f_2 are equivalent with respect to g if and only if $f_1 - f_2$ is a multiple of g in the polynomial ring $F[x]$. If $\deg(g) = n$, this equivalence relation clearly has s^n equivalence classes, where s denotes the order of the field F. Representatives for these s^n classes are given by

$$a_0 + a_1 x + \cdots + a_{n-1} x^{n-1}, \quad a_i \in F, \quad i = 0, 1, \ldots, n-1 . \tag{A.1}$$

The set of these s^n equivalence classes will be denoted by $F[x]/(g)$, and an element of this set will be denoted by $[f]$, $f \in F[x]$. Thus $[f]$ is the class of all polynomials in $F[x]$ that are equivalent to f with respect to g.

The binary operations on $F[x]$ induce binary operations, again denoted by $+$ and $*$, on $F[x]/(g)$. These can be defined for f_1, $f_2 \in F$ by

$$[f_1] + [f_2] = [f_1 + f_2]$$

and

$$[f_1] * [f_2] = [f_1 * f_2] .$$

It is easily verified that these operations are well-defined, in the sense that if $[f_1'] = [f_1]$ and $[f_2'] = [f_2]$ then $[f_1' + f_2'] = [f_1 + f_2]$ and $[f_1' * f_2'] = [f_1 * f_2]$.

We will soon see that if $g \in F[x]$ is chosen appropriately, then the triplet $(F[x]/(g), +, *)$ forms a Galois field. In order to make this more specific, we shall now introduce some additional concepts. An element $\alpha \in F$ is said to be a *root* of a nonzero polynomial $f \in F[x]$ if $f(\alpha) = 0$. It is easy to see that α is a root of f if and only if $x - \alpha$ divides $f(x)$, or, equivalently, $f(x) \equiv 0 \pmod{x - \alpha}$. If α is a root of f, then its multiplicity is defined to be the largest integer m for which $(x - \alpha)^m$ divides $f(x)$. By saying that a nonzero polynomial $f \in F[x]$ has l roots in F we mean that the sum of the multiplicities of the distinct roots of f in F is equal to l. With that terminology we have the following result, the proof of which is left to the reader.

Lemma A.6. *If $f \in F[x]$ and $\deg(f) = k$, then f has at most k roots in F.*

A nonzero polynomial $f \in F[x]$ is said to be *irreducible* over F if for any f_1, $f_2 \in F[x]$ with $f = f_1 f_2$ either $\deg(f_1) = 0$ or $\deg(f_2) = 0$. Otherwise f is said to be *reducible*. Clearly, if $\deg(f) \geq 2$ and f has a root in F, then f is reducible over F. The converse, however, is false. A polynomial may be reducible over a given field without having a root in it.

We now return to the question of when $(F[x]/(g), +, *)$ forms a Galois field.

Theorem A.7. *Let F be a Galois field and let $g \in F[x]$, $g \neq 0$. The triplet $(F[x]/(g), +, *)$ is a Galois field if and only if g is irreducible over F.*

Proof: If $(F[x]/(g), +, *)$ is a field, then it is a Galois field since $F[x]/(g)$ is clearly a finite set. Of the requirements for a field, properties 1 through 7 and 9 are satisfied for any choice of g. If g is irreducible, then for f_1, $f_2 \in F[x]$ with $[f_1] * [f_2] = [0]$, we can conclude (with the help of Problem A.2) that $[f_1] = [0]$ or $[f_2] = [0]$. Therefore, for any $[f] \in F[x]/(g)$, $[f] \neq [0]$, the elements

$$[f] * [f'] ,$$

for $[f'] \in F[x]/(g)$, are different for different elements $[f']$. Hence if $[f']$ runs through $F[x]/(g)$, so does $[f] * [f']$ and in particular there is a unique $[f']$ such that $[f] * [f'] = [1]$. Conversely, if $g = g_1 g_2$ is reducible with $\deg(g_1) < \deg(g)$, $\deg(g_2) < \deg(g)$, then $[g_1] * [g_2] = 0$, but $[g_1] \neq 0$, $[g_2] \neq 0$, which is impossible in a field. ∎

Thus, when we are interested in the construction of a Galois field with order $s = p^n$ and characteristic p, where p is a prime number and n is a positive integer, Theorem A.7 suggests the following strategy. Take F to be the field of residues modulo p, with the usual addition and multiplication modulo p. Select an irreducible polynomial $g \in F[x]$ of degree n. Then $F[x]/(g)$ provides a field with the desired order and characteristic.

There are various questions that remain to be considered. For example, does the required polynomial g in the previous paragraph always exist, for any values of p and n? If there is more than one such polynomial g, do the fields obtained by taking different polynomials g have different structures or are they in some sense the same? If there is such a polynomial g, how can we find it? Should we consider other methods of construction for Galois fields, in order to obtain fields with desirable additional properties? Before addressing these questions, we will first illustrate the concepts that have been introduced so far with a detailed example.

Example A.8. Let $F = \{0, 1, 2\}$ with the usual addition and multiplication modulo 3. Some elements of $F[x]$ are:

$$f_1(x) = 2x^5 - x^3 + 1 ,$$

$$\begin{aligned} f_2(x) &= x^3 + x^2 + 2x + 2 , \\ f_3(x) &= x^2 + x + 2 , \\ f_4(x) &= x^4 + x^3 + x + 2 . \end{aligned}$$

Then, for example, $\deg(f_1) = 5$, $\deg(f_2) = 3$, and

$$\begin{aligned} f_1(x) + f_2(x) &= 2x^5 + x^2 + 2x , \\ f_1(x) * f_2(x) &= 2x^8 + 2x^7 + x^4 + 2x^3 + x^2 + 2x + 2 . \end{aligned}$$

The polynomial f_1 has only one root in F, namely 2 with multiplicity 1; f_2 has three roots in F, namely 1 with multiplicity 1 and 2 with multiplicity 2; f_3 and f_4 have no roots at all in F. The factorization of these polynomials into irreducible polynomials is

$$\begin{aligned} f_1(x) &= (x+1)(2x^4 + x^3 + x^2 + 2x + 1) , \\ f_2(x) &= (x+1)^2(x+2) , \\ f_3(x) &= x^2 + x + 2 , \\ f_4(x) &= (x^2+1)(x^2 + x + 2) . \end{aligned}$$

Thus only $f_3(x)$ is an irreducible polynomial over F. Observe that

$$\begin{aligned} f_1(x) &= (2x^3 + x^2 + 1)(x^2 + x + 2) + 2x + 2 , \\ f_2(x) &= x(x^2 + x + 2) + 2 , \\ f_4(x) &= (x^2+1)(x^2 + x + 2) + 0 . \end{aligned}$$

Hence f_1, f_2 and f_4 correspond to three different elements in $F[x]/(f_3)$, namely

$$[f_1] = [2x+2] , \quad [f_2] = [2] , \quad [f_4] = [0] .$$

The Galois field $F[x]/(f_3)$ has order 9 and characteristic 3. The elements of the field, with the brackets deleted for convenience of notation, may be represented by 0, 1, 2, x, $x{+}1$, $x{+}2$, $2x$, $2x{+}1$ and $2x{+}2$. Notice that the successive powers of the element x in this field generate all the nonzero elements of the field. The successive powers of x are x, $x^2 = 2x+1$, $x^3 = 2x+2$, $x^4 = 2$, $x^5 = 2x$, $x^6 = x+2$, $x^7 = x+1$ and $x^8 = 1$. An element of a field with the property that its powers generate all the nonzero elements is called a *primitive element* of the field. Knowing a primitive element of a field can provide a considerable simplification of computations that have to be performed in that field. As an example, in $F[x]/(f_3)$ we have

$$(x+2)(2x+1)(x+1)(2x+2)^{-1} = x^6x^2x^7x^{-3} = x^{12} = x^4 = 2 .$$

The addition and multiplication tables for this field are given in Tables A.9 and A.10.

Table A.9. Addition table for a field of order 3^2.

$+$	0	1	2	x	$x+1$	$x+2$	$2x$	$2x+1$	$2x+2$
0	0	1	2	x	$x+1$	$x+2$	$2x$	$2x+1$	$2x+2$
1	1	2	0	$x+1$	$x+2$	x	$2x+1$	$2x+2$	$2x$
2	2	0	1	$x+2$	x	$x+1$	$2x+2$	$2x$	$2x+1$
x	x	$x+1$	$x+2$	$2x$	$2x+1$	$2x+2$	0	1	2
$x+1$	$x+1$	$x+2$	x	$2x+1$	$2x+2$	$2x$	1	2	0
$x+2$	$x+2$	x	$x+1$	$2x+2$	$2x$	$2x+1$	2	0	1
$2x$	$2x$	$2x+1$	$2x+2$	0	1	2	x	$x+1$	$x+2$
$2x+1$	$2x+1$	$2x+2$	$2x$	1	2	0	$x+1$	$x+2$	x
$2x+2$	$2x+2$	$2x$	$2x+1$	2	0	1	$x+2$	x	$x+1$

Table A.10. Multiplication table for a field of order 3^2.

$*$	0	1	2	x	$x+1$	$x+2$	$2x$	$2x+1$	$2x+2$
0	0	0	0	0	0	0	0	0	0
1	0	1	2	x	$x+1$	$x+2$	$2x$	$2x+1$	$2x+2$
2	0	2	1	$2x$	$2x+2$	$2x+1$	x	$x+2$	$x+1$
x	0	x	$2x$	$2x+1$	1	$x+1$	$x+2$	$2x+2$	2
$x+1$	0	$x+1$	$2x+2$	1	$x+2$	$2x$	2	x	$2x+1$
$x+2$	0	$x+2$	$2x+1$	$x+1$	$2x$	2	$2x+2$	1	x
$2x$	0	$2x$	x	$x+2$	2	$2x+2$	$2x+1$	$x+1$	1
$2x+1$	0	$2x+1$	$x+2$	$2x+2$	x	1	$x+1$	2	$2x$
$2x+2$	0	$2x+2$	$x+1$	2	$2x+1$	x	1	$2x$	$x+2$

If we had used the irreducible polynomial $2f_3(x)$ to construct a Galois field with order 3^2, we would obviously wind up with the same addition and multiplication tables as for the field obtained by using $f_3(x)$. More generally, we can always assume that the irreducible polynomial we use to generate the field is monic (i.e. has leading coefficient equal to unity). There are, however, other irreducible monic polynomials of degree 2 besides $f_3(x)$. From the tables in Lidl and Niederreiter (1986) we find that there are two other monic polynomials that we could have used to construct a field of order 3^2, namely

$$\begin{aligned} f_5(x) &= x^2+1 , \\ f_6(x) &= x^2+2x+2 . \end{aligned}$$

The *addition* tables for the two fields that would be obtained by using either of these polynomials would be the same as for the field obtained by using $f_3(x)$. However, that is not true for the *multiplication* tables. This is easily seen from Table A.11, where the powers of x in the three fields are compared.

Table A.11. Powers of x in three fields of order 3^2.

		Irreducible monic polynomial		
		$f_3(x)$	$f_5(x)$	$f_6(x)$
	1	x	x	x
	2	$2x+1$	2	$x+1$
Powers	3	$2x+2$	$2x$	$2x+1$
of	4	2	1	2
x	5	$2x$	x	$2x$
	6	$x+2$	2	$2x+2$
	7	$x+1$	$2x$	$x+2$
	8	1	1	1

Observe that x is not even a primitive element in the Galois field obtained by using $f_5(x)$. However, there are primitive elements in that field as well. The powers of $x+1$, for example, generate all the nonzero elements of that field. ∎

An element a of a field F is said to be a *k-th root of unity* if $a^k = 1$. Also, a is a *primitive* k-th root of unity if k is the smallest value of m such that a is an m-th root of unity. If a is a primitive k-th root of unity then

$$\{1, a, a^2, \ldots, a^{k-1}\}$$

are all the k-th roots of unity in F. If F has order p^n, the primitive elements of F are thus precisely the primitive (p^n-1)-th roots of unity in F; these are also called the *primitive roots* in F.

We will now introduce some concepts that will be useful for addressing the questions that were posed immediately before Example A.8. The first definition will help to decide when two fields are really the same.

Definition A.12. A one-to-one mapping ϕ from a field F_1 onto a field F_2 is said to be an *isomorphism* if

(1) $\phi(a+b) = \phi(a) + \phi(b)$, $a, b \in F_1$, and

(2) $\phi(ab) = \phi(a)\phi(b)$, $a, b \in F_1$.

If there exists an isomorphism between two fields, they are said to be *isomorphic*. For most practical purposes, isomorphic fields are the *same* field (and in particular their orders are the same).

Example A.13. Continuing Example A.8, let us take two of the finite fields of order 3^2, say those obtained by using $f_3(x) = x^2+x+2$ and $f_5(x) = x^2+1$, and see if they are isomorphic. Let us write $F_1 = F[x]/(f_3)$ and $F_2 = F[x]/(f_5)$, so that F_1 is the field of order 9 from Example A.8. If ϕ is an isomorphism from F_1 onto F_2 and $\alpha \in F_1$, then from condition (2) of Definition A.12 we see that $\phi(\alpha^i) = \phi(\alpha)^i$. So if α is a primitive element in F_1, $\phi(\alpha)$ should be a primitive element in F_2, and vice versa. Also, it follows immediately from condition (1) of Definition A.12 that $\phi(0) = 0$. Thus, if we know $\phi(\alpha)$, where α is a primitive element in F_1, then ϕ is completely determined since $\phi(0) = 0$ and $\phi(\alpha^i) = \phi(\alpha)^i$. Based on the observations in Example A.8 we might try $\phi(x) = x + 1$. With $\phi(0) = 0$ and $\phi(x^i) = (x+1)^i$ this leads to the following map

$$\phi(0) = 0,\ \phi(x) = x+1,\ \phi(2x+1) = 2x,\ \phi(2x+2) = 2x+1,\ \phi(2) = 2,$$
$$\phi(2x) = 2x+2,\ \phi(x+2) = x,\ \phi(x+1) = x+2,\ \phi(1) = 1\ .$$

It is left to the reader to verify that ϕ is indeed an isomorphism from F_1 onto F_2. ∎

Other important concepts are those of subfields and extension fields, although their relevance is not entirely apparent from the brief discussion in this appendix.

Definition A.14. A subset E of a field F is called a *subfield* of F if E is itself a field under the binary operations that are associated with F. We also express this by saying that F is an *extension field* of E.

If E is a Galois field of order s and F is a finite extension field of E, then it follows from arguments analogous to those in the proof of Theorem A.3 that the order of F must be of the form s^n, for some positive integer n. Moreover, since we may again represent every element of F by

$$\lambda_1\omega_1 + \cdots + \lambda_n\omega_n, \quad \lambda_i \in E, \quad i = 1, \ldots, n\ ,$$

for appropriately selected elements $\omega_i \in F$, $i = 1, \ldots, n$, we can think of the elements of F as an n-dimensional vector space over E.

Thus if E is a subfield of F, and the corresponding orders of these fields are p^m and p^n, respectively, with p a prime number, then m must divide n. Conversely, if F is a Galois field of order p^n, then there exists for every m that divides n a unique subfield of order p^m. In particular, F will have a unique subfield of p elements. This is the subfield that consists of the integral elements in F.

A.3. The Existence of Galois Fields

We now turn to some crucial properties of Galois fields, in order to answer our previously posed questions. Although the results are essential for an adequate understanding of Galois fields, the proofs can be omitted without jeopardizing the level of understanding that is required for this book, and we will not give the proofs here. Readers who wish to learn more may consult any of the books mentioned at the beginning of this appendix.

Theorem A.15. *Over $GF(p)$, $x^{p^n} - x$ factors into the product of all distinct monic irreducible polynomials whose degree divides n.*

For example, when $p = 2$, the factorizations of the first few values of $x^{2^n} - x$ are:

$$\begin{aligned} x^2 - x &= x(x+1) , \\ x^4 - x &= x(x+1)(x^2+x+1) , \\ x^8 - x &= x(x+1)(x^3+x+1)(x^3+x^2+1) , \\ x^{16} - x &= x(x+1)(x^2+x+1)(x^4+x+1)(x^4+x^3+1) \\ & \quad (x^4+x^3+x^2+x+1) . \end{aligned}$$

Therefore Theorem A.15 tells us that the complete list of monic irreducible polynomials over $GF(2)$ of degrees ≤ 4 is:

degree	polynomials
1	$x,\ x+1$
2	x^2+x+1
3	$x^3+x+1,\ x^3+x^2+1$
4	$x^4+x+1,\ x^4+x^3+1,\ x^4+x^3+x^2+x+1$

Computer algebra systems such as Macsyma, Magma, Maple or Mathematica make it easy to carry out such factorizations and hence to find irreducible polynomials of any degree.

Using Theorem A.15 it can be shown that the number of distinct monic irreducible polynomials of degree n over $GF(p)$ is given by the formula

$$\frac{1}{n} \sum_{d|n} \mu\left(\frac{n}{d}\right) p^d ,$$

where the sum extends over all numbers d in the range $1 \leq d \leq n$ that divide n, and μ is the Möbius function defined by $\mu(1) = 1,\ \mu(a) = (-1)^k$ if a is the product of k distinct primes, and $\mu(a) = 0$ if a is divisible by a square > 1.

In particular, there is always at least one such polynomial.

Corollary A.16. *For any prime number p and any positive integer n, there exists an irreducible polynomial of degree n over the field of residues modulo p.*

As a consequence of Theorem A.7 and Corollary A.16 we can conclude that a Galois field of order p^n exists for any prime number p and any positive integer n.

Theorem A.17. *Any two Galois fields with the same order are isomorphic.*

This result states that the structure of any two Galois fields of order p^n is essentially the same, and that we may therefore refer to *the* Galois field of order p^n. Generalizing the notation introduced in Section A.1, we will denote this field by $GF(p^n)$.

Theorem A.18. *Any Galois field contains a primitive element.*

The advantages of knowing a primitive element of a Galois field were pointed out in Example A.8. It was however also observed there that the choice of a primitive element in $GF(p^n)$ depends on which irreducible polynomial $f(x)$ is used to define the field. If $n \geq 2$ it is clear that an integral element can never be a primitive element, since the powers of an integral element generate only integral elements, and for $n \geq 2$, $GF(p^n)$ contains nonintegral elements. We might however hope that we can choose an irreducible polynomial over $GF(p)$ such that a known element of $GF(p^n)$, for example x, will be a primitive element. That is indeed always possible.

An irreducible polynomial $f(x)$ over $F = GF(p)$ of degree n is said to be *primitive* if x is a primitive element in the field $GF(p^n) = F[x]/(f(x))$. There is a simple test for being primitive. From Theorem A.15 we know that an irreducible polynomial $f(x)$ of degree n divides $x^{p^n-1} - 1$. Then $f(x)$ is primitive if it does not divide $x^m - 1$ for any $m < p^n - 1$. (Suppose $f(x)$ is not primitive. Then $[x]$ is not a primitive element in $F[x]/(f(x))$, and so two of the elements in $\{[x], [x]^2, \ldots, [x]^{p^n-1}\}$ are equal. Thus $[x]^m = 1$ for some $m < p^n - 1$, i.e. $x^m - 1 \equiv 0 \bmod f(x)$.)

From this it follows that the number of primitive irreducible polynomials over $GF(p)$ of degree n is given by the formula

$$\frac{1}{n}\phi(p^n - 1) ,$$

where ϕ is the Euler totient function, i.e. $\phi(a)$ is equal to the number of positive integers that do not exceed a and that are relatively prime to a. (For example, $\phi(1) = \phi(2) = 1$, $\phi(3) = \phi(4) = \phi(6) = 2$, $\phi(30) = 8$, $\phi(31) = 30$, $\phi(32) = 16$.) In particular, there is always at least one such polynomial. Computer algebra systems also make it very easy to find primitive polynomials.

Table A.19 gives a list of primitive irreducible polynomials for selected values of n and p. Thus if $GF(p^n)$ is obtained as $GF(p)[x]/(f)$ using the polynomial

f listed in the table, the element x of this field (or more precisely $[x]$) will be a primitive element. More extensive tables can be found for example in Chapter 10 of Lidl and Niederreiter (1986).

Table A.19. Selected primitive irreducible polynomials.

p	n	Primitive polynomial over $GF(p)$
2	2	x^2+x+1
2	3	x^3+x+1
2	4	x^4+x+1
2	5	x^5+x^2+1
2	6	x^6+x+1
2	7	x^7+x+1
2	8	$x^8+x^4+x^3+x^2+1$
2	9	x^9+x^4+1
3	2	x^2+x+2
3	3	x^3+2x+1
3	4	x^4+x+2
3	5	x^5+2x+1
3	6	x^6+x+2
5	2	x^2+x+2
5	3	x^3+3x+2
5	4	x^4+x^2+2x+2
5	5	x^5+4x+2
5	6	x^6+x^5+2
7	2	x^2+x+3
7	3	x^3+3x+2
7	4	x^4+5x^3+5x+5
7	5	x^5+x+4
11	2	x^2+x+7
11	3	x^3+x^2+3
11	4	x^4+4x^3+2
11	5	$x^5+x^3+x^2+9$
13	2	x^2+x+2
13	3	x^3+x^2+2
13	4	$x^4+x^3+3x^2+2$
13	5	x^5+x^3+x+2
17	2	x^2+x+3
17	3	x^3+x+3
17	4	x^4+4x^2+x+3
17	5	$x^5+x^4+x^3+3$

Primitive elements for the Galois $GF(p)$ have also been tabulated. A primitive element a for each prime number $p < 10000$ is given in Table 24.8 of Abramowitz and Stegun (1964). Table A.20 shows these elements for $p < 100$.

Table A.20. Examples of primitive elements a in $GF(p)$ for $p < 100$.

p	a	p	a
3	2	43	3
5	2	47	5
7	3	53	2
11	2	59	2
13	2	61	2
17	3	67	2
19	2	71	7
23	5	73	5
29	2	79	3
31	3	83	2
37	2	89	3
41	6	97	5

Notation for field elements. Suppose the field $GF(p^n)$ is defined using a primitive polynomial $f(x)$ over $GF(p)$ of degree n. Then $[x]$ is a primitive element of $GF(p^n)$. It is customary to denote $[x]$ by a Greek letter, such as α, ω or ξ, in order to better distinguish the nonintegral elements in the field. We will use α here.

Then we have two alternative ways to write the elements of $GF(p^n)$. We can write an element $\beta \in GF(p^n)$ as a polynomial in α of degree at most $n-1$:

$$\beta = b_0 + b_1\alpha + b_2\alpha^2 + \cdots + b_{n-1}\alpha^{n-1} ,$$

where the coefficients $b_0, \ldots, b_{n-1}$ are in $GF(p)$. Once some facility with Galois fields has been attained it is customary to replace this polynomial by the vector of its coefficients, and to simply write $\beta = b_0 b_1 \cdots b_{n-1}$.

Alternatively (from Theorem A.18) we may represent the elements of $GF(p^n)$ as

$$\alpha_0 = 0,\ \alpha_1 = \alpha,\ \alpha_2 = \alpha^2,\ \alpha_3 = \alpha^3,\ \ldots,\ \alpha_{p^n-1} = \alpha^{p^n-1} = 1.$$

These two representations are analogues of the two familiar ways of writing complex numbers, in rectangular coordinates (as $x + iy$) or polar coordinates (as $re^{i\theta}$). In Galois fields the "rectangular coordinates" or vector representation is most convenient for performing addition and subtraction, while the representation in "polar coordinates" as powers of a primitive element is better when performing multiplication, division, exponentiation, etc. References such as Lidl and Niederreiter (1986), Table A, provide logarithm and exponential tables that make it easy to switch between the two representations.

The reader will now appreciate the remark made at the beginning of this chapter about Galois fields making it possible to perform arithmetic on vectors!

The lowest degree monic polynomial over $GF(p)$ that has α^i as a root is called the *minimal polynomial* of α^i, and denoted by $M^{(i)}(x)$. The roots of $M^{(i)}(x)$ then consist of the set of field elements

$$\{\alpha^i, \alpha^{pi}, \alpha^{p^2 i}, \alpha^{p^3 i}, \ldots, \alpha^{p^{a-1} i}\} ,$$

where $p^a i \equiv i \pmod{p^n - 1}$, for some integer a. The set of exponents $C_i = \{i, pi, p^2 i, \ldots, p^{a-1} i\} \pmod{p^n - 1}$ is called a *cyclotomic coset* modulo $p^n - 1$, and so

$$M^{(i)}(x) = \prod_{j=0}^{a-1} (x - \alpha^{p^j i}) ,$$

and

$$x^{p^n - 1} - 1 = \prod M^{(i)}(x) , \tag{A.2}$$

where the product in (A.2) runs through one representative from each distinct nonzero cyclotomic coset modulo $p^n - 1$.

Example A.21. For $p = 2$, $n = 4$, suppose $GF(2^4)$ has been constructed using the primitive polynomial $f(x) = x^4 + x + 1$. The nonzero cyclotomic cosets modulo 15 are:

$$\begin{aligned} C_1 &= \{1, 2, 4, 8\} , \\ C_3 &= \{3, 6, 12, 9\} , \\ C_5 &= \{5, 10\} , \\ C_7 &= \{7, 14, 13, 11\} , \end{aligned}$$

the corresponding minimal polynomials are

$$\begin{aligned} M^{(1)}(x) &= f(x) = x^4 + x + 1 , \\ M^{(3)}(x) &= x^4 + x^3 + x^2 + x + 1 , \\ M^{(5)}(x) &= x^2 + x + 1 , \\ M^{(7)}(x) &= x^4 + x^3 + 1 , \end{aligned}$$

and we have

$$x^{15} - 1 = M^{(1)}(x) M^{(3)}(x) M^{(5)}(x) M^{(7)}(x) .$$

A.4. Quadratic Residues in Galois Fields

In this section we direct our attention to the topic of quadratic residues in a Galois field.

Definition A.22. The *quadratic residues* in the Galois field $GF(p^n)$ are the nonzero squares in the field, or, more precisely, the elements

$$\{\beta \in GF(p^n) : \beta = \gamma^2 \quad \text{for some nonzero} \quad \gamma \in GF(p^n)\} .$$

The quadratic residues are used in various constructions in this book, and are (not surprisingly) of importance for solving quadratic equations in the field. We first consider the number of quadratic residues in $GF(p^n)$.

Theorem A.23. *(i) If p is an odd prime number, then there are $(p^n - 1)/2$ quadratic residues in $GF(p^n)$. (ii) In $GF(2^n)$ every nonzero element is a quadratic residue.*

Proof: Let α be a primitive element of $GF(s)$, $s = p^n$, and assume that p is an odd prime number. The $(s-1)/2$ elements α^{2i}, $i = 1, 2, \ldots, (s-1)/2$, are certainly quadratic residues in $GF(s)$. We claim that there are no others. If there were another quadratic residue in this field, it would be of the form α^{2i-1}, $i \in \{1, 2, \ldots, (s-1)/2\}$. Assume then that for some $i \in \{1, 2, \ldots, (s-1)/2\}$, $\alpha^{2i-1} = \gamma^2$ for some $\gamma \in GF(s)$. As γ must also be a power of α, say α^j, this implies that $\alpha^{2i-1} = \alpha^{2j}$, or, equivalently $\alpha^{2i-2j-1} = 1$. Since α is a primitive element this implies $2i - 2j - 1 \equiv 0 \pmod{s-1}$. However, this is impossible since $2i - 2j - 1$ is odd and $s - 1$ is even.

If $s = 2^n$, then

$$\alpha^i = \alpha^i \alpha^{s-1} = \alpha^{s+i-1} .$$

Since either i or $s + i - 1$, $i = 1, 2, \ldots, s - 1$ is even, it is clear that in this case every nonzero element of $GF(s)$ is a quadratic residue. ∎

The concept of quadratic residue is thus of greatest interest for Galois fields of odd order.

Example A.24. The quadratic residues in $GF(7)$ are 1, 2 and 4. This follows since $1^2 = 6^2 = 1$, $2^2 = 5^2 = 4$ and $3^2 = 4^2 = 2$. The quadratic residues in $GF(9)$ when constructed by using a primitive polynomial are x^2, x^4, x^6 and x^8. With the primitive polynomial $f_3(x) = x^2 + x + 2$ of Example A.8, this means that the quadratic residues are given by 1, 2, $x + 2$ and $2x + 1$. ∎

Any element in $GF(s)$ that is not a quadratic residue is called a (*quadratic*) *nonresidue*. A convenient notation is to define

$$\chi(\beta) = \begin{cases} 1 & \text{if } \beta \text{ is a quadratic residue in } GF(s) , \\ 0 & \text{if } \beta = 0 , \\ -1 & \text{if } \beta \text{ is a nonzero quadratic nonresidue} . \end{cases} \tag{A.3}$$

It is easily seen that for any β and γ in $GF(s)$ we have $\chi(\beta\gamma) = \chi(\beta)\chi(\gamma)$. Thus the product of two elements in $GF(s)$ is a quadratic residue if and only if either both elements are quadratic residues or both are nonzero nonresidues. The reader will be asked to verify this as well as various other interesting properties of quadratic residues at the end of this appendix.

Finally, we turn to the problem of solving a quadratic equation

$$ay^2 + by + c = 0, \quad a, b, c \in GF(s), \quad a \neq 0 , \tag{A.4}$$

in the Galois field $GF(s)$. As in the field of the real numbers, such an equation may have two solutions, which are possibly equal, or no solution at all. Equivalently, if a polynomial of degree 2 over $GF(s)$ has one root in $GF(s)$, then it has also a second root in that field. If the order of the field is odd, it is easily seen that the answer as to when (A.4) has a solution is analogous to the answer in the field of the real numbers. A solution exists if and only if the discriminant D, defined by

$$D = b^2 - 4ac ,$$

is zero or a quadratic residue. If this is the case, then the solutions are given by

$$(-b \pm D^{1/2})/(2a) ,$$

where $D^{1/2}$ denotes an element of $GF(s)$ whose square equals D. Also, by 4 and 2 we mean the integral elements $1+1+1+1$ and $1+1$, respectively. The former will just be 1 if the characteristic of the field equals 3, for example.

If $GF(s)$ is a field of even order, and thus with characteristic 2, the previous expressions do not apply since the integral element 2 would be equal to 0 and would therefore not have a multiplicative inverse. In this case, a subset of $GF(2^n)$ that plays an important role in deciding whether (A.4) has a solution in $GF(2^n)$ is given by

$$G = \{\beta \in GF(2^n) : \beta = \gamma + \gamma^{-1} \quad \text{for some} \quad \gamma \in GF(2^n), \;\; \gamma \neq 0\} .$$

By taking $\gamma = 1$, it is clear that $0 \in G$. Also, if $\gamma + \gamma^{-1} = \delta + \delta^{-1}$, then $\gamma = \delta$ or $\gamma = \delta^{-1}$ (see Problem A.18). It follows that the cardinality of G is equal to $s/2$.

If $c = 0$, then the two solutions of (A.4) in $GF(2^n)$ are clearly given by $y = 0$ and $y = a^{-1}b$. If $c \neq 0$, then (A.4) has solutions in $GF(2^n)$ if and only if $b/(ac)^{1/2} \in G$. (Since $ac \neq 0$ and every nonzero element in $GF(2^n)$ is a quadratic residue, there exists a unique element in $GF(2^n)$ whose square is equal to ac — it is this element that we denote by $(ac)^{1/2}$.) If $b/(ac)^{1/2} \in G$ and thus $\gamma + \gamma^{-1} = b/(ac)^{1/2}$ for some $\gamma \in GF(2^n)$, $\gamma \neq 0$, then the solutions to (A.4) in $GF(2^n)$ are given by

$$y = c^{1/2}\gamma/a^{1/2}, \quad y = c^{1/2}/(\gamma a^{1/2}) .$$

These are distinct solutions unless $\gamma = 1$, or equivalently unless $b = 0$.

Example A.25. In the Galois field of order 9 constructed using the primitive polynomial $f_3(x) = x^2 + x + 2$ in Example A.8, consider the quadratic equation

$$\alpha y^2 + (2\alpha + 1)y + 2 = 0 . \tag{A.5}$$

Then $D = (2\alpha + 1)^2 - 2\alpha = \alpha + 2 = (\alpha + 1)^2$. Thus D is a quadratic residue and the solutions of (A.5) are given by $y = (-(2\alpha + 1) \pm (\alpha + 1))/(2\alpha))$, i.e. $y = 1$ and $y = 2\alpha + 2$. ■

Example A.26. In the Galois field of order 8 constructed using the primitive polynomial $f(x) = x^3 + x + 1$, consider the quadratic equation

$$\alpha y^2 + (\alpha^2 + 1)y + 1 = 0 . \tag{A.6}$$

Then $ac = \alpha$ and $(ac)^{1/2} = \alpha^4 = \alpha^2 + \alpha$. Hence

$$b/(ac)^{1/2} = (\alpha^2 + 1)/(\alpha^2 + \alpha) = \alpha^2 .$$

Since $\alpha + \alpha^{-1} = \alpha^2 + \alpha + 1$, $\alpha^2 + \alpha^{-2} = \alpha + 1$, $\alpha^3 + \alpha^{-3} = \alpha^2 + 1$, we see that

$$G = \{0, \alpha + 1, \alpha^2 + 1, \alpha^2 + \alpha + 1\} ,$$

and so $b/(ac)^{1/2} \notin G$. Hence (A.6) has no solution in this field. ■

A.5. Problems

A.1. Let $f(x)$ and $g(x)$ be polynomials in $F[x]$, not both zero, where F is a field. Show that there is a unique polynomial $d(x) \in F[x]$, called the *greatest common divisor* or g.c.d. of $f(x)$ and $g(x)$, with the following properties:

(i) the leading coefficient of $d(x)$ is 1;
(ii) $d(x)$ divides both $f(x)$ and $g(x)$;
(iii) every polynomial that divides $f(x)$ and $g(x)$ is a divisor of $d(x)$.

Show that $d(x)$ can be characterized as the lowest degree monic polynomial of the form

$$d(x) = s(x)f(x) + t(x)g(x), \quad \text{for} \quad s(x), t(x) \in F[x] . \tag{A.7}$$

A.2. The result in this problem is needed in the proof of Theorem A.7. Suppose F is a field, $g(x) \in F[x]$ is an irreducible polynomial, and $f_1(x), f_2(x) \in F[x]$. Show that if $g(x)$ divides $f_1(x)f_2(x)$ then either $g(x)$ divides $f_1(x)$ or $g(x)$ divides $f_2(x)$. (Hint: use (A.7).)

A.3. Prove Lemma A.5.

A.4. Find several finite fields such that the polynomial

$$f(x) = 1 + 2x^2 + x^3$$

is reducible over those fields. In each case provide possible roots of this polynomial in those fields.

A.5. Let $f(x) = 1 + x + x^3 + x^4$. Can you find a field F such that $f(x)$ is irreducible?

A.6. For each of the following polynomials decide whether it is irreducible over $GF(3)$: (i) $x^2 + 1$, (ii) $x^2 + x + 2$, (iii) $x^4 + 2x^2 + 1$, (iv) $x^2 + 2x + 2$.

A.7. Let $f \in GF(5)[x]$ be the polynomial

$$f(x) = x^7 + 3x^6 + 2x^4 + 4x + 2 .$$

Find a polynomial $g \in GF(5)[x]$ of degree less than 3 that is congruent to f modulo $h(x) = x^3 + 4x + 2$, i.e. such that

$$g(x) \equiv f(x) \pmod{h(x)} .$$

A.8. Construct a field of order 27.

A.9. Is 7 a primitive element of $GF(31)$?

A.10. Let F be a finite field of order s. Show that the number of primitive elements in F is equal to $\phi(s-1)$, where ϕ is Euler's totient function.

A.11. Show that $a^s = a$ for every $a \in GF(s)$.

A.12. If F has characteristic p, show that $(a+b)^{p^m} = a^{p^m} + b^{p^m}$ for all positive integers m and for all $a, b \in F$. (This identity is sometimes called the "Freshman's dream".)

A.13. Let $a \in GF(p^n)$. How many solutions (in the unknown x) are there to $x^p = a$?

A.14. Solve the following system of equations over $GF(7)$: $2x + y + z = 0$, $x + 2y + z = 4$, $3x + 3y + 4z = 2$.

A.15. Let a be a quadratic residue in $GF(s)$. In each of the following cases determine whether $-a$ is a quadratic residue:

(i) $s \equiv 0 \pmod 2$,

(ii) $s \equiv 1 \pmod 4$,

(iii) $s \equiv 3 \pmod 4$.

A.16. Show that if s is odd, then a^{-1} is a quadratic residue if and only if a is quadratic residue in $GF(s)$.

A.17. Let p be an odd prime and let α be a primitive element of $GF(p)$. If $GF(p)$ is extended to $GF(p^n)$, show that α is a quadratic residue or nonresidue of the extended field according as n is even or odd.

A.18. Show that if $\gamma + \gamma^{-1} = \delta + \delta^{-1}$, where $\gamma, \delta \in GF(s)$, then either $\gamma = \delta$ or $\gamma = \delta^{-1}$.

A.19. Show that the product of two quadratic residues, or the product of two nonresidues, is a quadratic residue, and the product of a quadratic residue and a nonresidue is a nonresidue. Hence show that, if s is odd, the function χ defined in (A.3) satisfies $\chi(\beta\gamma) = \chi(\beta)\chi(\gamma)$ for all $\beta, \gamma \in GF(s)$.

Bibliography

[Numbers in square brackets after the references specify the chapters which cite them, or, in some cases, chapters which are related to them, even though they may actually not be mentioned there.]

Abel, R. J. R., Brouwer, A. E., Colbourn, C. J., & Dinitz, J. H. (1996). Mutually orthogonal Latin squares (MOLS). Pages 111–142 of: Colbourn, C. J., & Dinitz, J. H. (eds), *The CRC Handbook of Combinatorial Designs*. CRC Press. [8,10].

Abel, R. J. R., & Cheng, Y. W. (1994). Some new MOLS of order $2^n p$ for p a prime power. *Australas. J. Combin.*, **10**, 175–186. [6].

Abel, R. J. R., Colbourn, C. J., & Dinitz, J. H. (1996). Incomplete MOLS. Pages 142–172 of: Colbourn, C. J., & Dinitz, J. H. (eds), *The CRC Handbook of Combinatorial Designs*. CRC Press. [8].

Abramowitz, M., & Stegun, I. A. (1964). *Handbook of Mathematical Functions*. Washington, DC: National Bureau of Standards. [11,A].

Abt, M., Mayer, R., & Pukelsheim, F. (1995). Improving manufacturing quality through planned experiments: pressure governor case study. *Surv. Math. Indust.*, **5**, 35–47. [11].

Abt, M., & Pukelsheim, F. (1995). Improving manufacturing quality through planned experiments: statistical methodology. *Surv. Math. Indust.*, **5**, 27–33. [11].

Addelman, S. (1962a). Orthogonal main-effect plans for asymmetrical factorial experiments. *Technometrics*, **4**, 21–46. [9].

Addelman, S. (1962b). Symmetrical and asymmetrical fractional factorial plans. *Technometrics*, **4**, 47–58.

Addelman, S. (1972). Recent developments in the design of factorial experiments. *J. Amer. Statist. Assoc.*, **67**, 103–111.

Addelman, S., & Kempthorne, O. (1961a). Some main-effect plans and orthogonal arrays of strength two. *Ann. Math. Statist.*, **32**, 1167–1176. [3,6,11,12].

Addelman, S., & Kempthorne, O. (1961b). Orthogonal Main-Effect Plans. Technical Report ARL 79, Aeronautical Research Lab., Wright-Patterson Air Force Base, Ohio, Nov. 1961. [9,11].

Adhikary, B., & Das, P. (1981). Construction and combinatorial properties of orthogonal arrays with variable number of symbols in rows. Pages 165–170 of: Bhaskara Rao, S. (ed), *Combinatorics and Graph Theory (Calcutta, 1980)*. New York: Springer-Verlag. Lecture Notes in Math. **885**.

Agaian, S. S. (1985). *Hadamard Matrices and Their Applications.* Springer Verlag. Lecture Notes in Math **1168**. [7].

Aggarwal, K. R., & Singh, T. P. (1981). Methods of construction of balanced arrays with application to factorial designs. *Calcutta Statist. Assoc. Bull.*, **30**(117-118), 89–93.

Agrawal, H. L. See Boob, B. S.

Agrawal, V. See also Dey, A.

Agrawal, V., & Dey, A. (1973). A note on orthogonal main-effect plans for asymmetrical factorials. *Sankhyā*, **B 44**, 278–282. [12].

Ahmad, M., Quddus, A., & Khokhar, I. (1992). Development of software for selection of orthogonal arrays: Application to oilfield scaling. Pages 341–348 of: *Proc. Third Islamic Countries Conf. Statistical Sci.*

Aigner, M. (1973). On the dual of tactical configurations and orthogonal arrays of index unity. *Sankhyā*, **A 35**, 221–228.

Albert, A. A. (1956). *Fundamental Concepts of Higher Algebra.* Chicago: University of Chicago Press. [A].

Assmus, Jr., E. F., & Key, J. D. (1992a). *Designs and Their Codes.* Cambridge Univ. Press. [5].

Assmus, Jr., E. F., & Key, J. D. (1992b). Hadamard matrices and their designs: a coding theoretic approach. *Trans. Amer. Math. Soc.*, **330**, 269–293. [7].

Atkinson, A. C., & Donev, A. N. (1992). *Optimum Experimental Designs.* Oxford Univ. Press. [7,12].

Atsumi, T. (1983). A study of orthogonal arrays from the point of view of design theory. *J. Combin. Theory*, **A 35**, 241–251. [10].

Baker, R.D., van Lint, J. H., & Wilson, R. M. (1983). On the Preparata and Goethals codes. *IEEE Trans. Inform. Theory*, **29**, 342–345. [5].

Banerjee, K. S. (1975). *Weighing Designs.* New York: Dekker. [7].

Bannai, E., & Ito, T. (1984). *Algebraic Combinatorics I: Association Schemes.* Menlo Park, CA: Benjamin/Cummings. [10].

Barg, A. (1998). Complexity issues in coding theory. Pages 649–754 of: *Pless and Huffman (1998).* [12].

Baumert, L. D., & McEliece, R. J. (1973). A note on the Griesmer bound. *IEEE Trans. Inform. Theory*, **19**, 134–135. [4].

Becker, R. A., Chambers, J. M., & Wilks, A. R. (1988). *The New S Language.* Pacific Grove, CA: Wadsworth.

Beckman, R. J. See McKay, M.D.

Beder, J. H. (1998). On Rao's inequalities for arrays of strength *d*. *Utilitas Math.*, **54**, 85–109. [2].

Bennett, C. H., Brassard, G., & Robert, J. M. (1986). How to reduce your enemy's information. *Lect. Notes Comput. Sci.*, **218**, 468–476. [10].

Berk, K. N., & Picard, R. R. (1991). Significance tests for saturated orthogonal arrays. *J. Qual. Techn.*, **23**, 79–89.

Berlekamp, E. R. (1968). *Algebraic Coding Theory*. New York: McGraw-Hill. [5].

Beth, T., Jungnickel, D., & Lenz, H. (1986). *Design Theory*. Cambridge Univ. Press. [6,7,8,10,12].

Bhagwandas. See Shrikhande, S. S.

Bhargava, V. K. See Gulliver, T. A.

Bierbrauer, J. See also Edel, Y.

Bierbrauer, J. (1993). Construction of orthogonal arrays. *J. Statist. Plann. Infer.*, **56**, 207–221.

Bierbrauer, J. (1995). Bounds on orthogonal arrays and resilient functions. *J. Combin. Des.*, **3**, 179–183. [10].

Bierbrauer, J., & Colbourn, C. J. (1996). Orthogonal arrays of strength more than two. Pages 179–182 of: Colbourn, C. J., & Dinitz, J. H. (eds), *The CRC Handbook of Combinatorial Designs*. CRC Press.

Bierbrauer, J., & Edel, Y. (1997). Extending and lengthening BCH-codes. *Finite Fields Applic.*, **3**, 314–333. [12].

Bierbrauer, J., Gopalakrishnan, K., & Stinson, D. R. (1996). Orthogonal arrays, resilient functions, error-correcting codes, and linear programming bounds. *SIAM J. Discrete Math.*, **9**, 424–452. [4, 10].

Bierbrauer, J., Gopalakrishnan, K., & Stinson, D. R. (1998). A note on the duality of linear programming bounds for orthogonal arrays and codes. *Bull. Inst. Combin. Applic.*, **22**, 17–24. [4].

Bisgaard, S. See Box, G. E. P.

Blake, I. F. See Safavi-Naini, R.

Blanchard, J. L. (1994). *The existence of orthogonal arrays of strength 3 with large order*. Preprint. [10].

Blanchard, J. L. (1995). A construction for orthogonal arrays with strength $t \geq 3$. *Discrete Math.*, **137**, 35–44. [10].

Blanchard, J. L. (1996). *The existence of orthogonal arrays of any strength with large order*. Preprint. [10].

Blokh, E. L., & Zyablov, V. V. (1974). Coding of generalized concatenated codes. *Problems of Inform. Transm.*, **10**(3), 218–222. [12].

Blokhuis, A. See also Bruen, A. A.

Blokhuis, A., Bruen, A. A., & Thas, J. A. (1990). Arcs in $PG(n,q)$, MDS-codes and three fundamental problems of B. Segre — some extensions. *Geom. Dedicata*, **35**, 1–11. [5].

Blum, J. R., Schatz, J. A., & Seiden, E. (1970). On 2-level orthogonal arrays of odd index. *J. Combin. Theory*, **9**, 239–243. [2].

Boob, B. S., & Agrawal, H. L. (1976). A note on the construction of mutually orthogonal Latin squares. *Biometrics*, **32**, 191–193.

Bose, R. C. (1938). On the application of the properties of Galois fields to the problem of construction of hyper-Graeco-Latin squares. *Sankhyā*, **3**, 323–338. [8].

Bose, R. C. (1947). Mathematical theory of the symmetrical factorial design. *Sankhyā*, **8**, 107–166. [5].

Bose, R. C. (1950a). A note on orthogonal arrays. *Ann. Math. Statist.*, **21**, 304–305.

Bose, R. C. (1950b). Mathematics of factorial designs. Pages 543–548 of: *Proc. Internat. Congress Math.* Vol. 1.

Bose, R. C. (1961). On some connections between the design of experiments and information theory. *Bull. Internat. Statist. Inst.*, **38**(4), 257–271. [4].

Bose, R. C., & Bush, K. A. (1952). Orthogonal arrays of strength two and three. *Ann. Math. Statist.*, **23**, 508–524. [2,3,6,5,12].

Bose, R. C., Chakravarti, I. M., & Knuth, D. E. (1960). On methods of constructing sets of mutually orthogonal Latin squares using a computer. *Technometrics*, **2**, 507–516.

Bose, R. C., Chakravarti, I. M., & Knuth, D. E. (1961). On methods of constructing sets of mutually orthogonal Latin squares using a computer: II. *Technometrics*, **3**, 111–117. (Errata: Ibid., **20** (1978), 219.) [8].

Bose, R. C., & Nair, K. R. (1939). Partially balanced incomplete block designs. *Sankhyā*, **4**, 337–373. [4].

Bose, R. C., & Ray-Chaudhuri, D. K. (1960a). On a class of error-correcting binary group codes. *Inform. Control*, **3**, 68–79. [5].

Bose, R. C., & Ray-Chaudhuri, D. K. (1960b). Further results on error correcting binary group codes. *Inform. Control*, **3**, 279–290. [5].

Bose, R. C., & Shrikhande, S. S. (1959). On the falsity of Euler's conjecture about the nonexistence of two orthogonal Latin squares of order $4t + 2$. *Proc. Nat. Acad. Sci. USA*, **45**, 734–737. [8].

Bose, R. C., & Shrikhande, S. S. (1960). On the construction of sets of mutually orthogonal Latin squares and the falsity of a conjecture of Euler. *Trans. Amer. Math. Soc.*, **95**, 191–209. [8].

Bose, R. C., Shrikhande, S. S., & Parker, E. T. (1960). Further results on the construction of mutually orthogonal Latin squares and the falsity of Euler's conjecture. *Canad. J. Math.*, **12**, 189–203. [8].

Bose, R. C., & Srivastava, J. N. (1964). On a bound useful in the theory of factorial designs and error correcting codes. *Ann. Math. Statist.*, **35**, 408–414. [5].

Box, G. E. P., Bisgaard, S., & Fung, C. A. (1988). An explanation and critique of Taguchi's contributions to quality engineering. *Quality and Reliability Engineering Internat.*, **4**, 123–131. [11].

Box, G. E. P., & Draper, N. R. (1987). *Empirical Model-Building and Response Surfaces.* New York: Wiley. [11].

Box, G. E. P., & Hunter, J. S. (1961a). The 2^{k-p} fractional factorial designs. Part I. *Technometrics*, **3**, 311–351. [11].

Box, G. E. P., & Hunter, J. S. (1961b). The 2^{k-p} fractional factorial designs. Part II. *Technometrics*, **3**, 449–458. [11].

Box, G. E. P., Hunter, W. G., & Hunter, J. S. (1978). *Statistics for Experimenters. An Introduction to Design, Data Analysis, and Model Building.* New York: Wiley. [11].

Box, G. E. P., & Meyer, R. D. (1986). Dispersion effects from fractional designs. *Technometrics*, **28**, 19–27. (Errata: Ibid. **29** (1987), 250.) [11].

Box, G. E. P., & Meyer, R. D. (1993). Finding the active factors in fractionated screening experiments. *J. Qual. Techn.*, **25**, 94–105.

Brassard, G. See Bennett, C. H.

Bremner, A. (1979). A Diophantine equation arising from tight 4-designs. *Osaka J. Math.*, **15**, 253–256. [2].

Brouwer, A. E. See also Abel, R. J. R.

Brouwer, A. E. (1979). The number of mutually orthogonal Latin squares — a table up to order 10000. Technical Report ZW 123/79, Mathematisch Centrum, Amsterdam. [8].

Brouwer, A. E. (1984). Four MOLS of order 10 with a hole of order 2. *J. Statist. Plann. Infer.*, **10**, 203–205. [8].

Brouwer, A. E. (1993). The linear programming bound for binary linear codes. *IEEE Trans. Inform. Theory*, **39**, 677–680. [4].

Brouwer, A. E. (1998). Bounds on the size of linear codes. Pages 295–461 of: *Pless and Huffman (1998)*. [4,5,12].

Brouwer, A. E. (1999). *Automated server for linear code bounds.* Published electronically at http://www.win.tue.nl/math/dw/voorlincod.html. [12].

Brouwer, A. E., Hämäläinen, H. O., Östergård, P. R. J., & Sloane, N. J. A. (1998). Bounds on mixed binary/ternary codes. *IEEE Trans. Inform. Theory*, **44**, 140–161. [9,12].

Brouwer, A. E., & van Rees, G. H. J. (1982). More mutually orthogonal Latin squares. *Discrete Math.*, **39**, 263–281.

Brouwer, A. E., & Verhoeff, T. (1993). An updated table of minimum distance bounds for binary linear codes. *IEEE Trans. Inform. Theory*, **39**, 662–677. [12].

Brown, J. W., Hedayat, A. S., & Parker, E. T. (1993). A pair of orthogonal Latin squares of order ten with four shared parallel transversals. *J. Combin. Inform. System Sci.*, **18**, 113–115. [8].

Bruck, R. H. (1963). Finite nets II. Uniqueness and embedding. *Pacific J. Math.*, **13**, 421–457. [8].

Bruen, A. A. See also Blokhuis, A.

Bruen, A. A., Thas, J. A., & Blokhuis, A. (1988). On M.D.S. codes, arcs in $PG(n, q)$ with q even, and a solution of three fundamental problems of B. Segre. *Inventiones math.*, **92**, 441–459. [5].

Buck, R. J. See Welch, W. J.

Buratti, M. (1998). Small quasimultiple affine and projective planes: some improved bounds. *J. Combin. Des.*, **6**, 337–345. [10].

Burman, J. P. See Plackett, R. L.

Bursztyn, D. See Steinberg, D. M.

Bush, K. A. See also Bose, R. C.

Bush, K. A. (1950). *Orthogonal arrays.* Ph.D. thesis, University of North Carolina, Chapel Hill. [1,3,5].

Bush, K. A. (1952a). A generalization of a theorem due to MacNeish. *Ann. Math. Statist.*, **23**, 293–295.

Bush, K. A. (1952b). Orthogonal arrays of index unity. *Ann. Math. Statist.*, **23**, 426–434. [2,3,4,5,12].

Butson, A. T. (1962). Generalized Hadamard matrices. *Proc. Amer. Math. Soc.*, **13**, 894–898. [6].

Butson, A. T. (1963). Relations among generalized Hadamard matrices, relative difference sets, and maximal length recurring sequences. *Canad. J. Math.*, **15**, 42–48. [6].

Calderbank, A. R. See also Hammons, Jr., A. R.

Calderbank, A. R., & Goethals, J.-M. (1984). Three-weight codes and association schemes. *Philips J. Res.*, **39**, 143–152. [10].

Calderbank, A. R., & Kantor, W. M. (1986). The geometry of two-weight codes. *Bull. London Math. Soc.*, **18**, 97–122. [5].

Cameron, P. J. (1973). Near-regularity conditions for designs. *Geom. Dedicata*, **2**, 213–223.

Cameron, P. J., & van Lint, J. H. (1991). *Designs, Graphs, Codes and their Links.* Cambridge, England: Cambridge Univ. Press. [4].

Camion, P., & Canteaut, A. (1996). Construction of t-resilient functions over a finite alphabet. Pages 283–293 of: *Advances in Cryptology, EUROCRYPT '96.* New York: Springer-Verlag. Lecture Notes in Comp. Sci. **1070**. [10].

Camion, P., Carlet, C., Charpin, P., & Sendrier, N. (1991). On correlation-immune functions. Pages 86–100 of: *Advances in Cryptology, CRYPTO '91.* New York: Springer-Verlag. Lecture Notes in Comp. Sci. **576**. [10].

Canteaut, A. See Camion, P.

Carlet, C. See Camion, P.

Carroll, R. J. See Davidian, M.

Casse, L. R. A. (1969). A solution to Beniamino Segre's "Problem $I_{r,q}$" for q even. *Atti Accad. Naz. Lincei, Rend. Cl. Sc. Fis. Mat. Natur.*, **46**, 13–20. [5].

Ceccherini, P. V., & Tallini, G. (1981). Codes, caps and linear spaces. Pages 72–80 of: Cameron, P. J., Hirschfeld, J. W. P., & Hughes, D. R. (eds), *Finite Geometries and Designs.* Cambridge, England: Cambridge Univ. Press. London Math. Soc. Lecture Note Series **49**. [5].

Chacko, A., Dey, A., & Ramakrishna, G. V. S. (1979). Orthogonal main-effect plans for asymmetrical factorials. *Technometrics*, **21**, 269–270. [12].

Chadjipantelis, T. See also Kounias, S.

Chadjipantelis, T., Kounias, S., & Moyssiadis, C. (1985). Construction of D-optimal designs for $N \equiv 2 \bmod 4$ using block-circulant matrices. *J. Combin. Theory*, **A 40**, 125–135. [7].

Chadjipantelis, T., Kounias, S., & Moyssiadis, C. (1987). The maximum determinant of 21×21 $(+1, -1)$-matrices and D-optimal designs. *J. Statist. Plann. Inf.*, **16**, 167–178. [7].

Chaffer, R. A. (1982). On constructions of orthogonal arrays and application to embedding theorems. *Discrete Math.*, **39**, 23–29.

Chakravarti, I. M. See also Bose, R. C., Morgan, J. P.

Chakravarti, I. M. (1956). Fractional replication in asymmetrical factorial designs and partially balanced arrays. *Sankhyā*, **17**, 143–164.

Chakravarti, I. M. (1961). On some methods of construction of partially balanced arrays. *Ann. Math. Statist.*, **32**, 1181–1185.

Chakravarti, I. M. (1963). Orthogonal and partially balanced arrays and their application in design of experiments. *Metrika*, **7**, 231–243.

Chakravarti, I. M. (1983). On Williams' balanced designs from Latin squares admitting certain sets of serial symmetries. Pages 97–105 of: Sen, P. K. (ed), *Contributions to Statistics: Essays in Honour of N. L. Johnson*. Amsterdam: North-Holland.

Chakravarti, I. M. (1990a). Families of codes with few distinct weights from singular and nonsingular Hermitian varieties and quadrics in projective geometries and Hadamard difference sets and designs associated with two-weight codes. Pages 35–50 of: Ray-Chaudhuri, D. K. (ed), *Coding Theory and Design Theory, Part I: Coding Theory*. New York: Springer Verlag.

Chakravarti, I. M. (1990b). Association schemes, orthogonal arrays and codes from nondegenerate quadrics and Hermitian varieties in finite projective geometries. *Calcutta Statist. Assoc. Bull.*, **40**(157–160), 89–96.

Chakravarty, R., & Dey, A. (1976). On the construction of balanced and orthogonal arrays. *Canad. J. Statist.*, **4**, 109–117. [2].

Chambers, J. M. See also Becker, R. A.

Chambers, J. M., & Hastie, T. J. (1992). *Statistical Models in S*. Pacific Grove, CA: Wadsworth.

Charpin, P. See Camion, P.

Chen, C. L. See Sloane, N. J. A.

Chen, H. See also Suen, C.-Y.

Chen, H., & Hedayat, A. S. (1996). 2^{n-l} designs with weak minimum aberration. *Ann. Statist.*, **24**, 2536–2548. [11].

Chen, J., & Lin, D. K. J. (1998). On the identifiability of a supersaturated design. *J. Statist. Plann. Inf.*, **72**, 99–107. [11].

Chen, J., Sun, D. X., & Wu, C. F. J. (1993). A catalogue of two-level and three-level fractional factorial designs with small runs. *Internat. Statist. Rev.*, **61**, 131–145.

Chen, M. H. See Wang, P. C.

Chen, V. C. P. (1995). Space filling measures for certain orthogonal array designs. Technical Report, Georgia Institute of Technology, School of Industrial and Systems Engineering, Georgia Institute of Technology, Atlanta. [11].

Chen, V. C. P. (1999). Application of orthogonal arrays and MARS to inventory forecasting stochastic dynamic programs. *Comput. Statist. Data Anal.* To appear. [11].

Chen, Y. See Sun, D. X.

Cheng, C.-S. (1980). Orthogonal arrays with variable numbers of symbols. *Ann. Statist.*, **8**, 447–453. [11].

Cheng, C.-S. (1989). Some orthogonal main-effect plans for asymmetrical factorials. *Technometrics*, **31**, 475–477. [12].

Cheng, C.-S. (1990). Construction of run orders of factorial designs. Pages 423–439 of: Ghosh, S. (ed), *Statistical Design and Analysis of Industrial Experiments*. New York: Dekker.

Cheng, C.-S. (1993). Neighbor-balanced orthogonal arrays. *J. Combin. Inform. System Sci.*, **18**, 61–67.

Cheng, C.-S. (1995). Some projection properties of orthogonal arrays. *Ann. Statist.*, **23**, 1223–1233.

Cheng, C.-S. (1997). $E(s^2)$-optimal supersaturated designs. *Statist. Sinica*, **7**, 929–939. [11,12].

Cheng, C.-S., & Mukerjee, R. (1998). Regular fractional factorial designs with minimum aberration and maximum estimation capacity. *Ann. Statist.*, **26**, 2289–2300.

Cheng, C.-S., & Steinberg, D. M. (1991). Trend robust two-level factorial designs. *Biometrika*, **78**, 325–336. [11].

Cheng, C.-S., Steinberg, D. M., & Sun, D. X. (1999). Minimum aberration and model robustness for two-level fractional factorial designs. *J. Royal Statist. Soc.*, **B 61**, 85–93. [11].

Cheng, Y., & Sloane, N. J. A. (1989). Codes from symmetry groups and a $[32, 17, 8]$ code. *SIAM J. Discrete Math.*, **2**, 28–37. [12].

Cheng, Y. W. See Abel, R. J. R.

Chipman, H., & Hamada, M. (1996). Bayesian analysis of ordered categorical data from industrial experiments. *Technometrics*, **38**, 1–10. [11].

Chopra, D. V. See also Srivastava, J. N.

Chopra, D. V. (1975a). Balanced optimal 2^8 fractional factorial designs of resolution V, $52 \leq N \leq 59$. Pages 91–100 of: Srivastava, J. N. (ed), *A Survey of Statistical Design and Linear Models*. Amsterdam: North-Holland.

Chopra, D. V. (1975b). Optimal balanced 2^8 fractional factorial designs of resolution V, with 60 to 65 runs. *Bull. Internat. Statist. Inst.*, **46**, 161–166.

Chopra, D. V. (1976). On the maximum number of constraints for some orthogonal arrays of strength 4. Pages 113–123 of: *Proc. 5th British Combinatorial Conf.* Winnipeg: Utilitas Math. Publ. Inc. [2].

Chopra, D. V. (1982). A note on balanced arrays of strength four. *Sankhyā*, **B 44**, 71–75.

Chopra, D. V. (1983a). A note on an upper bound for the constraints of balanced arrays with strength t. *Commun. Statist. Theory Meth.*, **12**, 1755–1759.

Chopra, D. V. (1983b). On the maximum constraints for some balanced arrays. *Gujarat Statist. Rev.*, **10**(1), 1–10.

Chopra, D. V., & Srivastava, J. N. (1973a). Optimal balanced 2^7 fractional factorial designs of resolution V, with $N \leq 42$. *Ann. Inst. Statist. Math.*, **25**, 587–604.

Chopra, D. V., & Srivastava, J. N. (1973b). Optimal balanced 2^7 fractional factorial designs of resolution V, $49 \leq N \leq 55$. *Commun. Statist.*, **2**, 59–84.

Chopra, D. V., & Srivastava, J. N. (1974). Optimal balanced 2^8 fractional factorial designs of resolution V, $37 \leq N \leq 51$. *Sankhyā*, **A 36**, 41–52.

Chopra, D. V., & Srivastava, J. N. (1975). Optimal balanced 2^7 fractional factorial designs of resolution V, $43 \leq N \leq 48$. *Sankhyā*, **B 37**, 429–447.

Chor, B., Goldreich, O., Håstad, J., Friedman, J., Rudich, S., & Smolensky, R. (1985). The bit extraction problem. Pages 396–407 of: *26-th IEEE Sympos. Foundations Comput. Sci.* [10].

Chowla, S., Erdös, P., & Straus, E. G. (1960). On the maximal number of pairwise orthogonal Latin squares of a given order. *Canad. J. Math.*, **12**, 204–208. [8].

Chvátal, V. (1983). *Linear Programming.* New York: W. H. Freeman. [4].

Cochran, W. G. See also Snedecor, G. W.

Cochran, W. G., & Cox, G. M. (1957). *Experimental Designs.* 2nd edn. New York: Wiley and Sons. [11].

Cohen, G., Honkala, I., Litsyn, S., & Lobstein, A. (1997). *Covering Codes.* Amsterdam: North-Holland. [11].

Colbourn, C. J. See also Abel, R. J. R., Bierbrauer, J.

Colbourn, C. J. (1996a). Transversal designs with block size eight and nine. *Europ. J. Combin.*, **17**, 1–14. [10].

Colbourn, C. J. (1996b). Orthogonal arrays of index more than one. Pages 172–178 of: Colbourn, C. J., & Dinitz, J. H. (eds), *The CRC Handbook of Combinatorial Designs.* CRC Press. [10].

Colbourn, C. J., & De Launey, W. (1996b). Difference matrices. Pages 287–297 of: Colbourn, C. J., & Dinitz, J. H. (eds), *The CRC Handbook of Combinatorial Designs.* CRC Press. [6].

Colbourn, C. J., & Dinitz, J. H. (eds). (1996a). *The CRC Handbook of Combinatorial Designs.* Boca Raton: CRC Press. [7,9,10].

Colbourn, C. J., & Dinitz, J. H. (1996b). Making the MOLS table. Pages 67–134 of: Wallis, W. D. (ed), *Computational and Constructive Design Theory.* Dordrecht: Kluwer.

Colbourn, C. J., Dinitz, J. H., & Stinson, D. R. (1999). Applications of Combinatorial Designs to Communications, Cryptography and Networking. In: *Proc. 17th British Combin. Conf.* To appear. [10].

Colbourn, C. J., & Kreher, D. L. (1996). Concerning difference matrices. *Designs, Codes, Crypt.*, **9**, 61–70. [6].

Collombier, D. (1996). Tableaux à fréquences marginales d'ordre 2 proportionnelles. *Linear Algebra Appl.*, **237/238**, 509–537.

Conover, W. J. See McKay, M. D.

Constantine, G. M. (1987). *Combinatorial Theory and Experimental Design.* New York: Wiley. [7].

Conti, F., Falkner, G., & Heise, W. (1981). Some nonexistence theorems on optimal codes. *Rend. Math.*, **1**, 341–346. [5].

Conway, J. H., & Sloane, N. J. A. (1998). *Sphere-Packings, Lattices and Groups.* 3rd edn. New York: Springer-Verlag. [5].

Cornell, J. A. See Khuri, A. I.

Coster, D. C. (1993). Trend-free run orders of mixed-level fractional factorial designs. *Ann. Statist.*, **21**, 2072–2086. [11].

Cotter, S. C. (1974). A general method of confounding for symmetrical factorial experiments. *J. Royal Statist. Soc.*, **B 36**, 267–276.

Cox, D. R. (1958). *Planning of Experiments.* New York: Wiley. [11].

Cox, G. M. See Cochran, W. G.

CPLEX. (1991). *CPLEX Organization, Inc.*, Incline Village, Nevada. [4].

Crampin, D. J., & Hilton, A. J. W. (1975). Remarks on Sade's disproof of the Euler conjecture with an application to Latin squares orthogonal to their transposes. *J. Combin. Theory*, **A 18**, 47–59. [8].

Currin, C., Mitchell, T. J., Morris, M. D., & Ylvisaker, D. (1991). Bayesian prediction of deterministic functions, with applications to the design and analysis of computer experiments. *J. Amer. Statist. Assoc.*, **86**, 953–963. [11].

Czitrom, V. (1989). Taguchi methods: Linear graphs of high resolution. *Commun. Statist. Theory Meth.*, **18**, 4583–4606.

Dalal, S. R., & Mallows, C. L. (1998). Factor-covering designs for testing software. *Technometrics*, **40**, 234–243. [11].

Das, M. N. See Jain, R. C.

Das, P. See Adhikary, B.

Davidian, M., & Carroll, R. J. (1987). Variance function estimation. *J. Amer. Statist. Assoc.*, **82**, 1079–1091. [11].

Dawson, J. (1985). A construction for the generalized Hadamard matrices $GF(4q; EA(q))$. *J. Statist. Plann. Inf.*, **11**, 103–110. [6].

De Launey, W. See also Colbourn, C. J.

De Launey, W. (1984). On the nonexistence of generalized Hadamard matrices. *J. Statist. Plann. Infer.*, **10**, 385–396.

De Launey, W. (1986). A survey of generalized Hadamard matrices and difference matrices $D(k, \lambda; G)$ with large k. *Utilitas Math.*, **30**, 5–29. [6].

De Launey, W. (1987). On difference matrices, transversal designs, resolvable transversal designs and large sets of mutually orthogonal F-squares. *J. Statist. Plann. Inf.*, **16**, 107–125. [6].

De Launey, W. (1989). GBRD's: Some new constructions for difference matrices, generalized Hadamard matrices and balanced generalized weighing matrices. *Graphs Combin.*, **5**, 125–135.

Delsarte, P. (1973). An algebraic approach to the association schemes of coding theory. *Philips Res. Reports Suppl.*, **10**. [4,10].

Delsarte, P., & Goethals, J.-M. (1975). Alternating bilinear forms over $GF(q)$. *J. Combin. Theory*, **A 19**, 26–50. [5].

Delsarte, P., & Levenshtein, V. I. (1998). Association schemes and coding theory. *IEEE Trans. Inform. Theory*, **44**, 2477–2504. [4].

Dembowski, P. (1968). *Finite Geometries.* New York: Springer-Verlag. [5].

Dénes, J., & Keedwell, A. D. (1974). *Latin squares and their applications.* New York: Academic Press. [2,8].

Dénes, J., & Keedwell, A. D. (eds). (1991). *Latin squares. New developments in the theory and applications.* Amsterdam: North-Holland. Annals of Discrete Math. **46**. [2,8].

Dey, A. See also Agrawal, V., Chacko, A., Chakravarty, R.

Dey, A. (1985). *Orthogonal Fractional Factorial Designs.* New York: Halsted Press. [9,11,12].

Dey, A. (1993). Some orthogonal arrays with variable symbols. *J. Combin. Inform. System Sci.*, **18**, 209–215.

Dey, A., & Agrawal, V. (1985). Orthogonal fractional plans for asymmetrical factorials derivable from orthogonal arrays. *Sankhyā*, **B 47**, 56–66.

Dey, A., & Midha, C. K. (1996). Construction of some asymmetrical orthogonal arrays. *Statist. Probab. Lett.*, **28**, 211–217. [9, 12].

Dey, A., & Ramakrishna, G. V. S. (1977). A note on orthogonal main-effect plans. *Technometrics*, **19**, 511–512. [12].

Diamond, W. J. (1981). *Practical Experimental Designs for Engineers and Scientists.* Belmont, CA: Lifetime Learning Publications. [7].

Dillon, J. F., Ferragut, E., & Gealy, M. (1999). *Linear frequency squares.* Preprint. [8].

Dinitz, J. H. See also Abel, R. J. R., Colbourn, C. J.

Dinitz, J. H., & Stinson, D. R. (eds). (1992). *Contemporary Design Theory: A Collection of Essays.* New York: Wiley. [8].

Djoković, D. Z. (1992). Construction of some new Hadamard matrices. *Bull. Austral. Math. Soc.* [7].

Djoković, D. Z. (1994a). Five new orders for Hadamard matrices of skew type. *Australas. J. Combin.* [7].

Djoković, D. Z. (1994b). Two Hadamard matrices of order 956 of Goethals-Seidel type. *Combinatorica*, **14**, 375–377. [7].

Dodunekov, S. M., & Manev, N. L. (1985). An improvement of the Griesmer bound for some small minimum distances. *Discrete Applied Math.*, **12**, 103–114. [4].

Donev, A. N. See Atkinson, A.C.

Doyen, J., & Rosa, A. (1980). An updated bibliography and survey of Steiner systems. *Ann. Discrete Math.*, **7**, 317–349. [12].

Drake, D. A. (1979). Partial λ-geometries and generalized Hadamard matrices over groups. *Canad. J. Math.*, **31**, 617–627. [6].

Draper, N. R. See Box, G. E. P.

Dube, S. N. See Sharma, H. C.

Duckworth, II, W. M. (1999). *Some binary maximin distance designs.* Preprint. [4].

Dulmage, A. L., Johnson, D. M., & Mendelsohn, N. S. (1961). Orthomorphisms of groups and orthogonal Latin squares: I. *Canad. J. Math.*, **13**, 356–372. [6].

Dunietz, I. S., Ehrlich, W. K., Iannino, A., Mallows, C. L., & Szablak, B. D. (1997). Applying design of experiments to software testing. Pages 205–215 of: *Proc. Internat. Conf. Software Engineering (ICSE '97).* Piscataway, NJ: IEEE Computer Society. [11].

Eades, P., & Hain, R. M. (1976). On circulant weighing matrices. *Ars Combin.*, **2**, 265–284. [7].

Eager, T. W. See Hunter, G. B.

Edel, Y. See also Bierbrauer, J.

Edel, Y., & Bierbrauer, J. (1998). Inverting construction Y_1. *IEEE Trans. Inform. Theory*, **44**, 1993. [12].

Ehlich, H. (1964a). Determinantenabschätzungen für Binäre Matrizen. *Math. Zeit*, **83**, 123–132. [7].

Ehlich, H. (1964b). Determinantenabschätzungen für Binäre Matrizen mit $n \equiv 3 \bmod 4$. *Math. Zeit*, **84**, 438–447. [7].

Ehlich, H., & Zeller, K. (1962). Binäre Matrizen. *Zeit. Angew. Math. Mech.*, **42**, 20–21. [7].

Ehrenfeld, S. (1955). On the efficiency of experimental designs. *Ann. Math. Statist.*, **26**, 247–255. [7].

Ehrlich, W. K. See Dunietz, I. S.

Engel, J. (1992). Modelling variation in industrial experiments. *Appl. Statist.*, **41**, 579–593. [11].

Engel, J., & Huele, A. F. (1996). A generalized linear modeling approach to robust design. *Technometrics*, **38**, 365–373. [11].

Erdélyi, A. (ed). (1953). *Higher Transcendental Functions.* New York: McGraw-Hill. 3 vols. [11].

Erdös, P. See Chowla, S.

Euler, L. (1782). Recherches sur une nouvelle espèce de quarrés magiques. *Verh. Zeeuwsch Gen. Wetensch. Vlissingen*, **9**, 85–239. [8].

Evans, A. B. (1993). Mutually orthogonal Latin squares based on linear groups. Pages 171–175 of: Jungnickel, D. et al. (ed), *Coding Theory, Design Theory, Group Theory.* New York: Wiley. [8].

Falkner, G. See also Conti, F.

Falkner, G., Kowol, B., Heise, W., & Zehendner, E. (1980). On the existence of cyclic optimal codes. *Atti Sem. Mat. Fis. Univ. Modena*, **28**, 326–341. [5].

Farmakis, N. See also Kounias, S.

Farmakis, N. (1991). Constructions of A-optimal weighing designs when $n = 19$. *J. Statist. Plann. Inf.*, **27**, 249–261. [7].

Federer, W. T. See also Mandeli, J. P., Raktoe, B. L.

Federer, W. T. (1977). On the existence and construction of a complete set of orthogonal $F(4t; 2t, 2t)$-squares designs. *Ann. Statist.*, **5**, 561–564. [8].

Federer, W. T., & Mandeli, J. P. (1986). Orthogonal F-rectangles, orthogonal arrays, and codes. *J. Combin. Theory*, **A 43**, 149–164.

Ferragut, E. See Dillon, J. F.

Finney, D. J. (1945). The fractional replication of factorial arrangements. *Ann. Eugen.*, **12**, 291–301. [3].

Finney, D. J. (1982). Some enumerations for the 6×6 Latin squares. *Utilitas Math.*, **21**, 137–153. [8, 12].

Fisher, R. A. (1942). The theory of confounding in factorial experiments in relation to the theory of groups. *Ann. Eugen.*, **11**, 341–353. [3,5].

Fisher, R. A. (1945). A system of confounding for factors with more than two alternatives giving completely orthogonal cubes and higher powers. *Ann. Eugen.*, **12**, 283–290.

Fisher, R. A., & Yates, F. (1963). *Statistical Tables for Biological, Agricultural and Medical Research.* 6th edn. New York: Hafner. [11].

Forney, Jr., G. D., Sloane, N. J. A., & Trott, M. D. (1993). The Nordstrom-Robinson code is the binary image of the Octacode. Pages 19–26 of: Calderbank, A. R., Forney, Jr., G. D., & Moayeri, N. (eds), *Proceedings DIMACS/IEEE Workshop on Coding and Quantization.* Amer. Math. Soc. [5,12].

Fourer, R., Gay, D. M., & Kernighan, B. W. (1993). *AMPL: A Modeling Language for Mathematical Programming.* San Francisco: Scientific Press. [4].

Franklin, M. F. (1984). Triples of almost orthogonal 10×10 Latin squares useful in experimental design. *Ars Combin.*, **17**, 141–146. [8].

Friedman, J. See also Chor, B.

Friedman, J. (1992). On the bit extraction problem. Pages 314–319 of: *Proc. 33rd IEEE Symp. Foundations Comp. Sci.* IEEE Press. [4,10].

Friedman, J. H. (1991). Multivariate adaptive regression splines. *Ann. Statist.*, **19**, 1–67. (Discussion: Ibid. **19**, 67–141.) [11].

Fuji-Hara, R. (1987). Some ideas for computer construction of orthogonal arrays. (Japanese). *Japan J. Appl. Statist.*, **16**, 141–149. [12].

Fuji-Hara, R., & Kamimura, S. (1993). Orthogonal arrays from Baer subplanes. *Utilitas Math.*, **43**, 65–70.

Fuji-Hara, R., Kuriki, S., & Miyake, M. (1996). Cyclic orthogonal and balanced arrays. *J. Statist. Plann. Inf.*, **56**, 171–180.

Fujii, Y. See also Namikawa, T., Yamamoto, S.

Fujii, Y. (1988). Nonexistence of a two-symbol orthogonal array of strength 8, 11 constraints and index 6. *TRU Math.*, **24**, 153–165.

Fujii, Y., Namikawa, T., & Yamamoto, S. (1987). On three-symbol orthogonal arrays. *Contributed papers, 46th Session ISI*, 131–132. [2,12].

Fujii, Y., Namikawa, T., & Yamamoto, S. (1988). Two-symbol orthogonal arrays of strength t and $t+3$ constraints. *TRU Math.*, **24**, 55–63.

Fujii, Y., Namikawa, T., & Yamamoto, S. (1989). Classification of two-symbol orthogonal arrays of strength t, $t+3$ constraints and index 4, II. *SUT J. Math.*, **25**, 161–177. [5].

Fung, C. A. See Box, G. E. P.

Gabidulin, E. M. (1985). Theory of codes with maximum rank distance (Russian). *Problemy Peredachi Informatsii*, **21**(1), 3–16. English translation in *Problems of Inform. Transmission*, **12** (1985), 1–12.

Gabidulin, E. M. (1992). Comments on 'maximum-rank array codes and their application to crisscross error correction'. *IEEE Trans. Inform. Theory*, **38**, 1183.

Galil, Z. See also Kiefer, J.

Galil, Z. (1985). Computing D-optimum weighing designs: where statistics, combinatorics, and computation meet. Pages 635–650 of: Le Cam, L. M., & Olshen, R. A. (eds), *Proceedings Berkeley Conference in Honor of Jerzy Neyman and Jack Kiefer*, vol. II. Monterey, CA: Wadsworth. [7].

Galil, Z., & Kiefer, J. (1980a). D-optimum weighing designs. *Ann. Statist.*, **8**, 1293–1306. [7].

Galil, Z., & Kiefer, J. (1980b). Time- and space-saving computer methods, related to Mitchell's DETMAX, for finding D-optimum designs. *Technometrics*, **22**, 301–313. [7].

Galil, Z., & Kiefer, J. (1980c). Optimum weighing designs. Pages 183–189 of: Matusita, K. (ed), *Recent Developments in Statistical Inference and Data Analysis*. Amsterdam: North-Holland. [7].

Galil, Z., & Kiefer, J. (1982a). On the characterization of D-optimum weighing designs for $n \equiv 3 \pmod 4$. Pages 1–35 of: Gupta, S. S., & Berger, J. O. (eds), *Statistical Decision Theory and Related Topics III*, vol. 1. New York: Academic Press. [7].

Galil, Z., & Kiefer, J. (1982b). Construction methods for D-optimum weighing designs when $n \equiv 3 \pmod 4$. *Ann. Statist.*, **10**, 502–510. [7].

Ganter, B., Mathon, R., & Rosa, A. (1978). A complete census of $(10,3,2)$-block designs and of Mendelsohn triple systems of order ten. I. Mendelsohn triple systems without repeated blocks. Pages 383–398 of: *Seventh Manitoba Conf. Numer. Math. Computing, Congress. Numer. XX.* Winnipeg: Utilitas Math. [8].

Ganter, B., Mathon, R., & Rosa, A. (1979). A complete census of $(10,3,2)$-block designs and of Mendelsohn triple systems of order ten. II. Mendelsohn triple systems with repeated blocks. Pages 181–204 of: *Eighth Manitoba Conf. Numer. Math. Computing, Congress. Numer. XXII.* Winnipeg: Utilitas Math. [8].

Gay, D. M. See Fourer, R.

Geally, M. See Dillon, J. F.

Georgiades, J. (1982). Cyclic $(q+1,k)$-codes of odd order q and even dimension k are not optimal. *Atti Sem. Mat. Fis. Univ. Modena*, **30**, 284–285. [5].

Georgiades, J., Heise, W., & Quattrochi, P. (1984). Cyclic MDS codes of length $n = 8$. *Atti Sem. Mat. Fis. Univ. Modena*, **32**, 82–87. [5].

Ghosh, S. See also Srivastava, J. N.

Ghosh, S., & Lagergren, E. S. (1990). Dispersion models and estimation of dispersion effects in replicated factorial experiments. *J. Statist. Plann. Inf.*, **26**, 253–262. [11].

Gilbert, E. N. (1952). A comparison of signalling alphabets. *Bell Syst. Tech. J.*, **31**, 504–522. [10].

Gill, P. S. (1986). Balanced incomplete arrays. *J. Statist. Plann. Inf.*, **14**, 179–185.

Godsil, C. D., & McKay, B. D. (1990). Asymptotic enumeration of Latin rectangles. *J. Combin. Theory*, **B 48**, 19–44. [8].

Goethals, J.-M. See also Calderbank, A. R., Delsarte, P.

Goethals, J.-M. (1974). Two dual families of nonlinear binary codes. *Elect. Lett.*, **10**, 471–472. [5].

Goethals, J.-M. (1976). Nonlinear codes defined by quadratic forms over $GF(2)$. *Inform. Control*, **31**, 43–74. [5].

Goethals, J.-M. (1979). Association schemes. Pages 243–283 of: Longo, G. (ed), *Algebraic Coding Theory and Applications*. New York: Springer-Verlag. CISM Courses and Lectures No. 258. [10].

Golay, M. J. E. (1949). Notes on digital coding. *Proc. IEEE*, **37**, 657. [4,5].

Goldreich, O. See Chor, B.

Gopalakrishnan, K. See Bierbrauer, J.

Goswami, K. K., & Pal, S. N. (1988). On nonexistence of minimal orthogonal arrays of strength two with varying numbers of symbols. *Indian Inst. Tech., Bombay*, 194–197.

Graybill, F. A. (1983). *Matrices with Applications in Statistics*. 2nd edn. Pacific Grove, CA: Wadsworth & Brooks/Cole. [11].

Griesmer, J. H. (1960). A bound for error-correcting codes. *IBM J. Res. Develop.*, **4**, 532–542. [4].

Gross, A. J. (1973). Some augmentations of Bose-Chaudhuri error correcting codes. Pages 209–216 of: Srivastava, J. N. (ed), *A Survey of Combinatorial Theory*. Amsterdam: North-Holland. [3,5].

Gulati, B. R. (1971a). On maximal (k,t) sets. *Ann. Inst. Statist. Math.*, **23**, 279–292. (Errata: Ibid., **23**, 527–529.).

Gulati, B. R. (1971b). Orthogonal arrays of strength five. *Trabajos Estadist. Investigacion Oper.*, **22**(3), 51–77. [2].

Gulati, B. R., & Kounias, E. G. (1970). On bounds useful in the theory of symmetrical factorial designs. *J. Royal Statist. Soc.*, **B 32**, 123–133. [5].

Gulati, B. R., & Kounias, E. G. (1973a). Maximal sets of points in finite projective spaces, no t linearly independent. *J. Combin. Theory*, **A 15**, 54–65. [5].

Gulati, B. R., & Kounias, E. G. (1973b). On three level symmetrical factorial designs and ternary group codes. *Sankhyā*, **A 35**, 377–392.

Gulliver, T. A., & Bhargava, V. K. (1992). Some best rate $1/p$ and rate $(p-1)/p$ systematic quasi-cyclic codes over $GF(3)$ and $GF(4)$. *IEEE Trans. Inform. Theory*, **38**, 1369–1374. [12].

Gulliver, T. A., & Östergård, P. (1998). Improvements to the bounds on quaternary linear codes of dimension 6. *Applic. Algebra in Eng., Commun. & Comput.*, **9**(2), 153–159. [5,12].

Gupta, V. K., & Nigam, A. K. (1987). Mixed orthogonal arrays for variance estimation with unequal numbers of primary selections per stratum. *Biometrika*, **74**, 735–742.

Hadamard, J. (1893). Résolution d'une question relative aux déterminants. *Bull. des Sciences Math.* (2), **17**, 240–246. [7].

Hain, R. M. See Eades, P.

Hall, Jr., M. (1961). Hadamard matrices of order 16. Research Summary No. 36–10, Jet Propulsion Lab., Pasadena, CA, Vol. 1, pp. 21–26. [7].

Hall, Jr., M. (1965). Hadamard matrices of order 20. Technical Report 32–761, Jet Propulsion Lab., Pasadena, CA. [7].

Hall, Jr., M. (1986). *Combinatorial Theory.* 2nd edn. New York: Wiley. [6,7,12].

Hamada, M. See also Chipman, H.

Hamada, M., & Wu, C. F. J. (1992). Analysis of designed experiments with complex aliasing. *J. Qual. Techn.*, **24**, 130–137. [11].

Hamada, M., & Wu, C. F. J. (1996). The treatment of related experimental factors by sliding levels. *J. Qual. Techn.*, **27**, 45–55. [11].

Hämäläinen, H. O. See Brouwer, A. E.

Hamming, R. W. (1950). Error-detecting and error correcting codes. *Bell Syst. Tech. J.*, **29**, 147–160. [3,4,5].

Hammons Jr., A. R., Kumar, P. V., Calderbank, A. R., Sloane, N. J. A., & Solé, P. (1994). The $\mathbb{Z}_4$-linearity of Kerdock, Preparata, Goethals and related codes. *IEEE Trans. Inform. Theory*, **40**, 301–319. [5,10,12].

Hanani, H. (1975). On transversal designs. Pages 43–53 of: Hall, Jr., M., & van Lint, J. H. (eds), *Combinatorics*. Dordrecht, Holland: Reidel. [10].

Hardin, R. H., & Sloane, N. J. A. (1992). Operating Manual for Gosset: A General-Purpose Program for Constructing Experimental Designs (Second Edition). Statistics Dept. Report 106, AT&T Bell Laboratories, Murray Hill, NJ. [12].

Hardin, R. H., & Sloane, N. J. A. (1993). A new approach to the construction of optimal designs. *J. Statist. Plann. Infer.*, **37**, 339–369. [7,12].

Hartley, H. O. See Pearson, E. S.

Harwit, M. See also Sloane, N. J. A.

Harwit, M., & Sloane, N. J. A. (1979). *Hadamard Transform Optics.* New York: Academic Press. [7].

Hashim, A. A., & Pozdniakov, V. S. (1976). Computerized search for linear binary codes. *Elect. Lett.*, **12**, 350–351. [12].

Håstad, J. See Chor, B.

Hastie, T. J. See Chambers, J. M.

Hedayat, A. S. See also Brown, J. W., Chen, H., Raktoe, B. L.

Hedayat, A. S. (1971). A set of three mutually orthogonal Latin squares of order 15. *Technometrics*, **13**, 696–698. [8].

Hedayat, A. S. (1990). New properties of orthogonal arrays and their statistical applications. Pages 407–422 of: Ghosh, S. (ed), *Statistical Design and Analysis of Industrial Experiments*. New York: Dekker.

Hedayat, A. S., Pu, K., & Stufken, J. (1992). On the construction of asymmetrical orthogonal arrays. *Ann. Statist.*, **20**, 2142–2152. [9,12].

Hedayat, A. S., Raghavarao, D., & Seiden, E. (1975). Further contributions to the theory of F-squares design. *Ann. Statist.*, **3**, 712–716. [8].

Hedayat, A. S., Seiden, E., & Stufken, J. (1997). On the maximal number of factors and the enumeration of 3-symbol orthogonal arrays of strength 3 and index 2. *J. Statist. Plann. Infer*, **58**, 43–63. [2,5,12].

Hedayat, A. S., & Shrikhande, S. S. (1971). Experimental designs and combinatorial systems associated with Latin squares and sets of mutually orthogonal Latin squares. *Sankhyā*, **A 33**, 423–432. [8].

Hedayat, A. S., & Stufken, J. (1986). Fractional factorial designs in the form of incomplete orthogonal arrays. Pages 101–115 of: McCulloch, C.E., *et al.* (eds), *Statistical Design: Theory and Practice*. Ithaca: Cornell University. [8].

Hedayat, A. S., & Stufken, J. (1988). Two-symbol orthogonal arrays. Pages 47–58 of: Y. Dodge, V. V. Fedorov, & Wynn, H. P. (eds), *Optimal Design and Analysis of Experiments*. Amsterdam: North-Holland. [2].

Hedayat, A. S., & Stufken, J. (1989a). On the maximum number of constraints in orthogonal arrays. *Ann. Statist.*, **17**, 448–451. [2].

Hedayat, A. S., & Stufken, J. (1989b). On the maximum number of factors in two construction methods for orthogonal arrays. Pages 33–40 of: Dodge, Y. (ed), *Statistical Data Analysis and Inference*. Amsterdam: North-Holland.

Hedayat, A. S., & Stufken, J. (1992). Some mathematical results on incomplete orthogonal arrays. *Sankhyā, Special Volume*, **A 54**, 197–202. [8].

Hedayat, A. S., & Stufken, J. (1999). Compound orthogonal arrays. *Technometrics*, **41**, 57–61. [10,11].

Hedayat, A. S., Stufken, J., & Su, G. (1996). On difference schemes and orthogonal arrays of strength t. *J. Statist. Plann. Inf.*, **56**, 307–324. [6].

Hedayat, A. S., Stufken, J., & Su, G. (1997). On the construction and existence of orthogonal arrays with three levels and indexes 1 and 2. *Ann. Statist.*, **25**, 2044–2053. [2,12].

Hedayat, A. S., & Wallis, W. D. (1978). Hadamard matrices and their applications. *Ann. Statist.*, **6**, 1184–1238. [7].

Heise, W. See also Conti, F., Falkner, G., Georgiades, J.

Heise, W. (1978). Es gibt keinen optimalen $(n+2,3)$-code einer ungeraden Ordnung n. *Math. Zeit.*, **164**, 67–68. [5].

Heise, W., & Kowol, B. (1980). Il n'y a pas de $[q+1, 2]$ MDS code cyclique d'ordre impair q. *J. Combin. Theory*, **A 29**, 243. [5].

Helgert, H. J., & Stinaff, R. D. (1973). Minimum-distance bounds for binary linear codes. *IEEE Trans. Inform. Theory*, **19**, 344–356. [12].

Hergert, F. B. (1990). On the Delsarte-Goethals codes and their formal duals. *Discrete Math.*, **83**, 249–263. [5].

Hill, R. (1973). On the largest size of cap in $S_{5,3}$. *Rend. Accad. Naz. Lincei*, **54**, 378–384. [5,12].

Hill, R. (1976). Caps and groups. Pages 384–394 of: Segre, B., *et al.* (eds), *Colloquio Internazionale sulle Teorie Combinatorie (Roma 1973)*. Atti dei Convegni Lincei, Vol. **17**. [5].

Hill, R. (1978a). Caps and codes. *Discrete Math.*, **22**, 111–137. [5].

Hill, R. (1978b). Some results concerning linear codes and $(k, 3)$-caps in three-dimensional Galois space. *Math. Proc. Camb. Phil. Soc.*, **84**, 191–205. [5].

Hill, R., & Newton, D. E. (1988). Some optimal ternary linear codes. *Ars Combin.*, **25 A**, 61–72. [12].

Hilton, A. J. W. See Crampin, D. J.

Hinkelmann, K., & Kempthorne, O. (1994). *Design and Analysis of Experiments. Volume I: Introduction to Experimental Design.* New York: Wiley. [8,11].

Hirschfeld, J. W. P. (1979). *Projective Geometries over Finite Fields.* Oxford Univ. Press. [5].

Hirschfeld, J. W. P. (1983). Maximal sets in finite projective spaces. Pages 55–76 of: Lloyd, E. K. (ed), *Surveys in Combinatorics.* Cambridge Univ. Press. Lecture Notes in Math. **82**. [5].

Hirschfeld, J. W. P. (1985). *Finite Projective Spaces of Three Dimensions.* Oxford Univ. Press. [5].

Hirschfeld, J. W. P., & Storme, L. (1998). The packing problem in statistics, coding theory and finite projective spaces. *J. Statist. Plann. Inf.*, **72**, 355–380. [5].

Hocquenghem, A. (1959). Codes correcteurs d'erreurs. *Chiffres*, **2**, 147–156. [5].

Hodi, F. S. See Hunter, G. B.

Honkala, I. See Cohen, G.

Horn, R. A., & Johnson, C. R. (1985). *Matrix Analysis.* Cambridge University Press. [7].

Hotelling, H. (1944). Some improvements in weighing and other experimental techniques. *Ann. Math. Statist.*, **15**, 297–306. [7].

Huele, A. F. See Engel, J.

Huffman, W. C. See Pless, V. S.

Hughes, D. R., & Piper, F. C. (1973). *Projective Planes.* New York: Springer-Verlag. [8].

Hughes, D. R., & Piper, F. C. (1985). *Design Theory.* Cambridge Univ. Press. [7].

Hunter, G. B., Hodi, F. S., & Eager, T. W. (1982). High-cycle fatigue of weld repaired cast Ti-6A1-4V. *Metallurgical Transactions*, **13A**, 1589–1594. [11].

Hunter, J. S. See Box, G. E. P.

Hunter, W. G. See Box, G. E. P.

Hyodo, Y. See Yamamoto, S.

Iannino, A. See Dunietz, I. S.

Ito, N., Leon, J. S., & Longyear, J. Q. (1981). Classification of 3-$(24, 12, 5)$ designs and 24-dimensional Hadamard matrices. *J. Combin. Theory*, **A 31**, 66–93. [7].

Ito, T. See Bannai, E.

Itoh, H., & Nakamichi, M. (1983). SbEC-DbED codes derived from experiments on a computer for semiconductor memory systems. *Electronics Comm. Japan*, **66-A**, 26–35. [12].

Jacobson, M. T., & Matthews, P. (1996). Generating uniformly distributed random Latin squares. *J. Combin. Des.*, **4**, 405–437. [8].

Jacobson, N. (1964). *Lectures in Abstract Algebra*. New York: Van Nostrand. 3 vols. [A].

Jacroux, M. (1990). Methods for constructing trend-resistant run orders of 2-level factorial experiments. Pages 441–457 of: Ghosh, S. (ed), *Statistical Design and Analysis of Industrial Experiments*. New York: Dekker.

Jacroux, M., Majumdar, D., & Shah, K. R. (1995). Efficient block designs in the presence of trends. *Statist. Sinica*, **5**, 605–615. [6].

Jacroux, M., & Notz, W. (1983). On the optimality of spring balance weighing designs. *Ann. Statist.*, **11**, 970–978. [7].

Jaffe, D. B. (1999). Optimal binary linear codes of length ≤ 30. *Discrete Math.* To appear. [12].

Jain, R. C., & Das, M. N. (1991). Efficiency balanced block designs using orthogonal arrays with variable symbols. *Calcutta Statist. Assoc. Bull.*, **41**, 173–178.

Jan, H. W. See Wang, P. C.

Jiang, S. (1979). A simple construction of a $2p \times 2p$ difference scheme with modulus p where p is an arbitrary odd prime. (Chinese). *Acta Math. Appl. Sinica*, **2**, 75–80. [6].

John, P. W. M. (1971). *Statistical Design and Analysis of Experiments*. New York: Macmillan. [11].

John, P. W. M. (1985). Non-isomorphism of two Latin square designs for sixteen varieties. *Commun. Statist. Theory Meth.*, **14**, 117–121. [8].

Johnson, C. R. See Horn, R. A.

Johnson, J. M. See Dulmage, A. L.

Johnson, M. E., Moore, L. M., & Ylvisaker, D. (1990). Minimax and maximin distance designs. *J. Statist. Plann. Inf.*, **26**, 131–148. [4,11].

Jungnickel, D. See also Beth, T.

Jungnickel, D. (1979). On difference matrices, resolvable transversal designs and generalized Hadamard matrices. *Math. Z.*, **167**, 49–60. [6].

Jungnickel, D. (1990). Latin squares, their geometries and their groups. A survey. Pages 166–225 of: Ray-Chaudhuri, D. (ed), *Coding Theory and Design Theory, Part II: Design Theory.* New York: Springer-Verlag. [8].

Jungnickel, D. (1992). Difference sets. Pages 241–324 of: *Dinitz and Stinson (1992).* [6,12].

Jurick, R. R. (1968). An algorithm for determining the largest maximally independent set of vectors from an r-dimensional vector space over a Galois field of n elements. Technical Report ASD-TR-68-40, Wright State Air Force Base, Ohio.

Kageyama, S. See also Mukerjee, R.

Kageyama, S. (1988). On orthogonal arrays attaining Rao's bounds. *J. Statist. Plann. Infer.*, **19**, 395–400. (Errata: Ibid. **24** (1990), 135.) [2].

Kamimura, S. See Fuji-Hara, R.

Kaneta, H. See Maruta, T.

Kang, S. M. See Welch, W. J.

Kantor, W. M. See also Calderbank, A. R.

Kantor, W. M. (1969). Automorphism groups of Hadamard matrices. *J. Combin. Theory*, **6**, 279–281. [7].

Karlin, M. (1969). New binary coding results by circulants. *IEEE Trans. Inform. Theory*, **15**, 81–92. [12].

Karson, M. J. See Mullett, G. M.

Keedwell, A. D. See Dénes, J.

Kempthorne, O. See also Addelman, S., Hinkelmann, K.

Kempthorne, O. (1947). A simple approach to confounding and fractional replication in factorial experiments. *Biometrika*, **34**, 252–272. [3,4].

Keny, R. See Lee, N. S.

Kerdock, A. M. (1972). A class of low-rate binary codes. *Inform. Control*, **20**, 182–187. [5].

Kernighan, B. W. See Fourer, R.

Key, J. D. See Assmus, Jr., E. J.

Khokhar, I. See Ahmad, M.

Khuri, A. I. See also Myers, R. H.

Khuri, A. I., & Cornell, J. A. (1987). *Response Surfaces: Designs and Analyses.* New York: Dekker. [7].

Kiefer, J. See also Galil, Z.

Kiefer, J. (1959). Optimum experimental designs. *J. Royal Statist. Soc.*, **B 21**, 272–319. [7].

Kiefer, J., & Galil, Z. (1980). Optimum weighing designs. Pages 183–189 of: Matusita, K. (ed), *Recent Developments in Statistical Inference and Data Analysis.* Amsterdam: North-Holland. [7].

Kikumasa, I. See Maruta, T.

Kimura, H. (1986). Hadamard matrices of order 28 with automorphism groups of order two. *J. Combin. Theory*, **A 43**, 98–102. [7].

Kimura, H. (1989). New Hadamard matrix of order 24. *Graphs Combin.*, **5**, 235–242. [7].

Kimura, H. (1994a). Classification of Hadamard matrices of order 28 with Hall sets. *Discrete Math.*, **128**, 257–268. [7].

Kimura, H. (1994b). Classification of Hadamard matrices of order 28. *Discrete Math.*, **133**, 171–180. [7].

Kimura, H., & Ohmori, H. (1986). Construction of Hadamard matrices of order 28. *Graphs Combin.*, **2**, 247–257. [7].

Knuth, D. E. See Bose, R. C.

Koehler, J. R., & Owen, A. B. (1996). Computer experiments. Pages 261–308 of: Ghosh, S., & Rao, C. R. (eds), *Handbook of Statistics*, vol. 13. Amsterdam: Elsevier.

Kolesova, G., Lam, C. W. H., & Thiel, L. (1990). On the number of 8×8 Latin squares. *J. Combin. Theory*, **A 54**, 143–148. [8].

Korchmáros, G. (1983). New examples of complete k-arcs in $PG(2, q)$. *Europ. J. Combin.*, **4**, 329–334. [5].

Kounias, E. G. See Gulati, B. R.

Kounias, S. See also Chadjipantelis, T., Moyssiadis, C.

Kounias, S., & Chadjipantelis, T. (1983). Some D-optimal weighing designs for $n \equiv 3 \pmod 4$. *J. Statist. Plann. Inf.*, **8**, 117–127. [7].

Kounias, S., & Farmakis, N. (1984). A construction of D-optimal weighing designs when $n \equiv 3 \pmod 4$. *J. Statist. Plann. Inf.*, **10**, 177–187. [7].

Kounias, S., & Petros, C. I. (1975). Orthogonal arrays of strength three and four with index unity. *Sankhyā*, **B 37**, 228–240. [2].

Kowol, B. See Falkner, G., Heise, W.

Kreher, D. L. See also Colbourn, C. J.

Kreher, D. L. (1990). Design theory toolchest — user manual and report. Pages 226–235 of: Ray-Chaudhuri, D. (ed), *Coding Theory and Design Theory, Part II: Design Theory*. New York: Springer-Verlag. [6].

Kreher, D. L., & Radziszowski, S. P. (1986). Finding simple t-designs by basis reduction. *Congress. Numer.*, **55**, 235–244. [6].

Kreher, D. L., & Radziszowski, S. P. (1987). New t-designs found by basis reduction. *Congress. Numer.*, **59**, 155–164. [6,12].

Kreher, D. L., & Radziszowski, S. P. (1990). Constructing $6 - (14, 7, 6)$ designs. *Contemp. Math.*, **111**, 137–151. [6,12].

Kreher, D. L., & Stinson, D. R. (1998). *Combinatorial Algorithms : Generation, Enumeration and Search*. Boca Raton: CRC Press. [6,12].

Kschischang, F. R., & Pasupathy, R. (1992). Some ternary and quaternary codes and associated sphere packings. *IEEE Trans. Inform. Theory*, **38**, 227–246. [5,12].

Kumar, P. V. See Hammons, Jr., A. R.

Kuriki, S. See also Fuji-Hara, R., Sato, M., Yamamoto, S.

Kuriki, S. (1993). On existence and construction of balanced arrays. *Discrete Math.*, **116**, 137–155.

Kuwada, M. See Yamamoto, S.

Lagergren, E. S. See Ghosh, S.

Laihonen, T. (1998). *Estimates on the covering radius when the dual distance is known.* Ph.D. thesis, Turku Centre for Computer Science, Turku, Finland. [4].

Laihonen, T., & Litsyn, S. (1998). On upper bounds for minimum distance and covering radius of non-binary codes. *Designs, Codes, Crypt.*, **14**, 71–80. [4].

Laihonen, T., & Litsyn, S. (1999). New bounds on covering radius as a function of dual distance. *SIAM J. Discrete Math.* To appear. [4].

Lam, C. W. H. See also Kolesova, G.

Lam, C. W. H., Thiel, L., & Swiercz, S. (1989). The non-existence of finite projective planes of order 10. *Canad. J. Math.*, **41**, 1117–1123. [8].

Lam, C. W. H., & Tonchev, V. D. (1996). Classification of affine resolvable $2-(27,9,4)$ designs. *J. Statist. Plann. Inf.*, **56**, 187–202. [5].

Laywine, C. F. (1989). A geometric construction for sets of mutually orthogonal frequency squares. *Utilitas Math.*, **35**, 95–102.

Laywine, C. F., & Mullen, G. L. (1990). Mutually orthogonal frequency squares with nonconstant frequency vectors. *Ars Combin.*, **29**, 259–264.

Laywine, C. F., & Mullen, G. L. (1992). Generalizations of Bose's equivalence between complete sets of mutually orthogonal Latin squares and affine planes. *J. Combin. Theory*, **A 61**, 13–35.

Laywine, C. F., & Mullen, G. L. (1993). Mutually orthogonal frequency hypercubes and affine geometries. Pages 183–193 of: Jungnickel, D. *et al.* (ed), *Coding Theory, Design Theory, Group Theory.* New York: Wiley.

Laywine, C. F., & Mullen, G. L. (1998). *Discrete Mathematics Using Latin Squares.* New York: Wiley. [8].

Laywine, C. F., Mullin, G. L., & Suchower, S. J. (1990). Orthogonal frequency squares of type $F(4t;t)$. *Utilitas Math.*, **37**, 207–213.

Lecointe, P. (1977). Algèbre des tableaux orthogonaux. *Comptes Rendus Acad. Sciences Paris*, **284**, 199–202 and 241–244.

Lee, F.-C. H. See Mandeli, J. P.

Lee, N. S., Phadke, M. S., & Keny, R. (1987). An expert system for experimental design: Automating the design of orthogonal array experiments. *ASQC Techn. Confer. Trans.*, 270–277.

Lempel, A. See Roth, R. M.

Lenz, H. See Beth, T.

Leon, J. S. See also Ito, N.

Leon, J. S. (1979). An algorithm for computing the automorphism group of a Hadamard matrix. *J. Combin. Theory*, **A 27**, 289–306. [7].

Levenshtein, V. L. See also P. Delsarte.

Levenshtein, V. L. (1992). Designs as maximum codes in polynomial metric spaces. *Acta Applicandae Mathematicae*, **29**, 1–82.

Levenshtein, V. L. (1995). Krawtchouk polynomials and universal bounds for codes and designs in Hamming spaces. *IEEE Trans. Inform. Theory*, **41**, 1303–1321. [4].

Levenshtein, V. L. (1997). Split orthogonal arrays and maximum independent resilient systems of functions. *Designs, Codes, Combin.*, **12**, 131–160.

Levenshtein, V. L. (1998). Universal bounds for codes and designs. Pages 499–648 of: *Pless and Huffman (1998)*. [4].

Levenshtein, V. L. (1999). Equivalence of Delsarte's bounds for codes and designs in symmetric association schemes, and some applications. *Discrete Math.*, **197/198**, 515–536. [4].

Lidl, R., & Niederreiter, H. (1983). *Finite Fields.* Reading, MA: Addison-Wesley. [A].

Lidl, R., & Niederreiter, H. (1986). *Introduction to Finite Fields and Their Applications.* Cambridge Univ. Press. [A].

Lin, D. K. J. See also Chen, J.

Lin, D. K. J. (1993). A new class of supersaturated designs. *Technometrics*, **35**, 28–31. [11].

Lin, W.-C., & Stufken, J. (1999). Varietal trials in the presence of trends. *J. Combin. Inform. System Sci.* To appear. [11].

Lindner, C. C. (1983). Quasigroup identities and orthogonal arrays. Pages 77–105 of: Lloyd, E. K. (ed), *Surveys in Combinatorics.* Cambridge Univ. Press. Lecture Notes Math., **82**.

Lindner, C. C., Mullin, R. C., & van Rees, G. H. J. (1987). Separable orthogonal arrays. *Utilitas Math.*, **31**, 25–32.

van Lint, J. H. See also Baker, R. D., Cameron, P. J.

van Lint, J. H. (1998). *Introduction to Coding Theory.* 3rd edn. New York: Springer-Verlag. [4,10].

Litsyn, S. See also Cohen, G., Laihonen, T.

Litsyn, S. (1998). An updated table of the best binary codes known. Pages 463–498 of: *Pless and Huffman (1998)*. [12].

Litsyn, S., Rains, E. M., & Sloane, N. J. A. (1999). *Table of nonlinear binary codes.* http://akalice.research.att.com/~njas/codes/And/. [12].

Liu, M.-S., & Yao, J.-S. (1989). Some orthogonal main-effect plans for asymmetrical factorials derivable from orthogonal arrays. *Chinese J. Math.*, **17**, 9–27.

Liu, Z.-W. (1977a). A difference scheme $D(14, 14, 7, 2)$ generating the orthogonal array $OA(98, 15, 7, 2)$. (Chinese). *Kexue Tongbao*, **22**, 31–32. [6].

Liu, Z.-W. (1977b). Construction of a class of difference schemes generating orthogonal arrays $(2p^2, 2p + 1, p, 2)$ for odd primes p. (Chinese). *Yingyong Shuxue Xuebao*, **3**, 35–45. [6].

Lloyd, S. P. (1957). Binary block coding. *Bell Syst. Tech. J.*, **36**, 517–535. [4].

Lobstein, A. See Cohen, G.

Longyear, J. Q. See Ito, N.

Lu, Y. See Zhang, Y.

Ma, F. S. See Wu, C.-F. J.

Mackenzie, C., & Seberry, J. (1984). Maximal ternary codes and Plotkin's bound. *Ars Combin.*, **17 A**, 251–270. [4].

MacNeish, H. F. (1922). Euler squares. *Ann. Math.*, **23**, 221–227. [8].

MacWilliams, F. J., Odlyzko, A. M., Sloane, N. J. A., & Ward, H. N. (1978). Self-dual codes over $GF(4)$. *J. Combin. Theory*, **A 25**, 288–318. [5].

MacWilliams, F. J., & Sloane, N. J. A. (1977). *The Theory of Error-Correcting Codes.* Amsterdam: North-Holland. [2,3,4,5,10,12].

MacWilliams, F. J., Sloane, N. J. A., & Thompson, J. G. (1973). On the existence of a projective plane of order 10. *J. Combin. Theory*, **A 14**, 66–78. [8].

Majumdar, D. See Jacroux, M.

Majumdar, K. N. (1953). On some theorems in combinatorics relating to incomplete block designs. *Ann. Math. Statist.*, **24**, 377–389.

Mallows, C. L. See also Dalal, S. R., Dunietz, I. S.

Mallows, C. L. (1997). Covering designs in random environments. Pages 235–246 of: Brillinger, D. R., Fernholz, L. T., & Morgenthaler, S. (eds), *The Practice of Data Analysis: Essays in Honor of John W. Tukey.* Princeton University Press. [11].

Mallows, C. L., Pless, V. S., & Sloane, N. J. A. (1976). Self-dual codes over $GF(3)$. *SIAM J. Appl. Math.*, **31**, 649–666. [5,12].

Mandeli, J. P. See also Federer, W. T.

Mandeli, J. P. (1995). Construction of asymmetrical orthogonal arrays having factors with a large non-prime power number of levels. *J. Statist. Plann. Inf.*, **47**, 377–391. [12].

Mandeli, J. P., Lee, F.-C. H., & Federer, W. T. (1981). On the construction of orthogonal F-squares of order n from an orthogonal array $(n, k, s, 2)$ and an $OL(s, t)$ set. *J. Statist. Plann. Inf.*, **5**, 267–272. [8].

Mandl, R. (1985). Orthogonal Latin squares: An application of experiment design to compiler testing. *Commun. ACM*, **28**, 1054–1058. [11].

Maneri, C., & Silverman, R. (1966). A vector-space packing problem. *J. Algebra*, **4**, 321–330.

Maneri, C., & Silverman, R. (1971). A combinatorial problem with applications to geometry. *J. Combin. Theory*, **A 11**, 118–121.

Manev, N. L. See Dodunekov, S. M.

Mao, S. S. See Wu, C.-F. J.

Martin, W. J. (1998). Mixed block designs. *J. Combin. Des.*, **6**, 151–163.

Martin, W. J. (1999). *Linear programming bounds for ordered orthogonal arrays and (T, M, S)-nets.* Preprint.

Martin, W. J., & Stinson, D. R. (1997a). *Generalized Rao bound for ordered orthogonal arrays and (T, M, S)-nets.* Preprint.

Martin, W. J., & Stinson, D. R. (1997b). *Association schemes for ordered orthogonal arrays and (T, M, S)-nets.* Preprint.

Maruta, K. (1990). On Singleton arrays and Cauchy matrices. *Discrete Math.*, **81**, 33–36. [5].

Maruta, K., Kikumasa, I., & Kaneta, H. (1988). Singleton arrays in coding theory. *Bull. Austral. Math. Soc.*, **37**, 333–335. [5].

Masuyama, M. (1957). On difference sets for constructing orthogonal arrays of index two and of strength two. *Rep. Statist. Appl. Res., JUSE*, **5**, 27–34.

Masuyama, M. (1959). On cyclic difference sets which generate orthogonal arrays. *Rep. Statist. Appl. Res., JUSE*, **6**, 4–10.

Masuyama, M. (1960). On the existence theorem of the cyclic difference sets for the orthogonal arrays $(9\lambda, 3\lambda + 1, 3, 2)$. *Rep. Statist. Appl. Res., JUSE*, **7**, 1–3.

Masuyama, M. (1961). On cyclic difference sets which generate orthogonal arrays II. *Rep. Statist. Appl. Res., JUSE*, **8**, 20–26.

Masuyama, M. (1969a). Construction of difference sets for $OA(2p^2, 2p+1, p, 2)$, p being an odd prime. *Rep. Statist. Appl. Res., JUSE*, **16**, 1–9.

Masuyama, M. (1969b). Cyclic generation of $OA(2s^m, 2(s^m - 1)/(s - 1) - 1, s, 2)$. *Rep. Statist. Appl. Res., JUSE*, **16**, 10–16. [6].

Mathon, R. See also Ganter, B.

Mathon, R., & Rosa, A. (1985). Tables of parameters of BIBDs with $r \leq 41$ including existence, enumeration and resolvability results. *Ann. Discrete Math.*, **26**, 275–308. [12].

Matthews, P. See M. T. Jacobson.

Maurin, F. (1985). On incomplete orthogonal arrays. *J. Combin. Theory*, **A 40**, 183–185. [8].

Maurin, F. (1996). Incomplete orthogonal arrays and idempotent orthogonal arrays. *Graphs Combin.*, **12**, 253–266. [8].

Mayer, R. See Abt, M.

McCarthy, P. J. (1969). Pseudo-replication: half samples. *Review Internat. Statist. Inst.*, **37**, 239–263. [7,11].

McCoy, N. H. 1948. *Rings and Fields.* Washington, D.C.: Math. Assoc. America. [A].

McCullagh, P., & Pregibon, D. (1987). k-Statistics and dispersion effects in regression. *Ann. Statist.*, **15**, 202–219. [11].

McEliece, R. J. See also Baumert, L. D.

McEliece, R. J. (1977). *The Theory of Information and Coding.* Reading, MA: Addison-Wesley. [4].

McEliece, R. J., Rodemich, E. R., Rumsey Jr., H. C., & Welch, L. R. (1977). New upper bounds on the rate of a code via the Delsarte-MacWilliams inequalities. *IEEE Trans. Inform. Theory*, **23**, 157–166. [4,10].

McKay, B. D. See also Godsil, C. D.

McKay, B. D., & Rogoyski, E. (1995). Latin squares of order ten. *Electron. J. Combin.*, **2**(N3). [8].

McKay, M. D., Beckman, R. J., & Conover, W. J. (1979). A comparison of three methods for selecting values of input variables in the analysis of output from a computer code. *Technometrics*, **21**, 239–245. [11].

Mead, R. (1988). *The Design of Experiments.* Cambridge Univ. Press. [11].

Mendelsohn, N. S. See Dulmage, A. L.

Meng, J. (1995). *Some new orthogonal arrays of strength 2.* M.Sc. thesis, Michigan Technological University, Houghton, MI. [6].

Meyer, R. D. See Box, G. E. P.

Midha, C. K. See Dey, A.

Miller, Jr., R. G. (1986). *Beyond ANOVA, Basics of Applied Statistics.* New York: Wiley. [11].

Mills, W. H. (1972). Three mutually orthogonal Latin squares. *J. Combin. Theory*, **13**, 79–82. [8].

Mitchell, T. J. See also Currin, C., Sacks, J., Welch, W. J.

Mitchell, T. J. (1974a). An algorithm for the construction of 'D-optimal' experimental designs. *Technometrics*, **16**, 203–210. [7].

Mitchell, T. J. (1974b). Computer construction of 'D-optimal' first-order designs. *Technometrics*, **16**, 211–220. [7].

Mitsuoka, M. See Yamamoto, S.

Miyakawa, M. (1988). Resampling plan using orthogonal array and its application to influence analysis. (Japanese). *Japanese J. Appl. Statist.*, **17**, 69–79.

Miyakawa, M. (1993). Aliasing pattern between main effect and interaction in L_{36} orthogonal array. (Japanese). *J. Japanese Soc. Quality Control*, **23**, 416–421.

Miyake, M. See Fuji-Hara, R.

Monroe, L., & Pless, V. S. (1993). Personal communication. [12].

Montgomery, D. C. See also Myers, R. H.

Montgomery, D. C. (1997). *Design and Analysis of Experiments.* 4th edn. New York: Wiley. [11].

Mood, A. M. (1946). On Hotelling's weighing problem. *Ann. Math. Statist.*, **17**, 432–446. [7].

Moore, L. M. See Johnson, M. E.

Morgan, J. P., & Chakravarti, I. M. (1988). Block designs for first and second order neighbor correlations. *Ann. Statist.*, **16**, 1206–1224. [6].

Morris, M. D. See also Currin, C., Welch, W. J.

Morris, M. D. (1991). Factorial sampling plans for preliminary computational experiments. *Technometrics*, **33**, 161–174. [11].

Moyssiadis, C. See also Chadjipantelis, T.

Moyssiadis, C., & Kounias, S. (1982). The exact D-optimal first order saturated design with 17 observations. *J. Statist. Plann. Inf.*, **7**, 13–27. [7].

Mukerjee, R. See also Cheng, C.-S.

Mukerjee, R. (1979). On a theorem of Bush on orthogonal arrays. *Calcutta Statist. Assoc. Bull.*, **28**, 169–170. [2].

Mukerjee, R. (1982). Universal optimality of fractional factorial plans derivable through orthogonal arrays. *Calcutta Statist. Assoc. Bull.*, **31**, 63–68. [11].

Mukerjee, R. (1995). On E-optimal fractions of symmetric and asymmetric factorials. *Statist. Sinica*, **5**, 515–533.

Mukerjee, R. (1998). On balanced orthogonal multi-arrays: Existence, construction and application to design of experiments. *J. Statist. Plann. Infer.*, **73**, 149–162. [10].

Mukerjee, R., & Kageyama, S. (1994). On existence of two symbol complete orthogonal arrays. *J. Combin. Theory*, **A 66**, 176–181. [2].

Mukerjee, R., & Wu, C.-F. J. (1995). On the existence of saturated and nearly saturated asymmetric orthogonal arrays. *Ann. Statist.*, **23**, 2102–2115. [9].

Mukhopadhyay, A. C. (1972). Construction of BIBD's from OA's and combinatorial arrangements analogous to OA's. *Calcutta Statist. Assoc. Bull.*, **21**, 45–50. [6].

Mukhopadhyay, A. C. (1978). Incomplete block designs through orthogonal and incomplete orthogonal arrays. *J. Statist. Plann. Inf.*, **2**, 189–196.

Mukhopadhyay, A. C. (1981). Construction of some series of orthogonal arrays. *Sankhyā*, **B 43**, 81–92. [3,6].

Mukhopadhyay, A. C. (1985). Orthogonal arrays and applications. Pages 523–527 of: Kotz, S., Johnson, N. L., & Read, C. B. (eds), *Encyclopedia of Statistical Sciences*, vol. 6. New York: Wiley.

Mullen, G. L. See also Laywine, C. F.

Mullen, G. L. (1988). Polynomial representation of complete sets of mutually orthogonal frequency squares of prime power order. *Discrete Math.*, **69**, 79–84.

Mullen, G. L. (1995). A candidate for the "Next Fermat Problem". *Math. Intelligencer*, **17**(3), 18–22. [8].

Muller, D. E. (1954). Application of Boolean algebra to switching circuit design and to error detection. *IRE Trans. Computers*, **3**, 6–12. [5].

Mullett, G. M., & Karson, M. J. (1986). Percentiles of LINMAP conjoint indices of fit for various orthogonal arrays: A simulation study. *J. Marketing Res.*, **23**, 286–290.

Mullin, R. C. See also Lindner, C. C.

Mullin, R. C., Schellenberg, P. J., Stinson, D. R., & Vanstone, S. A. (1978). On the existence of 7 and 8 mutually orthogonal Latin squares. Research Report 78–14, Dept. Combin. Optimiz., University of Waterloo.

Mullin, R. C., Schellenberg, P. J., Stinson, D. R., & Vanstone, S. A. (1980). Some results on the existence of squares. *Annals Discrete Math.*, **6**, 257–274.

Mullin, R. C., Schellenberg, P. J., van Rees, G. H. J., & Vanstone, S. A. (1980). On the construction of perpendicular arrays. *Utilitas Math.*, **18**, 141–160.

Myers, R. H. See also Vining, G. G.

Myers, R. H., Khuri, I., & Vining, G. G. (1992). Response surface alternatives to the Taguchi robust parameter design approach. *Amer. Statistician*, **46**, 131–139.

Myers, R. H., & Montgomery, D. C. (1995). *Response Surface Methodology : Process and Product in Optimization Using Designed Experiments.* New York: Wiley. [11].

Nair, K. R. See Bose, R. C.

Nair, V. N., & Pregibon, D. (1988). Analyzing dispersion effects from replicated factorial experiments. *Technometrics*, **30**, 247–257. (Errata: Ibid. **31** (1989), 133.) [11].

Nair, V. N. (ed.). (1992). Taguchi's parameter design: A panel discussion. *Technometrics*, **34**, 127–161. [11].

Nakamichi, M. See Itoh, H.

Namikawa, T. See also Fujii, Y., Yamamoto, S.

Namikawa, T., Fujii, Y., & Yamamoto, S. (1989). Computational study on the classification of two-symbol orthogonal arrays of strength t, $m = t + e$ constraints for $e \leq 3$. *SUT J. Math.*, **25**, 179–195. [5].

Newhart, D. W. (1988). On minimum weight codewords in QR codes. *J. Combin. Theory*, **A 48**, 104–119. [5].

Newton, D. E. See Hill, R.

Nguyen, N.-K. (1996). A note on the construction of near-orthogonal arrays with mixed levels and economic run size. *Technometrics*, **38**, 279–283.

Niederreiter, H. See also R. Lidl.

Niederreiter, H. (1987). Point sets and sequences with small discrepancy. *Monatsh. Math.*, **104**, 273–337. [10].

Niederreiter, H. (1992a). *Random Number Generation and Quasi-Monte Carlo Methods.* Philadelphia: SIAM. [10].

Niederreiter, H. (1992b). Orthogonal arrays and other combinatorial aspects in the theory of uniform point distributions in unit cubes. *Discrete Math.*, **106/107**, 361–367.

Nigam, A. K. See also Gupta, V. K.

Nigam, A. K. (1985). Main-effect orthogonal plans from regular group divisible designs. *Sankhyā*, **B 47**, 365–371.

Nishida, F. See Tanaka, H.

Noda, R. (1979). On orthogonal arrays of strength 4 achieving Rao's bound. *J. London Math. Soc.*, **19**, 385–390. [2].

Noda, R. (1986). On orthogonal arrays of strength 3 and 5 achieving Rao's bound. *Graphs Combin.*, **2**, 277–282. [2,12].

Nordstrom, A. W., & Robinson, J. P. (1967). An optimum nonlinear code. *Inform. Control*, **11**, 613–616. [5,12].

Notz, W. See Jacroux, M.

Odlyzko, A. M. See MacWilliams, F. J.

Ohmori, H. See Kimura, H.

Östergård, P. R. J. See Brouwer, A. E., Gulliver, T. A.

Owen, A. B. See also Koehler, J. R.

Owen, A. B. (1992). Orthogonal arrays for computer experiments, integration and visualization. *Statist. Sinica*, **2**, 439–452. (Errata: Ibid. **3** (1993) 261.) [11].

Owen, A. B. (1994). Lattice sampling revisited: Monte Carlo variance of means over randomized orthogonal arrays. *Ann. Statist.*, **22**, 930–945. [11].

Pal, S. N. See Goswami, K. K.

Paley, R. E. A. C. (1933). On orthogonal matrices. *J. Math. Phys.*, **12**, 311–320. [7].

Pang, S. See Zhang, Y.

Parker, E. T. See also Bose, R. C., Brown, J. W.

Parker, E. T. (1958). Construction of some sets of pairwise orthogonal Latin squares. (Abstract). *Amer. Math. Soc. Notices*, **5**, 815. [8].

Parker, E. T. (1959a). Construction of some sets of mutually orthogonal Latin squares. *Proc. Amer. Math. Soc.*, **10**, 946–951. [8].

Parker, E. T. (1959b). Orthogonal Latin squares. *Proc. Nat. Acad. Sci. U.S.A.*, **45**, 859–862. [8].

Parker, E. T. (1962a). Nonextendibility conditions on mutually orthogonal Latin squares. *Proc. Amer. Math. Soc.*, **13**, 219–221. [8].

Parker, E. T. (1962b). On orthogonal Latin squares. *Proceedings Symposium in Pure Mathematics*, **6**, 43–46.

Parker, E. T. (1963). Computer investigations of orthogonal Latin squares of order 10. *Proceedings Symposium in Applied Mathematics*, **15**, 73–81.

Parker, E. T. (1988). The almost orthogonal 6-squares. *Congress. Numer.*, **62**, 85–86. [8].

Parker, E. T. (1991). Attempts for orthogonal latin 10-squares. *Abstracts Amer. Math. Soc.*, **12**(#9IT-05-27). [8].

Parker, E. T., & Somer, L. (1988). A partial generalization of Mann's theorem concerning orthogonal Latin squares. *Canad. Math. Bull.*, **31**, 409–413. [8].

Pasupathy, R. See Kschischang, F. R.

Peace, G. S. (1993). *Taguchi Methods: A Hands-on Approach.* Reading: Addison-Wesley.

Pearson, E. S., & Hartley, H. O. (1966). *Biometrika Tables for Statisticians.* 3rd edn. 2 vols. Cambridge Univ. Press. [11].

Peters, M., & Roth, R. (1987). Four pairwise orthogonal Latin squares of order 24. *J. Combin. Theory*, **A 44**, 152–153. [8].

Peterson, W. W., & Weldon, Jr., E. J. (1972). *Error-Correcting Codes.* 2nd edn. Cambridge, MA: MIT Press. [5].

Petros, C. I. See Kounias, S.

Phadke, M. S. See Lee, N. S.

Picard, R. R. See Berk, K. N.

Piepel, G. F. (1997). Survey of software with mixture experiment capabilities. *J. Qual. Techn.*, **29**, 76–85. [11].

Piper, F. C. See Hughes, D. R.

Piret, P. (1980). Good linear codes of length 27 and 28. *IEEE Trans. Inform. Theory*, **26**, 227. [12].

Plackett, R. L. (1946). Some generalizations in the multifactorial design. *Biometrika*, **33**, 328–332. [11].

Plackett, R. L., & Burman, J. P. (1946). The design of optimum multifactorial experiments. *Biometrika*, **33**, 305–325. [2,7].

Pless, V. S. See also Mallows, C. L., Monroe, L.

Pless, V. S. (1972). Symmetry codes over $GF(3)$ and new five-designs. *J. Combin. Theory*, **12**, 119–142. [12].

Pless, V. S., & Huffman, W. C. (eds). (1998). *Handbook of Coding Theory.* Amsterdam: Elsevier. [2,4].

Plotkin, M. (1960). Binary codes with specified minimal distances. *IEEE Trans. Inform. Theory*, **6**, 445–450. [4].

Pozdniakov, V. S. See Hashim, A. A.

Pregibon, D. See McCullagh, P., Nair, V. N.

Preparata, F. P. (1968). A class of optimum nonlinear double-error correcting codes. *Inform. Control*, **13**, 378–400. [5].

Pu, K. See Hedayat, A. S.

Pukelsheim, F. See also Abt, M.

Pukelsheim, F. (1988). Analysis of variability by analysis of variance. Pages 281–292 of: Dodge, Y., Fedorov, V. V., & Wynn, H. P. (eds), *Optimal Design and Analysis of Experiments.* Amsterdam: North-Holland. [11].

Quattrochi, P. See Georgiades, J.

Quddus, A. See Ahmad, M.

Qvist, B. (1952). Some remarks concerning curves of the second degree in a finite plane. *Ann. Acad. Sci. Fenn.*, **134**, 1–27. [5].

Radziszowski, S. P. See Kreher, D. L.

Rafter, J. A., & Seiden, E. (1974). Contributions to the theory and construction of balanced arrays. *Ann. Statist.*, **2**, 1256–1273.

Raghavarao, D. See also Hedayat, A. S.

Raghavarao, D. (1971). *Constructions and Combinatorial Problems in Design of Experiments.* New York: Wiley. [3,7,8].

Rains, E. M. See also Litsyn, S.

Rains, E. M. (1999). Nonbinary quantum codes. *IEEE Trans. Inform. Theory.* To appear. [4].

Rains, E. M., & Sloane, N. J. A. (1998). Self-dual codes. Pages 177–294 of: *Pless and Huffman (1998).* [5].

Raktoe, B. L., Hedayat, A. S., & Federer, W. T. (1981). *Factorial Designs*. New York: Wiley. [10,11].

Ramakrishna, G. V. S. See Chacko, A., Dey, A.

Rao, C. R. (1945). Finite geometries and certain derived results in number theory. *Proc. Nat. Inst. Sci. India*, **11**, 136–149.

Rao, C. R. (1945–1946). On the most efficient designs in weighing. *Sankhyā*, **7**, 440. [7].

Rao, C. R. (1946a). Hypercubes of strength *d* leading to confounded designs in factorial experiments. *Bull. Calcutta Math. Soc.*, **38**, 67–78. [1,3,4,5].

Rao, C. R. (1946b). Difference sets and combinatorial arrangements derivable from finite geometries. *Proc. Nat. Inst. Sci. India*, **12**, 123–135.

Rao, C. R. (1947). Factorial experiments derivable from combinatorial arrangements of arrays. *J. Royal Statist. Soc. (Suppl.)*, **9**, 128–139. [1,2,3,4,5,11].

Rao, C. R. (1949). On a class of arrangements. *Proc. Edinburgh Math. Soc.*, **8**, 119–125. [1,3,4,5].

Rao, C. R. (1950). The theory of fractional replications in factorial experiments. *Sankhya*, **10**, 81–86.

Rao, C. R. (1951). A simplified approach to factorial experiments and the punched card technique in the construction and analysis of designs. *Bull. Inst. Internat. Statist.*, **23**, 1–28.

Rao, C. R. (1961a). Combinatorial arrangements analogous to orthogonal arrays. *Sankhyā*, **A 23**, 283–286. [6].

Rao, C. R. (1961b). A combinatorial assignment problem. *Nature*, **191**, 100.

Rao, C. R. (1969). Cyclic generation of linear subspaces in finite geometries. Pages 515–535 of: R. C. Bose, T. A. Bowling (ed), *Combinatorial Mathematics and its Applications*. Chapel Hill: Univ. of North Carolina Press.

Rao, C. R. (1973). Some combinatorial problems of arrays and applications to design of experiments. Pages 349–359 of: Srivastava, J. N. (ed), *A Survey of Combinatorial Theory*. Amsterdam: North-Holland. [6,9].

Ray-Chaudhuri, D. K. See also Bose, R. C.

Ray-Chaudhuri, D. K., & Singhi, N. M. (1988). On existence and number of orthogonal arrays. *J. Combin. Theory*, **A 47**, 28–36. (Errata: Ibid., **A 66** (1994), 327–328.) [10].

Reddy, S. M. See Sloane, N. J. A.

Reed, I. S. (1954). A class of multiple-error-correcting codes and the decoding scheme. *IEEE Trans. Inform. Theory*, **4**, 38–49. [5].

Reed, I. S., & Solomon, G. (1960). Polynomial codes over certain finite fields. *J. Soc. Indust. Appl. Math.*, **8**, 300–304. [3,4,5,12].

van Rees, G. H. J. See Brouwer, A. E., Lindner, C. C., Mullin, R. C., Schellenberg, P. J.

Robert, J. M. See Bennett, C. H.

Robinson, J. P. See Nordstrom, A. W.

Rodemich, E. R. See McEliece, R. J.

Rogoyski, E. See McKay, B. D.

Rosa, A. See Doyen, J., Ganter, B., Mathon, R.

Rosenbaum, P. R. (1994). Dispersion effects from fractional factorials in Taguchi's method of quality design. *J. Royal Statist. Soc.*, **B 56**, 641–652. [11].

Rosenbaum, P. R. (1996). Some useful compound dispersion experiments in quality design. *Technometrics*, **38**, 354–364. [10,11].

Rosenbaum, P. R. (1997). *Blocking in compound dispersion experiments.* Preprint. [10,11].

Rosenberg, S. J. (1995). A large index theorem for orthogonal arrays, with bounds. *Discrete Math.*, **137**, 315–318.

Roth, R. See Peters, M.

Roth, R. M. See also Seroussi, G.

Roth, R. M. (1991a). Maximal-rank array codes and their application to crisscross error correction. *IEEE Trans. Inform. Theory*, **37**, 328–339. (Errata: Ibid. **38** (1992) 1183.) [5].

Roth, R. M., & Lempel, A. (1989a). A construction of non-Reed-Solomon type MDS codes. *IEEE Trans. Inform. Theory*, **35**, 655–657. [5].

Roth, R. M., & Lempel, A. (1989b). On MDS codes via Cauchy matrices. *IEEE Trans. Inform. Theory*, **35**, 1314–1319. [5].

Roth, R. M., & Seroussi, G. (1985). On generator matrices of MDS codes. *IEEE Trans. Inform. Theory*, **31**, 826–830. [5].

Roth, R. M., & Seroussi, G. (1986). On cyclic MDS codes of length q over $GF(q)$. *IEEE Trans. Inform. Theory*, **32**, 284–285. [5].

Rudich, S. See Chor, B.

Rumsey Jr., H. J. See McEliece, R. J.

Ryser, H. J. (1963). *Combinatorial Mathematics.* Carus Monograph **14**, Math. Assoc. Amer. [8].

Sacks, J. See also Welch, W. J.

Sacks, J., Schiller, S. B., & Welch, W. J. (1989). Designs for computer experiments. *Technometrics*, **31**, 41–47. [11].

Sacks, J., Welch, W. J., Mitchell, T. J., & Wynn, H. P. (1989). Design and analysis of computer experiments (with discussion). *Statist. Sci.*, **4**, 409–435. [11].

Sade, A. (1960). Produit direct-singulier de quasigroupes orthogonaux et anti-abéliens. *Ann. Soc. Sci. Bruxelles*, **I 74**, 91–99. [8].

Safadi, R. B. See also Wang, R. H.

Safadi, R. B., & Wang, R. H. (1991). The use of genetic algorithms in the construction of mixed multilevel orthogonal arrays. Pages 322–325 of: Keramidas, E. M., & Kaufman, S. M. (eds), *Computing Science and Statistics: Proceedings 23rd Symposium on the Interface.* Interface Foundation of North America. [12].

Safavi-Naini, R., & Blake, I. F. (1979). Generalized t-designs and orthogonal arrays. *Ars Combin.*, **7**, 135–151.

Sathe, Y. S., & Shenoy, R. G. (1991). Further results on construction methods for some A- and D-optimal weighing designs when $N \equiv 3 \pmod 4$. *J. Statist. Plann. Inf.*, **28**, 339–352. [7].

Sato, M. See also Yamamoto, S.

Sato, M., Kuriki, S., & Yamamoto, S. (1985). The nonexistence of a two-symbol orthogonal array of strength 4, 7 constraints and index 7. *TRU Math.*, **21**, 181–193.

Sawade, K. (1985). A Hadamard matrix of order 268. *Graphs Combin.*, **1**, 185–187. [7].

Schatz, J. A. See Blum, J. R.

Schellenberg, P. J. See also Mullin, R. C.

Schellenberg, P. J., van Rees, G. H. J., & Vanstone, S. A. (1978). Four pairwise orthogonal Latin squares of order 15. *Ars Combin.*, **6**, 141–150. [6].

Schiller, S. B. See Sacks, J.

Schrijver, A. (1986). *Theory of Linear and Integer Programming.* New York: Wiley. [4].

Searle, S. R. (1971). *Linear Models.* New York: Wiley. [9].

Seberry, J. See also Mackenzie, C., Wallis, J. S. (The early works of J. Seberry were published under the name J. S. Wallis.).

Seberry, J. (1980). A construction for generalized Hadamard matrices. *J. Statist. Plann. Inf.*, **4**, 365–368. [6].

Seberry, J., & Yamada, M. (1992). Hadamard matrices, sequences, and block designs. Pages 431–560 of: *Dinitz and Stinson (1992).* [7,9].

Segre, B. (1955). Curve razionali normali e k-archi negli spazi finiti. *Ann. Mat. Pura Appl.*, **39**, 357–379. [5].

Seiden, E. See also Blum, J. R., Hedayat, A. S., Rafter, J. A.

Seiden, E. (1954). On the problem of construction of orthogonal arrays. *Ann. Math. Statist.*, **25**, 151–156. [2,6].

Seiden, E. (1955a). On the maximum number of constraints of an orthogonal array. *Ann. Math. Statist.*, **26**, 132–135. (Errata: Ibid., **27** (1956), 204). [2,5,12].

Seiden, E. (1955b). Further remarks on the maximum number of constraints of an orthogonal array. *Ann. Math. Statist.*, **26**, 759–763. [2,5,12].

Seiden, E. (1986). On the existence, construction and enumeration of some 2-symbol orthogonal arrays. Pages 127–138 of: McCulloch, C.E., *et al.* (eds), *Statistical Design: Theory and Practice.* Ithaca: Cornell University. [8].

Seiden, E., & Zemach, R. (1966). On orthogonal arrays. *Ann. Math. Statist.*, **37**, 1355–1370. [2,6,12].

Sendrier, N. See Camion, P.

Seroussi, G. See also Roth, R. M.

Seroussi, G., & Roth, R. M. (1986). On MDS extensions of generalized Reed-Solomon codes. *IEEE Trans. Inform. Theory*, **32**, 349–354. [5].

Shah, K. R. See also Jacroux, M.

Shah, K. R., & Sinha, B. K. (1989). *Theory of Optimal Designs.* New York: Springer-Verlag. Lecture Notes in Statistics **54**. [7].

Shannon, C. E. (1948). A mathematical theory of communication. *Bell. Syst. Tech. J.*, **27**, 379–423 and 623–656. [4,10].

Shannon, C. E. (1992). *Collected Papers.* New York: IEEE Press. Edited by Sloane, N. J. A. and Wyner, A. D. [4,10].

Sharma, H. C., & Dube, S. N. (1986). Structure of PD-cycles and MOLS through them. *J. Statist. Plann. Inf.*, **14**, 221–232.

Shearer, J. B. (1988). Personal communication. [12].

Shenoy, R. G. See Sathe, Y. S.

Sherwood, G. B. (1994). Effective testing of factor combinations. In: *Proc. 3rd Internat. Conf. Software Testing, Analysis and Review (Washington, D.C., May 1994).* Jacksonville, FL.: Software Quality Engineering. [11].

Shirakura, T. See also Yamamoto, S.

Shirakura, T. (1976). Optimal balanced fractional 2^m factorial designs of resolution VII, $6 \le m \le 8$. *Ann. Statist.*, **4**, 515–531.

Shoemaker, A. C., Tsui, K. L., & Wu, C.-F. J. (1991). Economical experimentation methods for robust design. *Technometrics*, **33**, 415–427. (Errata: Ibid. **34** (1992) 502.) [11].

Shrikhande, S. S. See also Bose, R. C., Hedayat, A. S.

Shrikhande, S. S. (1952). On the dual of some balanced incomplete block designs. *Biometrics*, **8**, 66–72. [4].

Shrikhande, S. S. (1953). The non-existence of certain affine resolvable balanced incomplete block designs. *Canad. J. Math.*, **5**, 413–420.

Shrikhande, S. S. (1961). A note on mutually orthogonal Latin squares. *Sankhyā*, **A 23**, 115–116. (Errata: Ibid., 426.) [8].

Shrikhande, S. S. (1964). Generalized Hadamard matrices and orthogonal arrays of strength two. *Canad. J. Math.*, **16**, 736–740. [6].

Shrikhande, S. S., & Bhagwandas. (1965). Duals of incomplete block designs. *J. Indian Statist. Assoc. Bull.*, **3**, 30–37.

Shrikhande, S. S., & Bhagwandas. (1969a). On embedding of orthogonal arrays of strength two. Pages 256–273 of: Bose, R. C., & Dowling, T. A. (eds), *Combinatorial Mathematics and Its Applications.* Chapel Hill, NC: Univ. of North Carolina Press.

Shrikhande, S. S., & Bhagwandas. (1969b). A note on embedding for Hadamard matrices. Pages 673–688 of: Bose, R. C., *et al.* (eds), *Essays in Probability and Statistics, S. N. Roy Memorial Volume.* Chapel Hill, NC: Univ. of North Carolina Press.

Shrikhande, S. S., & Singhi, N. M. (1979a). A note on embedding of orthogonal arrays of strength two. *J. Statist. Plann. Inf.*, **3**, 267–271. [8].

Shrikhande, S. S., & Singhi, N. M. (1979b). Embedding of orthogonal arrays of strength two and deficiency greater than two. *J. Statist. Plann. Inf.*, **3**, 367–379. [8].

Sierksma, G. (1996). *Linear Integer Programming: Theory and Practice.* New York: Dekker. [4].

Silverman, J. H. (1986). *The Arithmetic of Elliptic Curves.* New York: Springer-Verlag. [5].

Silverman, R. See Maneri, C.

Simonnard, M. (1966). *Linear Programming.* Englewood Cliffs, NJ: Prentice-Hall. [4].

Singh, T. P. See Aggarwal, K. R.

Singhi, N. M. See Ray-Chaudhuri, D. K., Shrikhande, S. S.

Singleton, R. C. (1964). Maximum distance q-nary codes. *IEEE Trans. Inform. Theory*, **10**, 116–118. [4,5].

Sinha, B. K. See Shah, K. R.

Sitter, R. R (1993). Balanced repeated replications based on orthogonal multi-arrays. *Biometrika*, **80**, 211–221. [10,11].

Sloane, N. J. A. See also Brouwer, A. E., Cheng, Y., Conway, J. H., Forney, Jr., G. D., Hammons, Jr., A. R., Hardin, R. H., Harwit, M., Litsyn, S., MacWilliams, F. J., Mallows, C. L., Rains, E. M.

Sloane, N. J. A. (1973). Covering arrays and intersecting codes. *J. Combin. Des.*, **1**, 51–63. [11].

Sloane, N. J. A. (1975). An introduction to association schemes and coding theory. Pages 225–260 of: Askey, R. A. (ed), *Theory and Applications of Special Functions.* New York: Academic Press. [10].

Sloane, N. J. A. (1999). *The On-Line Encyclopedia of Integer Sequences.* Published electronically at http://www.research.att.com/~njas/sequences/. [8].

Sloane, N. J. A., & Harwit, M. (1976). Masks for Hadamard transform optics, and weighing designs. *Applied Optics*, **15**, 107–114. [7].

Sloane, N. J. A., Reddy, S. M., & Chen, C. L. (1972). New binary codes. *IEEE Trans. Inform. Theory*, **18**, 503–510. [10].

Sloane, N. J. A., & Stufken, J. (1996). A linear programming bound for orthogonal arrays with mixed levels. *J. Statist. Plann. Infer.*, **56**, 295–305. [9,12].

Sloane, N. J. A., & Whitehead, D. S. (1970). A new family of single-error correcting codes. *IEEE Trans. Inform. Theory*, **16**, 717–719. [10].

Sloane, N. J. A., & Young, J. (1999). *Linear programming bounds for orthogonal arrays.* In preparation. [4,9].

Smolensky, R. See Chor, B.

Snedecor, G. W., & Cochran, W. G. (1989). *Statistical Methods.* 8th edn. Ames, Iowa: Iowa State Univ. Press. [11].

Solé, P. See Hammons, Jr., A. R.

Solomon, G. See Reed, I. S.

Somer, L. See Parker, E. T.

Srivastava, J. N. See also Bose, R. C., Chopra, D. V.

Srivastava, J. N. (1972). Some general existence conditions for balanced arrays of strength t and 2 symbols. *J. Combin. Theory*, **A 13**, 198–206.

Srivastava, J. N. (1976). Smaller sized factor screening designs through the use of linear search models. Pages 139–162 of: *Proc. 9th Internat. Biometric Conf. (Boston, 1976)*, vol. 1. Raleigh, NC: Biometric Society.

Srivastava, J. N. (1978). A review of some recent work on discrete optimal factorial designs for statisticians and experimenters. Pages 267–329 of: Krishnaiah, P. R. (ed), *Developments in Statistics*, vol. 1. New York: Academic Press.

Srivastava, J. N. (1987). On the inadequacy of the customary orthogonal arrays in quality control and general scientific experimentation, and the need of probing designs of higher revealing power. *Commun. Statist. Theory Meth.*, **16**, 2901–2941.

Srivastava, J. N. (1990). Modern factorial design theory for experimenters and statisticians. Pages 311–406 of: Ghosh, S. (ed), *Statistical Design and Analysis of Industrial Experiments.* New York: Dekker.

Srivastava, J. N. (1994). A new mathematical space with application to statistical experimental design-I. Preliminary results. *J. Statist. Plann. Inf.*, **40**, 113–125.

Srivastava, J. N. (1996). A critique of some aspects of experimental design. Pages 309–341 of: Ghosh, S., & Rao, C. R. (eds), *Handbook of Statistics*, vol. 13. Amsterdam: Elsevier.

Srivastava, J. N., & Chopra, D. V. (1973). Balanced arrays and orthogonal arrays. Pages 411–428 of: Srivastava, J. N. (ed), *A Survey of Combinatorial Theory.* Amsterdam: North-Holland. [10].

Srivastava, J. N., & Ghosh, S. (1977). On the existence of search designs with continuous factors. *Ann. Inst. Statist. Math.*, **A 29**, 301–306. [11].

Srivastava, J. N., & Throop, D. (1990). Orthogonal arrays obtainable as solutions to linear equations over finite fields. *Linear Algebra Appl.*, **127**, 283–300. [10].

Stegun, I. A. See Abramowitz, M.

Steinberg, D. M. See also Cheng, C.-S.

Steinberg, D. M. (1996). Robust design: experiments for improving quality. Pages 199–240 of: Ghosh, S., & Rao, C. R. (eds), *Handbook of Statistics*, vol. 13. Amsterdam: Elsevier.

Steinberg, D. M., & Bursztyn, D. (1994). Dispersion effects in robust-design experiments with noise factors. *J. Qual. Techn.*, **26**, 21–20. [11].

Stinaff, R. D. See Helgert, H. J.

Stinson, D. R. See also Bierbrauer, J., Colbourn, C. J., Dinitz, J. H., Kreher, D. L., Martin, W. J., Mullin, R. C.

Stinson, D. R. (1984). A short proof of the nonexistence of a pair of orthogonal Latin squares of order six. *J. Combin. Theory*, **A 36**, 373–376. [8].

Stinson, D. R. (1993). Resilient functions and large sets of orthogonal arrays. *Congress. Numer.*, **92**, 105–110. [10].

Stinson, D. R. (1994). Combinatorial techniques for universal hashing. *J. Computer System Sci.*, **48**, 337–346.

Stinson, D. R. (1997). On some methods for unconditionally secure key distribution and broadcast encryption. *Designs, Codes, Crypt.*, **12**, 215–243. [10].

Storme, L. See also Hirschfeld, J. W. P.

Storme, L. (1992). Completeness of normal rational curves. *J. Algebraic Combin.*, **1**, 197–202. [5].

Storme, L. (1993). k-Arcs in $PG(n, q)$ and linear M.D.S. codes. *Med. Konink. Acad. Wet. Belg.*, **55**, 87–126. [5].

Storme, L., & Szönyi, T. (1993a). Caps in $PG(n, q)$, q even, $n \geq 3$. *Geom. Dedicata*, **45**, 163–169. [5].

Storme, L., & Szönyi, T. (1993b). Note on a problem of Kaneta concerning arcs in $PG(3, q)$. *Atti Sem. Math. Fis. Univ. Modena*, **41**, 409–416. [5].

Storme, L., & Thas, J. A. (1991). Generalized Reed-Solomon codes and normal rational curves: an improvement of results by Seroussi and Roth. Pages 369–389 of: Hirschfeld, J. W. P., Hughes, D. R., & Thas, J. A. (eds), *Advances in Finite Geometries and Designs (Isle of Thorns, 1990)*. Oxford Univ. Press. [5].

Storme, L., & Thas, J. A. (1992). Complete k-arcs in $PG(n, q)$, q even. *Discrete Math.*, **106/107**, 455–469. [5].

Storme, L., & Thas, J. A. (1993). M.D.S. codes and arcs in $PG(n, q)$ with q even: an improvement of the bounds of Bruen, Thas, and Blokhuis. *J. Combin. Theory*, **A 62**, 139–154. [5].

Storme, L., & Thas, J. A. (1994). k-Arcs and dual k-arcs. *Discrete Math.*, **125**, 357–370. [5].

Straus, E. G. See Chowla, S.

Street, A. P. See also Wallis, W. D.

Street, A. P., & Street, D. J. (1987). *Combinatorics of Experimental Design*. Oxford Univ. Press. [8].

Street, D. J. See also Street, A. P.

Street, D. J. (1979). Generalized Hadamard matrices, orthogonal arrays and F-squares. *Ars Combin.*, **8**, 131–141. [6,8].

Stufken, J. See also Hedayat, A. S., Lin, W.-C., Sloane, N. J. A.

Stufken, J. (1991). Some families of optimal and efficient repeated measurements designs. *J. Statist. Plann. Infer.*, **27**, 75–83. [6].

Su, G. See Hedayat, A. S.

Suchower, S. J. See also Laywine, C. F.

Suchower, S. J. (1993). Polynomial representations of complete sets of mutually orthogonal frequency squares of prime power order. *J. Combin. Theory*, **A 62**, 46–65. [8].

Suchower, S. J. (1994). Nonisomorphic complete sets of F-Rectangles with varying numbers of symbols. *Contemporary Math.*, **168**, 353–362. [8].

Suchower, S. J. (1995). Nonisomorphic complete sets of F-Rectangles with prime power dimensions. *Designs, Codes, Crypt.*, **5**, 155–174. [8].

Suen, C.-Y. (1989a). Some resolvable orthogonal arrays with two symbols. *Commun. Statist. Theory Meth.*, **18**, 3875–3881.

Suen, C.-Y. (1989b). A class of orthogonal main-effect plans. *J. Statist. Plann. Inf.*, **21**, 391–394. [12].

Suen, C.-Y., Chen, H., & Wu, C.-F. J. (1997). Some identities on q^{n-m} designs with application to minimum aberration designs. *Ann. Statist.*, **25**, 1176–1188. [11].

Sun, D. X. See also Chen, J., Cheng, C.-S.

Sun, D. X. (1993). *Estimation capacity and related topics in experimental designs.* Ph.D. thesis, University of Waterloo.

Sun, D. X., & Wu, C.-F. J. (1993). Statistical properties of Hadamard matrices of order 16. Pages 169–179 of: Kuo, W. (ed), *Quality through Engineering Design.* Amsterdam: Elsevier.

Sun, D. X., & Wu, C.-F. J. (1994). Interaction graphs for three-level fractional factorial designs. *J. Qual. Techn.*, **26**, 297–307.

Sun, D. X., Wu, C.-F. J., & Chen, Y. (1997). Optimal blocking schemes for 2^n and 2^{n-p} designs. *Technometrics*, **39**, 298–307.

Swiercz, S. See Lam, C. W. H.

Szablak, B. D. See Dunietz, I. S.

Szegö, G. (1967). *Orthogonal Polynomials.* 3rd edn. Providence, RI: Amer. Math. Soc. [4,11].

Szönyi, T. See Storme, L.

Taguchi, G. (1986). *Introduction to Quality Engineering: Designing Quality into Products and Processes.* Tokyo: Asian Productivity Organization. [11].

Taguchi, G. (1987). *System of Experimental Design: Engineering Methods to Optimize Quality and Minimize Costs.* White Plains, NY:UNIPUB, and Dearborn, MI: American Supplier Institute, Inc. [11,12].

Taguchi, G., & Wu, Y. (1985). *Introduction to off-line quality control.* Romulus, MI: Central Japan Quality Control Association, ASI Inc. [1].

Tallini, G. See Ceccherini, P. V.

Tanaka, H., & Nishida, F. (1970). A construction of a polynomial code. *Electron. Commun. Japan*, **53-A**(8), 24–31. [3,5].

Tang, B. (1993). Orthogonal array-based Latin hypercubes. *J. Amer. Statist. Assoc.*, **88**, 1392–1397.

Tang, B. (1994). A theorem for selecting OA-based Latin hypercubes using a distance criterion. *Commun. Statist. Theory Meth.*, **23**, 2047–2058.

Tang, B., & Wu, C.-F. J. (1996). Characterization of minimum aberration 2^{n-k} designs in terms of their complementary designs. *Ann. Statist.*, **24**, 2549–2559.

Tang, B., & Wu, C.-F. J. (1997). A method for constructing super-saturated designs and its Es^2 optimality. *Canad. J. Statist.*, **25**, 191–201. [11,12].

Tarry, G. (1900). Le problème des 36 officiers (i). *Comptes Rendus Assoc. Franc. Avanc. Sci.*, **1**, 122–123. [8].

Tarry, G. (1901). Le problème des 36 officiers (ii). *Comptes Rendus Assoc. Franc. Avanc. Sci.*, **2**, 177–203. [8].

Thas, J. A. See also Blokhuis, A., Bruen, A. A., Storme, L.

Thas, J. A. (1968). Normal rational curves and k-arcs in Galois spaces. *Rend. Mat.*, **1**, 331–334. [5].

Thas, J. A. (1969). Connection between the Grassmanian $G_{k-1;n}$ and the set of the k-arcs of the Galois space $S_{n,q}$. *Rend. Mat.*, **2**, 121–134. [5].

Thiel, L. See Kolesova, G., Lam, C. W. H.

Thompson, J. G. See MacWilliams, F. J.

Throop, D. See Srivastava, J. N.

Tietäväinen, A. (1990). An upper bound on the covering radius as a function of the dual distance. *IEEE Trans. Inform. Theory*, **36**, 1472–1474. [4].

Tietäväinen, A. (1991). Covering radius and dual distance. *Designs, Codes, Crypt.*, **1**, 31–46. [4].

Tits, J. (1960/1961). Les groupes simples de Suzuki et de Ree. *Séminaire Bourbaki*, **13**(210), 1–18. [5].

Tits, J. (1962). Ovoïdes et groupes de Suzuki. *Archiv. Math.*, **13**, 187–198. [5].

Todd, J. A. (1933). A combinatorial problem. *J. Math. Phys.*, **12**, 321–333. [7].

Tonchev, V. D. See also Lam, C. W. H.

Tonchev, V. D. (1983). Hadamard matrices of order 28 with automorphisms of order 13. *J. Combin. Theory*, **A 35**, 43–57. [7].

Tonchev, V. D. (1985). Hadamard matrices of order 28 with automorphisms of order 7. *J. Combin. Theory*, **A 40**, 62–81. [7].

Tonchev, V. D. (1988). *Combinatorial Configurations, Designs, Codes, Graphs.* New York: Wiley.

Tonchev, V. D. (1989). Self-orthogonal designs and extremal doubly even codes. *J. Combin. Theory*, **A 52**, 197–205. [7].

Trott, M. D. See Forney, Jr., G. D.

Tsfasman, M. A., & Vlăduţ, S. G. (1991). *Algebraic-Geometric Codes.* Dordrecht, Holland: Kluwer. [4].

Tsui, K. L. See Shoemaker, A. C.

van der Waerden, B. L. (1950). *Modern Algebra.* New York: Ungar. 3rd ed., 2 vols. [A].

van der Waerden, B. L. (1993). *Algebra.* New York: Springer-Verlag. [A].

Vanstone, S. A. See Mullin, R. C., Schellenberg, P. J.

Varshamov, R. R. (1957). Estimate of the number of signals in error correcting codes. (Russian). *Dokl. Akad. Nauk SSSR*, **17**, 739–741. [10].

Verhoeff, T. See also Brouwer, A. E.

Verhoeff, T. (1987). An updated table of minimum-distance bounds for binary linear codes. *IEEE Trans. Inform. Theory*, **33**, 665–680. [12].

Vining, G. G. See also Myers, R. H.

Vining, G. G., & Myers, R. H. (1990). Combining Taguchi and response surface philosophies: a dual response approach. *J. Qual. Techn.*, **22**, 38–45. [11].

Vlăduţ, S. G. See Tsfasman, M. A.

Wagner, T. J. (1965). A remark concerning the minimum distance of group codes. *IEEE Trans. Inform. Theory*, **11**, 458. [5,12].

Wallis, J. S. See also Seberry, J., Wallis, W. D. (The early works of J. Seberry were published under the name J. S. Wallis.).

Wallis, J. S. (1976). On the existence of Hadamard matrices. *J. Combin. Theory*, **A 21**, 188–195. [7].

Wallis, W. D. See also Hedayat, A. S.

Wallis, W. D., Street, A. P., & Wallis, J. S. (1972). *Combinatorics: Room Squares, Sum-Free Sets, Hadamard Matrices.* New York: Springer-Verlag. Lecture Notes in Math. **292**. [7].

Wang, J. C. (1996a). Mixed difference matrices and the construction of orthogonal arrays. *Statist. Probab. Lett.*, **28**, 121–126. [9,12].

Wang, J. C. (1996b). *A recursive construction of orthogonal arrays.* Preprint. [12].

Wang, J. C., & Wu, C.-F. J. (1989). An approach to the construction of asymmetrical orthogonal arrays. Technical Report 89-01, Institute for Improvement in Quality and Productivity, University of Waterloo, 1989. [12].

Wang, J. C., & Wu, C.-F. J. (1991). An approach to the construction of asymmetrical orthogonal arrays. *J. Amer. Statist. Assoc.*, **86**, 450–456. [9,12].

Wang, J.-C., & Wu, C.-F. J. (1992). Nearly orthogonal arrays with mixed levels and small runs. *Technometrics*, **34**, 409–422. [12].

Wang, P. C. (1988). A simple method for analyzing binary data from orthogonal arrays. *J. Qual. Techn.*, **20**, 230–232.

Wang, P. C. (1989). Tests for dispersion effects from orthogonal arrays. *Comput. Statist. Data Anal.*, **8**, 109–117.

Wang, P. C., Chen, M. H., & Wei, S. C. (1995). To analyze binary data from mixed-level orthogonal arrays. *Comput. Statist. Data Anal.*, **20**, 689–695.

Wang, P. C., & Jan, H. W. (1995). Designing two-level factorial experiments using orthogonal arrays when the run order is important. *The Statistician*, **44**, 379–388.

Wang, R. See Wu, C.-F. J.

Wang, R. H. See also Safadi, R. B.

Wang, R. H., & Safadi, R. B. (1990). Generating mixed multilevel orthogonal arrays by simulated annealing. Pages 557–560 of: Page, C., & LePage, R. (eds), *Computing Science and Statistics: Proceedings 22nd Symposium on the Interface.* New York: Springer-Verlag. [12].

Ward, H. N. See also MacWilliams, F. J.

Ward, H. N. (1990). Quadratic residue codes of length 27. *IEEE Trans. Inform. Theory*, **36**, 950–953. [5].

Wei, S. C. See Wang, P. C.

Welch, L. R. See McEliece, R. J.

Welch, W. J. See also Sacks, J.

Welch, W. J., Buck, R. J., Sacks, J., Wynn, H. P., Mitchell, T. J., & Morris, M. D. (1992). Screening, predicting, and computer experiments. *Technometrics*, **34**, 15–25. [11].

Welch, W. J., Yu, T.-K., Kang, S. M., & Sacks, J. (1990). Computer experiments for quality control by parameter design. *J. Qual. Techn.*, **22**, 15–22. [11].

Weldon Jr., E. J. See Peterson, W. W.

White, D. (1978). On computer generation of balanced arrays. *J. Statist. Plann. Inf.*, **2**, 252–275. [12].

Whitehead, D. S. See Sloane, N. J. A.

Wilks, A. R. See Becker, R. A.

Williamson, J. (1944). Hadamard's determinant theorem and the sum of four squares. *Duke Math. J.*, **11**, 65–81. [7].

Williamson, J. (1946). Determinants whose elements are 0 and 1. *Amer. Math. Monthly*, **53**, 427–434. [7].

Wilson, R. M. See also Baker, R. D.

Wilson, R. M. (1974). Concerning the number of mutually orthogonal Latin squares. *Discrete Math.*, **9**, 181–198.

Wolf, J. K. (1969). Adding two information symbols to certain nonbinary BCH codes and some applications. *Bell Syst. Tech. J.*, **48**, 2405–2424. [3,5].

Wu, C.-F. J. See also Chen, J., Hamada, M., Mukerjee, R., Shoemaker, A. C., Sun, D. X., Suen, C.-Y., Tang, B., Wang, J. C.

Wu, C.-F. J. (1989). Construction of $2^m 4^n$ designs via a grouping scheme. *Ann. Statist.*, **17**, 1880–1885. [9].

Wu, C.-F. J. (1991). Balanced repeated replications based on mixed orthogonal arrays. *Biometrika*, **78**, 181–188. [11].

Wu, C.-F. J. (1993). Construction of supersaturated designs through partially aliased interactions. *Biometrika*, **80**, 661–669. [11,12].

Wu, C.-F. J., Mao, S. S., & Ma, F. S. (1990). SEL: A search method based on orthogonal arrays. Pages 279–310 of: Ghosh, S. (ed), *Statistical Design and Analysis of Industrial Experiments*. New York: Dekker.

Wu, C.-F. J., Zhang, R. C., & Wang, R. (1992). Construction of asymmetrical orthogonal arrays of the type $OA(s^k, s^m(s^{r_1})^{n_1} \cdots (s^{r_t})^{n_t})$. *Statist. Sinica*, **2**, 203–219. [9,12].

Wu, Y. See Taguchi, G.

Wynn, H. P. See Sacks, J., Welch, W. J.

Xiang, K. F. (1983). The difference set table for $\lambda = 2$. (Chinese). *Acta Math. Appl. Sinica*, **6**, 160–166. [6].

Xu, C. X. (1979). Construction of orthogonal arrays $L_{2p^u}(p^{1+\sum_{i=1}^{u-1} 2p^i})$ with odd prime p. (Chinese). *Acta Math. Appl. Sinica*, **2**, 92–97. [6].

Yamada, M. See also Seberry, J.

Yamada, M. (1988). On a series of Hadamard matrices of order 2^t and the maximal excess of Hadamard matrices of order 2^{2t}. *Graphs Combin.*, **4**, 297–301. [7].

Yamamoto, S. See also Fujii, Y., Namikawa, T., Sato, M.

Yamamoto, S., Fujii, Y., Hyodo, Y., & Yumiba, H. (1992a). Classification of two-symbol orthogonal arrays of strength 2, size 16, 15 (maximal) constraints and index 4. *SUT J. Math.*, **28**, 47–59. [5,7].

Yamamoto, S., Fujii, Y., Hyodo, Y., & Yumiba, H. (1992b). Classification of two-symbol orthogonal arrays of strength 2, size 20 and 19 (maximal) constraints. *SUT J. Math.*, **28**, 191–209. [5,7].

Yamamoto, S., Fujii, Y., Hyodo, Y., & Yumiba, H. (1993). Connections between the numbers of nonisomorphic solutions of Hadamard matrices, orthogonal arrays and balanced incomplete block designs. *SUT J. Math.*, **29**, 143–165. [7].

Yamamoto, S., Fujii, Y., Hyodo, Y., & Yumiba, H. (1994a). Classification of fractional 2^5 and 2^6 factorial designs having 16 and 20 runs derivable from saturated orthogonal arrays of strength two. *SUT J. Math.*, **30**, 35–50. [5].

Yamamoto, S., Fujii, Y., Hyodo, Y., & Yumiba, H. (1994b). Saturated two-symbol orthogonal arrays of strength two and Hadamard three designs. *SUT J. Math.*, **30**, 51–64.

Yamamoto, S., Fujii, Y., Hyodo, Y., & Yumiba, H. (1995). Classification of orthogonal 24-run 2^5 factorial designs derivable from saturated two-symbol orthogonal arrays of strength 2, size 24 and index 6. *SUT J. Math.*, **31**, 39–53. [5].

Yamamoto, S., Fujii, Y., & Mitsuoka, M. (1993). Three-symbol orthogonal arrays of strength 2 and index 2 having maximal constraints: Computational study. *J. Combin. Inform. System Sci.*, **18**, 329–342.

Yamamoto, S., Fujii, Y., Namikawa, T., & Mitsuoka, M. (1991). Three-symbol orthogonal arrays of strength t having $t+2$ constraints. *SUT J. Math.*, **27**, 93–111.

Yamamoto, S., Kuriki, S., & Sato, M. (1984). On existence and construction of some 2-symbol orthogonal arrays. *TRU Math.*, **20**, 317–331. [2].

Yamamoto, S., Namikawa, T., & Fujii, Y. (1988). Classification of two-symbol orthogonal arrays of strength t, $t+3$ constraints and index 4. *TRU Math.*, **24**, 167–184. [5].

Yamamoto, S., Shirakura, T., & Kuwada, M. (1975). Balanced arrays of strength $2t$ and balanced fractional 2^m factorial designs. *Ann. Inst. Statist. Math.*, **27**, 143–157.

Yao, J.-S. See Liu, M.-S.

Yates, F. See R. A. Fisher.

Ylvisaker, D. See Currin, C., Johnson, E. M.

Yoshizawa, M. (1987). On schematic orthogonal arrays $OA(q^t, k, q, t)$. *Graphs Combin.*, **3**, 81–89. [10].

Young, J. See Sloane, N. J. A.

Yu, T.-K. See Welch, W. J.

Yumiba, H. See Yamamoto, S.

Zehendner, E. See Falkner, G.

Zeller, K. See Ehlich, H.

Zemach, R. See Seiden, E.

Zemroch, P. J. (1988). Strategies for generating blocked fractional replicate designs by computer. *Comput. Statist. Quart.*, **4**, 43–57.

Zhang, R. C. See Wu, C.-F. J.

Zhang, Y., Lu, Y., & Pang, S. (1999). Orthogonal arrays obtained by decomposition of projection matrices. *Statist. Sinica.* To appear. [9,12].

Zhu, L. (1984). Six pairwise orthogonal Latin squares of order 69. *J. Austral. Math. Soc.*, **A 37**, 1–3. [8].

Zyablov, V. V. See Blokh, E. L.

Author Index

Subject Index

Springer Series in Statistics

(continued from p. ii)

Kotz/Johnson (Eds.): Breakthroughs in Statistics Volume II.
Kotz/Johnson (Eds.): Breakthroughs in Statistics Volume III.
Kres: Statistical Tables for Multivariate Analysis.
Küchler/Sørensen: Exponential Families of Stochastic Processes.
Le Cam: Asymptotic Methods in Statistical Decision Theory.
Le Cam/Yang: Asymptotics in Statistics: Some Basic Concepts.
Longford: Models for Uncertainty in Educational Testing.
Manoukian: Modern Concepts and Theorems of Mathematical Statistics.
Miller, Jr.: Simultaneous Statistical Inference, 2nd edition.
Mosteller/Wallace: Applied Bayesian and Classical Inference: The Case of the Federalist Papers.
Parzen/Tanabe/Kitagawa: Selected Papers of Hirotugu Akaike.
Pollard: Convergence of Stochastic Processes.
Pratt/Gibbons: Concepts of Nonparametric Theory.
Ramsay/Silverman: Functional Data Analysis.
Read/Cressie: Goodness-of-Fit Statistics for Discrete Multivariate Data.
Reinsel: Elements of Multivariate Time Series Analysis, 2nd edition.
Reiss: A Course on Point Processes.
Reiss: Approximate Distributions of Order Statistics: With Applications to Non-parametric Statistics.
Rieder: Robust Asymptotic Statistics.
Rosenbaum: Observational Studies.
Ross: Nonlinear Estimation.
Sachs: Applied Statistics: A Handbook of Techniques, 2nd edition.
Särndal/Swensson/Wretman: Model Assisted Survey Sampling.
Schervish: Theory of Statistics.
Seneta: Non-Negative Matrices and Markov Chains, 2nd edition.
Shao/Tu: The Jackknife and Bootstrap.
Siegmund: Sequential Analysis: Tests and Confidence Intervals.
Simonoff: Smoothing Methods in Statistics.
Small: The Statistical Theory of Shape.
Tanner: Tools for Statistical Inference: Methods for the Exploration of Posterior Distributions and Likelihood Functions, 3rd edition.
Tong: The Multivariate Normal Distribution.
van der Vaart/Wellner: Weak Convergence and Empirical Processes: With Applications to Statistics.
Vapnik: Estimation of Dependences Based on Empirical Data.
Weerahandi: Exact Statistical Methods for Data Analysis.
West/Harrison: Bayesian Forecasting and Dynamic Models, 2nd edition.
Wolter: Introduction to Variance Estimation.
Yaglom: Correlation Theory of Stationary and Related Random Functions I: Basic Results.
Yaglom: Correlation Theory of Stationary and Related Random Functions II: Supplementary Notes and References.

www.ingramcontent.com/pod-product-compliance
Lightning Source LLC
Chambersburg PA
CBHW071729290626
47387CB00004B/608
* 9 7 8 0 3 8 7 9 8 7 6 6 8 *